Study Gui

for

Biology Life on Earth

Ninth Edition

Biology Life on Earth with Physiology

Ninth Edition

Teresa Audesirk
Gerald Audesirk
Professors Emeriti, University of Colorado, Denver

Bruce E. Byers
University of Massachusetts, Amherst

Lisa K. Bonneau, Ph.D.
Metropolitan Community College, Blue River

Clifton Cooper
Linn-Benton Community College

Benjamin Cummings

Boston Columbus Indianapolis New York San Francisco Upper Saddle River
Amsterdam Cape Town Dubai London Madrid Milan Munich Paris Montréal Toronto
Delhi Mexico City São Paulo Sydney Hong Kong Seoul Singapore Taipei Tokyo

Vice President/Editor-in-Chief: Beth Wilbur
Senior Acquisitions Editor: Star MacKenzie
Project Editor: Nina Lewallen Hufford
Editorial Assistant: Frances Sink
Managing Editor, Production: Mike Early
Production Project Manager: Jane Brundage
Compositor: S4Carlisle Publishing Services, Carla Kipper

Cover Production: Seventeenth Street Studios
Photo Researcher: Yvonne Gerin
Manufacturing Buyer: Michael Penne
Executive Marketing Manager: Lauren Harp

Cover Photo Credit: Daniel J. Cox/Corbis

ISBN 10-digit   0-321-61179-9
        13-digit   978-0-321-61179-6

Benjamin Cummings
is an imprint of

www.pearsonhighered.com

1 2 3 4 5 6 7 8 9 10—EB—14 13 12 11 10

PHOTO CREDITS

Chapter 4 Page 59, micrograph of plastid: Biophoto Associates/Photo Researchers. **Chapter 19** Page 235, E. coli conjugating by means of F pilus: Dennis Kunkel/Phototake NYC. **Chapter 20** Page 246, giant kelp forest: Flip Nicklin/Minden Pictures. **Chapter 22** Page 265, SEM of mushroom gill: S Lowry/ University of Ulster/Getty Images. **Chapter 24** Page 286, crocodile hatching: Mark Deeble & Victoria Stone/Photolibrary.com. **Chapter 25** Page 297, trial-and-error learning in toad: Caroll Boltin. **Chapter 27** Page 323, false-eyed frogs: Zig Leszczynski/Animals Animals/EarthScenes; peacock moth: Thomas Marent/Minden Pictures; swallowtail caterpillar: Jeff Lepore/Photo Researchers; poison dart frog: Thomas Marent/Minden Pictures. **Chapter 29** Page 346, American cactus: Teresa and Gerald Audesirk; Euphorbia: Tom McHugh/Photo Researchers. Page 347, shortgrass prairie: Teresa and Gerald Audesirk. **Chapter 31** Page 369, desert pupfish: Tom McHugh/Photo Researchers; hummingbird: kwan tse/ Shutterstock; iguana: Ivan Cholakov Gostock-dot-ne/Shutterstock. **Chapter 32** Page 381, elephantiasis: World Health Organization/Image Library, Special Programme for Research and Training in Tropical Diseases/Photo by Chandran. **Chapter 34** Page 406, Kwashiorkor child: Howard Brinton/General Board of Global Ministries. Page 407, child with rickets: Biophoto Associates/Photo Researchers. **Chapter 37** Page 449, cicada molting: Q2 Kent Wood/Photo Researchers. Page 450, gigantism/dwarfism: Image courtesy of RINGLING BROS. AND BARNUM & BAILEY® THE GREATEST SHOW ON EARTH®. **Chapter 40** Page 491, osteoporosis victim: Yoav Levy/Phototake NYC. **Chapter 43** Page 539, guttation on a strawberry leaf: Noah Elhardt/Wikimedia Commons. Page 541, parenchyma cells, collenchyma cells, pear "stone cells" sclereids: George Wilder/Visuals Unlimited. **Chapter 44** Page 552, dandelion flower: Teresa and Gerald Audesirk; red maple seeds: Stephen Maka/CORBIS- NY. Page 553, flower with anthers: Teresa and Gerald Audesirk. **Chapter 45** Page 562, twining morning glory: Adam Jones/Visuals Unlimited/Getty Images. Page 563, abscission layer: Biophoto Associates/Photo Researchers.

CONTENTS

CHAPTER 1 AN INTRODUCTION TO LIFE ON EARTH

OUTLINE

Section 1.1 How Do Scientists Study Life?

Matter has different levels of organization, ranging from the **atom** (the lowest level) to the **biosphere** (the highest level, **Figure 1-1**). The **cell** is the smallest unit of life (**Figure 1-2**).

Scientific inquiry is based on the principles of (1) **natural causality**, (2) natural laws that are uniform in space and time, and (3) the common perception of natural events.

Scientific inquiry occurs by use of the **scientific method**, which consists of **observation**, **question**, **hypothesis**, **prediction**, **experiment**, and **conclusion** (**Figure 1-4**). **Controls** are used to isolate the effect of a single **variable** on an experimental observation.

Experimental results are useful only if they are communicated to other scientists.

A **scientific theory** is a broad explanation of a natural phenomenon, supported by extensive and reproducible observations. A scientific theory is formed through **inductive reasoning**, which can then be used to support **deductive reasoning**. All scientific theories have the potential to be disproved.

Section 1.2 Evolution: The Unifying Theory of Biology

Evolution accounts for both the diversity and the similarities of life. Evolution states that modern organisms descended, with modifications, from preexisting life forms.

Evolution occurs through the processes of (1) genetic variation among **population** members, (2) inheritance by offspring of genetic variations from their parents, and (3) **natural selection**.

Natural selection occurs when organisms with favorable **adaptations** to their environment survive and pass on these favorable **genes** to their offspring. Thus, favorable genes are preserved within a population.

Section 1.3 What Are the Characteristics of Living Things?

All living things exhibit seven characteristics. They (1) are structurally complex and composed of cells (**Figure 1-8**), (2) maintain internal **homeostasis** (**Figure 1-9**), (3) respond to environmental stimuli, (4) acquire and use **nutrients** and **energy** (**Figure 1-10**), (5) grow, (6) reproduce, and (7) can evolve.

Section 1.4 How Do Scientists Categorize the Diversity of Life?

All living things can be grouped into one of three **domains**: Bacteria, Archaea, and Eukarya. The Eukarya can be further subdivided into three **kingdoms** (Fungi, Plantae, and Animalia) as well as the unicellular protists (**Figure 1–11**, **Table 1–1**). Kingdoms are further subdivided into phyla, classes, orders, families, genera, and **species**. Species are named using the binomial system.

Bacteria and Archaea lack a nucleus (i.e., are **prokaryotic**), lack membrane-bound organelles, and are typically **unicellular**. Eukarya have a nucleus (i.e., are eukaryotic) and can be unicellular or **multicellular**.

Photosynthetic organisms can make their own energy (i.e., are **autotrophs**), whereas those organisms that cannot are **heterotrophs**.

FLASH CARDS

To use the flash cards, tear the page from the book and cut along the dashed lines. The key term appears on one side of the flash card, and its definition appears on the opposite side.

adaptation	cell
atom	cell theory
autotroph	chromosome
binomial system	community
biodiversity	conclusion
biosphere	control

The smallest unit of life, consisting, at a minimum, of an outer membrane that encloses a watery medium containing organic molecules, including genetic material composed of DNA.

A trait that increases the ability of an individual to survive and reproduce compared to individuals without the trait.

Scientific theory stating that every living organism is made up of one or more cells, cells are the functional units of all organisms, and all cells arise from preexisting cells.

The smallest particle of an element that retains the properties of the element.

A DNA double helix together with proteins that help to organize and regulate the use of the DNA.

"Self-feeder"; normally, a photosynthetic organism; a producer.

All the interacting populations within an ecosystem.

The method of naming organisms by genus and species, often called the scientific name, usually using Latin words or words derived from Latin.

In the scientific method, a decision about the validity of a hypothesis, made on the basis of experiments or observations.

The diversity of living organisms; often measured as the variety of different species in an individual ecosystem or in the entire biosphere.

That portion of an experiment in which all possible variables are held constant; in contrast to the "experimental" portion, in which a particular variable is altered.

That part of Earth inhabited by living organisms; includes both living and nonliving components.

cytoplasm	energy
deductive reasoning	eukaryotic
deoxyribonucleic acid (DNA)	evolution
domain	experiment
ecosystem	gene
element	heterotroph

The capacity to do work.

All of the material contained within the plasma membrane of a cell, exclusive of the nucleus.

Referring to cells of organisms of the domain Eukarya (plants, animals, fungi, and protists). Eukaryotic cells have genetic material enclosed within a membrane-bound nucleus and contain other organelles.

The process of generating hypotheses about the results of a specific experiment or the nature of a specific observation.

The descent of modern organisms, with modification, from preexisting life-forms; a change in the genetic makeup (the proportions of different genotypes) of a population from one generation to the next.

A molecule composed of deoxyribose nucleotides; contains the genetic information of all living cells.

In the scientific method, the use of carefully controlled observations or manipulations to test the predictions generated by a hypothesis.

The broadest category for classifying organisms; organisms are classified into three domains: Bacteria, Archaea, and Eukarya.

The unit of heredity; a segment of DNA located at a particular place on a chromosome that encodes the information for the amino acid sequence of a protein and, hence, particular traits.

All the organisms and their nonliving environment within a defined area.

Literally, "other-feeder"; an organism that eats other organisms; a consumer.

A substance that cannot be broken down, or converted, to a simpler substance by ordinary chemical means.

homeostasis	mutation
hypothesis	natural causality
inductive reasoning	natural selection
kingdom	nucleus
molecule	nutrient
multicellular	observation

A change in the DNA sequence of a gene.

The maintenance of a relatively constant internal environment that is required for the optimal functioning of cells.

The scientific principle that natural events occur as a result of preceding natural causes.

In the scientific method, a supposition based on previous observations that is offered as an explanation for an observed phenomenon and is used as the basis for further observations or experiments.

The unequal survival and reproduction of organisms due to environmental forces, resulting in the preservation of favorable adaptations. Usually, natural selection refers specifically to differential survival and reproduction on the basis of genetic differences among individuals.

The process of creating a generalization based on many specific observations that support the generalization, coupled with an absence of observations that contradict it.

The organelle of eukaryotic cells that contains the cell's genetic material.

The second broadest taxonomic category, contained within a domain and consisting of related phyla or divisions.

A substance acquired from the environment and needed for the survival, growth, and development of an organism.

A particle composed of one or more atoms held together by chemical bonds; the smallest particle of a compound that displays all the properties of that compound.

In the scientific method, the recognition of and a statement about a specific phenomenon, usually leading to the formulation of a question about the phenomenon.

Many-celled; most members of the kingdoms Fungi, Plantae, and Animalia are multicellular, with intimate cooperation among cells.

organ	plasma membrane
organ system	population
organelle	prediction
organic molecule	prokaryotic
organism	question
photosynthesis	scientific method

The outer membrane of a cell, composed of a bilayer of phospholipids in which proteins are embedded.

A structure (such as the liver, kidney, or skin) composed of two or more distinct tissue types that function together.

All the members of a particular species within an ecosystem, found in the same time and place and actually or potentially interbreeding.

Two or more organs that work together to perform a specific function; for example, the digestive system.

In the scientific method, a statement describing an expected observation or the expected outcome of an experiment, assuming that a specific hypothesis is true.

A membrane-enclosed structure found inside a eukaryotic cell that performs a specific function.

Referring to cells of the domains Bacteria or Archaea. Prokaryotic cells have genetic material that is not enclosed in a membrane-bound nucleus; they also lack other membrane-bound organelles.

A molecule that contains both carbon and hydrogen.

In the scientific method, a statement that identifies a particular aspect of an observation that a scientist wishes to explain.

An individual living thing.

A rigorous procedure for making observations of specific phenomena and searching for the order underlying those phenomena.

The series of chemical reactions in which the energy of light is used to synthesize high-energy organic molecules, usually carbohydrates, from low-energy inorganic molecules, usually carbon dioxide and water.

scientific theory	tissue
scientific theory of evolution	unicellular
species	variable

A group of (normally similar) cells that together carry out a specific function, for example, muscle; a tissue may also include extracellular material produced by its cells.

A general explanation of natural phenomena developed through extensive and reproducible observations; more general and reliable than a hypothesis.

Single-celled; most members of the domains Bacteria and Archaea and the kingdom Protista are unicellular.

The theory that modern organisms descended, with modification, from preexisting life-forms.

A factor in a scientific experiment that is deliberately manipulated in order to test a hypothesis.

The basic unit of taxonomic classification, consisting of a population or series of populations of closely related and similar organisms. In sexually reproducing organisms, a species can be defined as a population or series of populations of organisms that interbreed freely with one another under natural conditions but that do not interbreed with members of other species.

SELF TEST

1. Science assumes that natural laws (such as the law of gravity)
   a. apply uniformly through space and through time.
   b. apply in the laboratory but not in nature.
   c. change with time.
   d. differ depending on the location of the observer.

2. Medical researchers are testing a new drug to see whether it will lower blood pressure in humans. In addition to 20 experimental test subjects who receive injections of the actual drug, the researchers also inject a neutral saline solution into 20 other people. No one is told who has received the real drug. Why do the researchers do this?
   a. The real drug is very expensive, and the researchers only have enough of it to give to 20 people.
   b. The 20 people who receive the neutral saline injection will provide a reference value for blood pressure against which the effectiveness of the real drug, given to the other 20 people, may be judged.
   c. To increase the overall sample size, so that experimental sampling error will be reduced.
   d. The people who receive the neutral saline injection will serve as "backups" for the study, in case the real drug accidentally sickens or even kills the test subjects who receive it.

3. A group of researchers developed a hypothesis and tested it by designing an experiment. The results of the experiment did not support the original hypothesis. What should the researchers do next?
   a. Continue developing experiments to test the hypothesis.
   b. Throw out the results that did not support their hypothesis.
   c. Reject the original hypothesis and develop a new hypothesis.
   d. Conduct an experiment without a control.

4. Plants have numerous small openings called stomata in their leaves, which they use to exchange gases with the atmosphere. They are able to open or close stomata as they need. Specifically, they open the stomata to allow entry of carbon dioxide from the atmosphere into the interior of their leaves, where they carry out photosynthesis. They also close stomata to limit their rate of water loss to the atmosphere. Suppose that a researcher wants to study the possible effect of ozone air pollution on the number of stomata that a plant has. Which statement below represents a hypothesis dealing with this situation?
   a. Plants growing in polluted areas appear to have stunted growth and damaged leaves.
   b. Plants grown with high exposure to ozone had 10 stomata per square millimeter of leaf, whereas plants grown with no exposure to ozone had 15 stomata per square millimeter of leaf.
   c. Plants are grown inside two enclosures, one in which they are exposed to clean air and one in which they are exposed to ozone-polluted air.
   d. Because ozone can damage living tissue, plants grown in high-ozone locations will have fewer stomata, as a way of limiting harmful ozone penetration into the interior of their leaves.

5. A control is needed in an experiment to
   a. provide a comparison with the experimental results obtained when changing the variable.
   b. keep scientists from pursuing unethical questions and practices.
   c. increase the complexity of the experiment.
   d. duplicate the results of the experiment.

6. In a famous experiment that helped disprove the possibility of spontaneous generation, Francesco Redi hypothesized that contrary to popular belief, maggots did not appear spontaneously but came from flies. He predicted that if flies were kept away from meat, then maggots would not appear. To test his hypothesis, Redi placed identical pieces of meat in two identical jars and placed the two jars on the same windowsill. He placed a lid on one jar and left the other uncovered for several days. In the experiment the control is the
   a. jar.
   b. meat in the jar.
   c. time left on the windowsill.
   d. meat in the jar with the lid.

7. In his study of widowbird mating habits, Malte Andersson made experimental changes to the tail lengths of male widowbirds, and tracked their subsequent mating success. He used two control groups of birds, one of which had no changes made, and another of which had the tail feathers cut off and then glued back into place. Also, he had two experimental groups of birds, one of which had the tail feathers cut to much shorter lengths, and another onto which Andersson glued feather extensions to make the tails much longer. He then counted how many nests the female birds built in each

male's territory. In this experiment, what is being tested?

a. the effect of male tail length on female mate selection
b. how effectively males can drive other males out of their territory
c. effectiveness of different types of glue in reattaching feathers that have been cut off
d. whether females build larger nests when they mate with males that have longer tails

8. Simple experiments generally isolate and test one _____ at a time.

a. hypothesis
b. observation
c. control
d. variable

9. What is true about biodiversity and extinction rates?

a. Although extinction rates have risen in some areas because of human activities, they have fallen in others, so the overall global extinction rate has only changed slightly as a result of human activities.
b. Species within communities seldom have a strong effect on other species in the same communities, so if one species becomes extinct, it will have little effect on others.
c. Most species exist in the temperate (midlatitude) and polar regions, so these are the areas where most recent extinctions have occurred.
d. Because of human activities, global extinction rates are now many times higher than they were before humans became a major force on the planet.

10. The three natural processes that form the basis for evolution are

a. adaptation, natural selection, and inheritance.
b. predation, genetic variation, and natural selection.
c. mutation, genetic variation, and adaptation.
d. fossils, natural selection, and adaptation.
e. genetic variation, inheritance, and natural selection.

11. Which statement is TRUE?

a. Eukaryotic cells are simpler than prokaryotic cells.
b. Heterotroph means "self-feeder."
c. Mutations are accidental changes in genes.
d. A scientific theory is similar to an educated guess.
e. Genes are proteins that produce DNA.

12. The diversity of life is mainly due to

a. atoms.
b. genetic variation.

c. prokaryotic cells.
d. organ systems.

13. Natural selection would be best illustrated by which of the following?

a. a bacterial cell in the human body that dies when a person takes antibiotics
b. a bacterial cell in the human body with a genetic variation that allows it to survive when the person takes antibiotics
c. a bacterial cell in the human body that mutates when a person takes antibiotics
d. immune cells in the human body that learn from their first encounter with a particular kind of bacteria, and fight it much more effectively the next time a person becomes infected with the same type of bacteria

14. A fundamental characteristic of life on Earth is that

a. living things have a complex, organized structure based on inorganic molecules.
b. living things passively acquire materials and energy from their environment and convert them into different forms.
c. living things grow and reproduce.
d. living things reproduce using information stored in RNA.
e. energy cycles constantly between different living organisms, and does not need to be replaced.

15. You leave your house on a cold morning and walk down the street. To your shock, you find a man lying on the sidewalk, possibly injured or dead. You shout at and shake the person, but he does not move or attempt to speak. He is also very cold to the touch. Strictly from these two observations, two of the ordinary aspects of life that you tentatively conclude are no longer present are

a. growth and response to environmental stimuli.
b. response to environmental stimuli and maintenance of homeostasis.
c. maintenance of homeostasis and a complex structure composed of cells.
d. a complex structure composed of cells and utilization of materials from the environment.
e. utilization of materials from the environment and growth.

16. Identify a characteristic that is representative of some but not all life-forms.

a. change over time in order to adapt to changing environments
b. acquisition and use of energy
c. maintenance of a constant internal environment despite changing external environments
d. movement over great distances

17. The organic complexity and organization of living organisms depends on the periodic capture of energy from the environment. Ultimately, the source of this energy is
    a. metabolism.
    b. photosynthesis.
    c. the sun.
    d. other life-forms.

18. Energy, such as gasoline for your car, is required for organisms to survive, and even thrive, in the face of diverse environments. An autotrophic organism would
    a. be one that derived its energy from internalizing the cellular matter of other organisms (i.e., eats others).
    b. be one that derived its energy from a renewable external energy source such as sunlight (i.e., photosynthetic organisms).
    c. include cucumbers growing in your garden.
    d. includes both b and c.

19. The process of homeostasis is
    a. the mechanism by which living organisms evolve.
    b. how cells produce energy.
    c. how organisms maintain a constant internal environment.
    d. the basis for the passing of inherited traits from one generation to the next.

20. The complexity of living systems is
    a. highly organized, compared to the nonliving aspects of the environment such as the rocks, air, water, and sunlight.
    b. equal in all life-forms.
    c. the single defining characteristic of living things.
    d. found only in humans.

21. Your textbook lists seven characteristics that living organisms possess as a group. However, if you could distill these seven characteristics down to two general descriptive properties exclusive to all living organisms, what would they be?
    a. macromolecular complexity and multiple levels of organization
    b. atoms and elements
    c. cellular structure and RNA-based inheritance
    d. chemical relationships and diversity

22. In general, plants do not
    a. sense or respond to their environments.
    b. eat their food.
    c. produce reproductive structures.
    d. have any visible structures inside of their cells.

23. If an organism is multicelled, has cells with well-defined nuclei, and is an autotroph, it could be a member of the kingdom _____, and is very unlikely to be a member of the kingdom _____, instead.
    a. Plantae; Animalia
    b. Animalia; Fungi
    c. Fungi; Protista
    d. Protista; Plantae

24. What is the correct format for the binomial name of the grizzly bear?
    a. Ursus Arctos
    b. Ursus arctos
    c. *Ursus Arctos*
    d. *Ursus arctos*

25. A scientist examines an organism and finds that it is eukaryotic, heterotrophic, and multicellular, and it absorbs nutrients. She concludes that the organism is most likely a member of the kingdom
    a. Bacteria.   d. Fungi.
    b. Protista.   e. Animalia.
    c. Plantae.

26. You discover a new type of organism in the back of your fridge. Luckily, your roommate is a biology major and takes you to the lab where he works. You put a small piece of the fuzzy critter under the microscope and see that it is made of very simple single cells without a nucleus. What type of organism is this MOST likely to be?
    a. Bacterium   c. Protistan
    b. Animal   d. Fungus

27. Organisms that can make their own food are called _____; organisms that must obtain energy from molecules made by other organisms are called _____.
    a. herbivores; carnivores
    b. photosynthetic; herbivores
    c. heterotrophs; autotrophs
    d. autotrophs; heterotrophs

28. Which is almost certainly not a eukaryotic cell?
    a. a cell that is joined with many others of the same species, which take on various specialized roles and develop distinct types of tissues and organs
    b. a cell that is most similar to some of the earliest types of life-forms to inhabit the earth
    c. a cell with many distinct internal structures, plainly visible using an ordinary light microscope
    d. a cell that is approximately 100 micrometers in diameter

29. An understanding of basic biological concepts
  a. permits a deeper, and sometimes profound, appreciation of the world around us.
  b. provides just a set of facts and ideas about the world around us.
  c. is necessary only for biology majors.
  d. often decreases our appreciation (i.e., effectively dehumanizes) of the world around us.

ANSWER KEY

1. a		16. d
2. b		17. c
3. c		18. d
4. d		19. c
5. a		20. a
6. d		21. a
7. a		22. b
8. d		23. a
9. d		24. d
10. e		25. d
11. c		26. a
12. b		27. d
13. b		28. b
14. c		29. a
15. b		

CHAPTER 2 ATOMS, MOLECULES, AND LIFE

OUTLINE

Section 2.1 What Are Atoms?

Atoms are the fundamental structural units of matter. Each atom has a central atomic nucleus, which contains positively charged **protons** and uncharged **neutrons**. Orbiting the atomic nucleus are negatively charged **electrons** (**Figure 2-1**).

Atoms are grouped into 92 naturally occurring types (or **elements**) based on the number of protons in the nucleus (i.e., the **atomic number**). Certain elements are essential in life forms (**Table 2-1**). Atoms of the same element may have different numbers of neutrons and are referred to as **isotopes**.

Electrons orbit the nucleus within **electron shells**, each corresponding to a higher energy level. The innermost shell can hold two electrons, whereas higher-energy shells of the atoms of most biologically relevant elements can hold up to eight (**Figure 2-2**).

Exciting an electron causes it to jump to a higher-energy shell, which releases energy when it falls back to its original shell position (**Figure 2-3**).

Section 2.2 How Do Atoms Interact to Form Molecules?

Molecules are composed of two or more atoms of the same or different elements, whereas a **compound** is made up of atoms of different elements.

Stable (or inert) atoms have their outermost electron shells filled to capacity. An atom is reactive when its outermost electron shell is not full and will attempt to gain stability by gaining, losing, or sharing electrons with other atoms to fill its outer shell (i.e., it will form **chemical bonds**). Chemical bonds are formed (or broken) as a result of chemical reactions that form new substances. The major types of chemical bonds are **ionic bonds**, **covalent bonds**, and **hydrogen bonds** (**Table 2-2**).

Free radicals can damage biological molecules, by either stealing electrons from them or by donating electrons to them (**Figure 2-4**). Free radicals are neutralized by some important dietary **antioxidants** such as vitamins C and E.

Ionic bonds occur when one atom donates an electron to another atom, each filling its outermost electron shell as a result. Doing so causes the donating atom to become positively charged, while the receiving atom becomes negatively charged. These **ions** are then held together by their electrical attraction to each other (**Figure 2-5**).

Covalent bonds occur when atoms share electrons among their outermost shells (**Figure 2-6**). Depending on the atoms involved, covalent bonds can involve equal sharing of electrons (i.e., **nonpolar**, **Figure 2-6a**) or unequal sharing of electrons (i.e., **polar**, **Figure 2-6b**). Most biological molecules utilize covalent bonds.

Hydrogen bonds occur when the negatively charged end of one polar molecule is electrically attracted to the positively charged end (containing hydrogen atoms) of another (**Figure 2-7**).

Section 2.3 Why Is Water So Important to Life?

Water acts as an effective **solvent** that can dissolve molecules held together by ionic bonds (**Figure 2-9**) and polar covalent bonds. These molecules are called **hydrophilic**. Molecules with nonpolar covalent bonds typically do not dissolve in water and are called **hydrophobic**.

Polar water molecules have high **cohesion** due to the hydrogen bonds that form among them, resulting in high **surface tension** at the water's surface. Cohesion allows plants to transport substances through their tissues.

When water dissociates, hydrogen ions and hydroxide ions are released in equal proportions, forming a neutral **solution** (**Figure 2-11**). Water can react with **acids** to form an excess of hydrogen ions in solution (an **acidic** solution) or can react with **bases** to form an excess of hydroxide ions in solution (a **basic** solution). The degree of acidity is represented on a **pH scale** (**Figure 2-12**).

A **buffer** is a substance that accepts or releases hydrogen ions in response to changes in hydrogen ion concentrations, thus stabilizing pH levels.

Water moderates the effects of temperature changes on the body because it requires a lot of energy to heat water (a high **specific heat**), evaporate water (a high **heat of vaporization**), and freeze water (a high **heat of fusion**).

Water forms ice that becomes less dense when it solidifies (**Figure 2-14**). This ensures that ice will form at the surface of a body of water, forming an insulating layer that delays continued freezing of the water beneath it.

FLASH CARDS

To use the flash cards, tear the page from the book and cut along the dashed lines. The key term appears on one side of the flash card, and its definition appears on the opposite side.

acid	atomic nucleus
acidic	atomic number
adhesion	base
antioxidant	basic
atom	buffer
atomic mass	chemical bond

The central part of an atom that contains protons and neutrons.

A substance that releases hydrogen ions (H^+) into solution; a solution with a pH less than 7.

The number of protons in the nuclei of all atoms of a particular element.

Referring to a solution with an H^+ concentration exceeding that of OH^-; referring to a substance that releases H^+.

A substance capable of combining with and neutralizing H^+ ions in a solution; a solution with a pH greater than 7.

The tendency of polar molecules (such as water) to adhere to polar surfaces (such as glass).

Referring to a solution with an H^+ concentration less than that of OH^-; referring to a substance that combines with H^+.

Any molecule that reacts with free radicals, neutralizing their ability to damage biological molecules. Vitamins C and E are examples of dietary antioxidants.

A compound that minimizes changes in pH by reversibly taking up or releasing H^+ ions.

The smallest particle of an element that retains the properties of the element.

An attraction between two atoms or molecules that tends to hold them together. Types of bonds include covalent, ionic, and hydrogen.

The total mass of all the protons, neutrons, and electrons within an atom.

chemical reaction

electron shell

cohesion

element

compound

free radical

covalent bond

heat of fusion

dissolve

heat of vaporization

electron

hydrogen bond

A region in an atom within which electrons orbit; each shell corresponds to a fixed energy level at a given distance from the nucleus.

The process that forms and breaks chemical bonds that hold atoms together.

A substance that cannot be broken down, or converted, to a simpler substance by ordinary chemical means.

The tendency of the molecules of a substance to stick together.

A molecule containing an atom with an unpaired electron, which makes it highly unstable and reactive with nearby molecules. By removing an electron from the molecule it attacks, it creates a new free radical and begins a chain reaction that can lead to the destruction of biological molecules crucial to life.

A substance whose molecules are formed by different types of atoms; can be broken into its constituent elements by chemical means.

The energy that must be removed from a compound to transform it from a liquid into a solid at its freezing temperature.

A chemical bond between atoms in which electrons are shared.

The energy that must be supplied to a compound to transform it from a liquid into a gas at its boiling temperature.

To disperse completely and uniformly; usually referring to molecules of one substance, called a solute, becoming evenly dispersed in another substance, called a solvent.

The weak attraction between a hydrogen atom that bears a partial positive charge (due to polar covalent bonding with another atom) and another atom (oxygen, nitrogen, or fluorine) that bears a partial negative charge; hydrogen bonds may form between atoms of a single molecule or of different molecules.

A subatomic particle, found in an electron shell outside the nucleus of an atom, that bears a unit of negative charge and very little mass.

hydrophilic

molecule

hydrophobic

neutron

hydrophobic interaction

nonpolar covalent bond

ion

pH scale

ionic bond

polar covalent bond

isotope

proton

A particle composed of one or more atoms held together by chemical bonds; the smallest particle of a compound that displays all the properties of that compound.

Pertaining to molecules that dissolve readily in water, or to molecules that form hydrogen bonds with water; polar.

A subatomic particle that is found in the nuclei of atoms, bears no charge, and has a mass approximately equal to that of a proton.

Pertaining to molecules that do not dissolve in water or form hydrogen bonds with water; nonpolar.

A covalent bond with equal sharing of electrons.

The tendency for hydrophobic molecules to cluster together when immersed in water.

A scale, with values from 0 to 14, used for measuring the relative acidity of a solution; at pH 7 a solution is neutral, pH 0 to 7 is acidic, and pH 7 to 14 is basic; each unit on the scale represents a tenfold change in H^+ concentration.

A charged atom or molecule; an atom or molecule that has either an excess of electrons (and, hence, is negatively charged) or has lost electrons (and is positively charged).

A covalent bond with unequal sharing of electrons, such that one atom is relatively negative and the other is relatively positive.

A chemical bond formed by the electrical attraction between positively and negatively charged ions.

A subatomic particle that is found in the nuclei of atoms; it bears a unit of positive charge, and has a relatively large mass, roughly equal to the mass of the neutron.

One of several forms of a single element, the nuclei of which contain the same number of protons but different numbers of neutrons.

radioactive

specific heat

solution

surface tension

solvent

The amount of energy required to raise the temperature of 1 gram of a substance by 1°C.

Pertaining to an atom with an unstable nucleus that spontaneously disintegrates, with the emission of radiation.

The property of a liquid to resist penetration by objects at its interface with the air, due to cohesion between molecules of the liquid.

A solvent containing one or more dissolved substances.

A liquid capable of dissolving (uniformly dispersing) other substances in itself.

SELF TEST

1. The basic structural units of chemistry and life are
   a. atoms.
   b. electrons.
   c. protons.
   d. neutrons.
   e. molecules.

2. Which of the following list of terms is in the correct order of size, from smallest to largest?
   a. electron, proton, atomic nucleus, electron shell, atom, molecule
   b. proton, electron, atomic nucleus, electron shell, atom, molecule
   c. molecule, atom, electron shell, atomic nucleus, proton, electron
   d. atomic nucleus, electron, proton, electron shell, atom, molecule
   e. electron, proton, atomic nucleus, electron shell, molecule, atom

3. A typical oxygen atom has 8 protons, 8 neutrons, and 8 electrons. The fundamental fact that makes it an atom of oxygen, rather than an atom of some other element, is the fact that it has
   a. 8 protons.
   b. 8 neutrons.
   c. 8 electrons.
   d. a total of 16 heavy particles (the protons and neutrons).

4. What is the best description of the arrangement of subatomic particles within a typical atom?
   a. Neutrons occupy the neutrally charged nucleus, whereas the positively charged protons orbit close to the nucleus and the negatively charged electrons orbit farther away from the nucleus.
   b. The massive, positively charged protons and neutral neutrons occupy the central nucleus, whereas the lighter, negatively charged electrons orbit the nucleus at varying distances from it.
   c. The positive protons and negative electrons are bonded to each other in the nucleus, whereas the neutrons are repelled by both and are forced to orbit at a considerable distance from the nucleus.
   d. The nucleus always contains neutrons, but will also contain protons whenever the atom becomes a positively charged ion, and will contain electrons instead whenever the atom becomes a negatively charged ion.

5. The atomic number of carbon is 6. A carbon atom has _____ protons and _____ electrons.
   a. 3; 3
   b. 6; 6
   c. 6; 12
   d. 6; 3

6. Consider the chapter's discussion of electron shells, and how many electrons it takes to fill each of them. For a neutrally charged atom of sulfur, how many electrons would it have in its outer shell?
   a. 2
   b. 4
   c. 6
   d. 8
   e. 10

Table 2-1   **Common Elements in Living Organisms**

Element	Atomic Number[1]	Atomic Mass[2]	% by Weight in the Human Body[3]
Oxygen (O)	8	16	65
Carbon (C)	6	12	18.5
Hydrogen (H)	1	1	9.5
Nitrogen (N)	7	14	3.0
Calcium (Ca)	20	40	1.5
Phosphorus (P)	15	31	1.0
Potassium (K)	19	39	0.35
Sulfur (S)	16	32	0.25
Sodium (Na)	11	23	0.15
Chlorine (Cl)	17	35	0.15
Magnesium (Mg)	12	24	0.05
Iron (Fe)	26	56	Trace
Fluorine (F)	9	19	Trace
Zinc (Zn)	30	65	Trace

[1]Atomic number: number of protons in the atomic nucleus.
[2]Atomic weight: total mass of protons, neutrons, and electrons (negligible); numbers are rounded.
[3]Approximate percentage of this element, by weight, in the human body.

7. Isotopes are
   a. atoms with equal numbers of protons and electrons.
   b. atoms with unequal numbers of protons and electrons.
   c. multiple atoms with the same number of protons but different numbers of neutrons.
   d. atoms that are of the same element and that have the same weight, but different levels of electrical charge.

8. Elements are
   a. composed of molecules.
   b. unique forms of matter.
   c. found only in living matter.
   d. found only in nonliving matter.

9. If a single carbon atom has formed four covalent bonds with four surrounding hydrogen atoms, how many electrons, in total, is it sharing with the hydrogen atoms?
   a. 0
   b. 2
   c. 4
   d. 8

10. Nonpolar covalent bonds are different from polar covalent bonds because
   a. electrons are shared unequally in nonpolar covalent bonds and are shared equally in polar covalent bonds.

b. electrons are shared equally in nonpolar co-valent bonds and are shared unequally in polar covalent bonds.

c. electrons are lost in nonpolar covalent bonds and are gained in polar covalent bonds.

d. electrons are shared in nonpolar covalent bonds and are lost or gained in polar cova-lent bonds.

11. Oxygen atoms have an atomic number of 8. Neon atoms have 10 electrons. Which answer predicts whether or not these atoms are generally reactive (i.e., can form chemical bonds with other atoms)? (Note that this question is not asking whether oxygen can react with neon.)

a. Both oxygen and neon are not reactive because their outermost electron orbitals are filled.

b. Both oxygen and neon are reactive because their outermost electron orbitals are not filled.

c. Oxygen is reactive because its outermost elec-tron shell contains 6 electrons (is not filled); neon is not reactive because its outermost electron shell contains 8 electrons (is filled).

d. Oxygen is not reactive because its outer-most electron shell contains 8 electrons (is filled); neon is reactive because its outer-most electron shell contains 2 electrons (is not filled).

12. Imagine that you wanted to make a time capsule in which you would seal important artifacts from your life (pictures, poems, a lock of your baby hair, etc.), to be opened by your heirs 1,000 years from now. To prevent these artifacts from decaying, you want to fill the capsule with a gas that would be *least* reactive. Which of these gases would you choose: oxygen gas (O_2), car-bon dioxide (CO_2), argon gas (Ar), or hydrogen gas (H_2)? (The atomic numbers of the atoms in these molecules are oxygen = 8, carbon = 6, hy-drogen = 1, and argon = 18).

a. oxygen gas   c. argon gas
b. carbon dioxide d. hydrogen gas

13. What happens when an atom ionizes?
a. It shares one or more electrons with another atom.
b. It emits energy as it loses extra neutrons.
c. It gives up or takes up one or more electrons.
d. It shares a hydrogen atom with another atom.

14. Which is an important source of damaging free radicals?
a. oxygen-consuming metabolic processes
b. chocolate
c. vegetables, such as spinach
d. fruits, such as oranges
e. multivitamin pills

15. The chemical bonding properties of an atom are determined most directly by the
a. nucleus.
b. number of protons.
c. number of neutrons.
d. outer shell of electrons.

16. Which of the following is NOT a characteristic of covalent bonds?
a. result when atoms gain or lose one or more electrons
b. result when atoms share one or more electrons
c. are interactions between the outermost electron shells of atoms
d. are stronger than ionic bonds in water

17. Neutrally charged (nonionized) atoms of potassium (K) possess 19 electrons, whereas chlorine (Cl) atoms possess 17 electrons, sodium (Na) atoms possess 11 electrons, calcium (Ca) atoms possess 20 electrons, and magnesium (Mg) atoms possess 12 electrons. With which other element is potassium most likely to form ionic bonds?
a. chlorine   c. calcium
b. sodium   d. magnesium

18. What allows one atom to physically interact with a second atom?
a. properties of both nuclei
b. properties of the electrons
c. electron shells of both atoms
d. external energy sources

19. Generally speaking, the number of electrons in the outer electron shell of an isolated atom (one that is not bonded to another atom) is determined by
a. the number of neutrons in the nucleus.
b. the number of electrons contained within the inner electron shells.
c. the total number of heavy particles (protons and neutrons) present in the nucleus.
d. the number of protons in the nucleus.

20. An atom other than hydrogen has a single elec-tron in its outermost shell. The MOST likely outcome for this atom in terms of chemical bonding is that the atom will
a. gain electrons to fill its outermost shell, becoming a negatively charged ion.
b. lose its outermost electron, becoming an ion with a charge of $+1$.
c. share its single electron with another atom.
d. share enough electrons to fill its outermost shell.

21. In liquid water, which kinds of chemical bonds do we see?
    a. Hydrogen bonds form within individual water molecules, whereas nonpolar covalent bonds form between different water molecules.
    b. Nonpolar covalent bonds form within individual water molecules, whereas polar covalent bonds form between different water molecules.
    c. Polar covalent bonds form within individual water molecules, whereas ionic bonds form between different water molecules.
    d. Ionic bonds form within individual water molecules, whereas hydrogen bonds form between different water molecules.
    e. Polar covalent bonds form within individual water molecules, whereas hydrogen bonds form between different water molecules.

22. Why are hydrophobic molecules, such as fats and oils, unable to dissolve in watery solutions?
    a. Water cannot interact with molecules that have polar covalent bonds, as do fats and oils.
    b. Water cannot interact with molecules with ionic bonds, such as fats and oils.
    c. Water cannot interact with hydrophobic molecules, such as fats and oils, because hydrophobic molecules form hydrogen bonds with each other, excluding the water.
    d. Water molecules form hydrogen bonds with each other, excluding the hydrophobic molecules.

23. You are waiting backstage for your cue to come onstage when you notice that you are breathing rapidly and beginning to feel light-headed. As you try to control your anxiety and slow your breathing, you think about what you learned in your biology class this week and realize that your hyperventilation is changing your blood pH. Explain. Reminder: Blood pH is maintained by carbonate buffer, which is related to the amount of CO_2 you breathe in or out. One way that your body controls the amount of carbonate in your blood is to change the rate of breathing. When you breathe out, you remove and lower the amount of carbonic acid in the blood. As a result, the number of H^+ ions in the blood decreases. Take your time on this one; relating pH to changes in H^+ concentration can be confusing.
    a. Rapid breathing decreases my blood's pH, making it more basic.
    b. Rapid breathing decreases my blood's pH, making it more acidic.
    c. Rapid breathing increases the pH of my blood, making it more basic.
    d. Rapid breathing increases the pH of my blood, making it more acidic.

24. Water's ability to act as a "universal" solvent is due to
    a. the fact that there is so much of it in the world around us and in our own bodies.
    b. its natural ability to interact with polar molecules such as ions and proteins.
    c. the nature of oxygen, which pulls the hydrogen electrons a little closer to it than they are to the two hydrogen atoms.
    d. both b and c.

25. Water greatly resists increases in temperature because it takes great energy to break the
    a. ionic bonds that hold water molecules together.
    b. covalent bonds that hold water molecules together.
    c. huge number of hydrogen bonds that hold different water molecules together.
    d. pair of hydrogen bonds that holds the oxygen atom to the two hydrogen atoms, inside each water molecule.

26. A large, complex organic molecule contains a number of polar covalent bonds. Would you expect this molecule to be soluble in water? Why?
    a. No, I would not expect it to be soluble because it does not contain any ionic bonds.
    b. No, I would not expect it to be soluble because it contains polar covalent bonds.
    c. No, I would not expect it to be soluble because it is a large molecule.
    d. Yes, I would expect it to be soluble because it should be hydrophilic.

27. Which substance should be the most resistant to dissolving in water?
    a. pentane, a hydrocarbon with 5 carbon atoms and 12 hydrogen atoms
    b. ethanol, consisting of carbon, hydrogen, and oxygen, with one hydrogen bonded to the oxygen
    c. crystals of magnesium chloride, where positively charged magnesium ions are bonded to negatively charged chlorine ions
    d. proteins, which include many hydrogen atoms bonded to nitrogen atoms

28. One solution has a pH of 2 and another has a pH of 3. What would be the relative difference in ion concentration?
    a. twofold difference in concentration
    b. fivefold difference in concentration
    c. tenfold difference in concentration
    d. It is impossible to predict the exact difference in concentration.

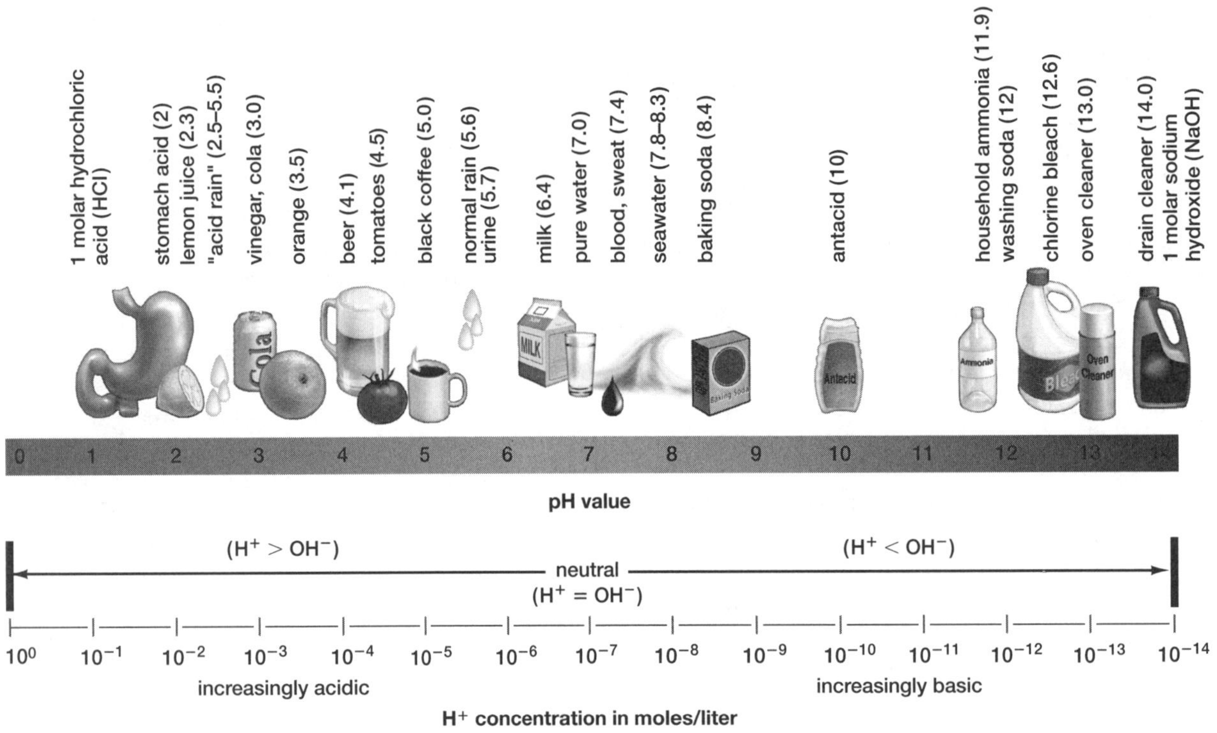

pH value

(H$^+$ > OH$^-$)   (H$^+$ < OH$^-$)

neutral
(H$^+$ = OH$^-$)

| 10^0 | 10^{-1} | 10^{-2} | 10^{-3} | 10^{-4} | 10^{-5} | 10^{-6} | 10^{-7} | 10^{-8} | 10^{-9} | 10^{-10} | 10^{-11} | 10^{-12} | 10^{-13} | 10^{-14} |

increasingly acidic   increasingly basic

H$^+$ concentration in moles/liter

29. Examine the pH scale. What is true about the hydrogen ion (H$^+$) concentration in stomach acid, compared to the H$^+$ concentration in pure water?
   a. Stomach acid has 5 times the concentration of H$^+$ ions that pure water does.
   b. Stomach acid has 50 times the concentration of H$^+$ ions that pure water does.
   c. Stomach acid has 100,000 times the concentration of H$^+$ ions that pure water does.
   d. Stomach acid has 1/50 as high a concentration of H$^+$ ions as pure water does.
   e. Stomach acid has 1/100,000 as high a concentration of H$^+$ ions as pure water does.

30. The high specific heat of water means that living systems are
   a. more dense than nonliving systems.
   b. more resistant to changes in temperature.
   c. made of solid water.
   d. unable to cool themselves by evaporation of water.

ANSWER KEY

1. a	16. a
2. a	17. a
3. a	18. b
4. b	19. d
5. b	20. b
6. c	21. e
7. c	22. d
8. b	23. c
9. d	24. d
10. b	25. c
11. c	26. d
12. c	27. a
13. c	28. c
14. a	29. c
15. d	30. b

CHAPTER 3 BIOLOGICAL MOLECULES

OUTLINE

Section 3.1 Why Is Carbon So Important in Biological Molecules?

Organic molecules are composed of both carbon and hydrogen atoms. All other molecules are considered **inorganic**.

Because of the versatility of the carbon atoms present in them, organic molecules have a tremendous variety of structures and functions. A carbon atom may form up to four covalent bonds with other atoms and can thus assume many sizes and complex shapes.

Functional groups of atoms can attach an organic molecule to a carbon backbone and affect its reactivity. Different functional groups impart different chemical properties to organic molecules (**Table 3-1**).

Section 3.2 How Are Organic Molecules Synthesized?

Organic molecules are composed of small subunits (**monomers**) bonded together to form longer chains (**polymers**).

Monomers are joined together to form a polymer by **dehydration synthesis**, whereas polymers are split into individual monomers by **hydrolysis** (**Figures 3-1** and **3-2**).

Most biological molecules fall into one of four different categories: carbohydrates, lipids, proteins, or nucleic acids (**Table 3-2**).

Section 3.3 What Are Carbohydrates?

Carbohydrates are made of carbon, hydrogen, and oxygen in a 1:2:1 ratio. All carbohydrates are composed of one or more **sugar** monomers. Small sugar molecules such as **glucose** dissolve easily in water (**Figure 3-3**).

Monosaccharides are carbohydrates composed of a single sugar molecule, with glucose being the most common form (**Figure 3-4**). Other monosaccharides include *fructose, galactose, ribose,* and *deoxyribose* (**Figures 3-5** and **3-6**).

Disaccharides are carbohydrates composed of two sugar monomers joined by dehydration synthesis (**Figure 3-7**). **Sucrose, lactose,** and **maltose** are common disaccharides.

Polysaccharides are carbohydrates composed of many sugar monomers. **Starch** is an energy-storage polysaccharide formed by plants (**Figure 3-8**), whereas **glycogen** is the animal equivalent. **Cellulose** is an important structural carbohydrate in plants (**Figure 3-9**), whereas the exoskeleton of some invertebrate animals is composed of **chitin** (**Figure 3-10**).

Section 3.4 What Are Lipids?

Lipids contain long regions made almost entirely of carbon and hydrogen that are joined by nonpolar covalent bonds. Lipids are hydrophobic and can act as energy-storage molecules, contribute to cell membrane structure, waterproof plant and animal surfaces, and form hormones.

Lipids fall into one of three categories: (1) **fats, oils,** and **waxes;** (2) **phospholipids;** and (3) **steroids**.

Fats, oils, and waxes are lipids that have three structural features in common: (1) They contain only carbon, hydrogen, and oxygen; (2) they contain at least one **fatty acid** subunit; and (3) they usually do not form ring structures.

Fats and oils (also called **triglycerides**) form by the dehydration synthesis of three fatty acid molecules and one **glycerol** molecule (**Figure 3-12**). These lipids contain a large amount of chemical energy and are used as a long-term energy-storage molecule in animals. Fats are solid at room temperature because their fatty acids are **saturated** with hydrogen atoms and thus lack double bonds between carbon atoms. Oils are liquid at room temperature because their fatty acids are **unsaturated** with hydrogen atoms, meaning there are double bonds between some carbon atoms (**Figure 3-13**).

Waxes are not a food source. Their function is to form waterproof coverings over various plant and animal structures.

Phospholipids are lipids composed of a phosphate "head," glycerol backbone, and two fatty acid "tails" (**Figure 3-14**). The phosphate head is polar and is thus hydrophilic, whereas the fatty acid tails are not. Phospholipids form the bulk of the cell plasma membrane.

Steroids are composed of four carbon rings fused together, with various functional groups attached (**Figure 3-15**). Steroids form cholesterol as well as some hormones.

Section 3.5 What Are Proteins?

Because of the diversity of their structures, **proteins** can perform many diverse functions (**Table 3-3**). Proteins are composed of one or more **amino acid** monomers (**Figures 3-17** and **3-18**), which are joined by **peptide bonds** during dehydration synthesis (**Figure 3-19**).

Proteins can have up to four structural levels. The simplest structure, a chain of amino acids, is called the **primary structure**. The amino acid chain can be twisted into a coiled **helix** or a **pleated sheet**, resulting in a **secondary structure**. The helix can be folded upon itself in three dimensions, resulting in a **tertiary structure**. Multiple tertiary structure polypeptides can be joined to form one large protein molecule called a **quaternary structure** (**Figures 3-20** and **3-21**).

The function of a protein is dependent on its three-dimensional shape, so if the shape is altered (i.e., **denatured**), its original function is disrupted.

Section 3.6 What Are Nucleotides and Nucleic Acids?

Nucleic acids are composed of chains of **nucleotide** monomers (**Figure 3-22**), each consisting of a phosphate, a five-carbon sugar, and one of several nitrogen-containing **bases**.

Nucleotides perform a variety of useful functions. Cyclic nucleotides play a role in intercellular communication, **adenosine triphosphate** (or **ATP**, **Figure 3-23**) molecules carry energy from place to place within a cell, and coenzymes facilitate cellular metabolism.

Deoxyribonucleic acid (or **DNA**) is a double-stranded polymer composed of nucleotides containing the sugar deoxyribose. DNA spells out the genetic information used to construct the proteins of each organism (**Figure 3-24**).

Ribonucleic acid (or **RNA**) is a single-stranded polymer composed of nucleotides containing the sugar ribose. RNA carries the code from DNA into the cell cytoplasm and directs protein synthesis.

FLASH CARDS

To use the flash cards, tear the page from the book and cut along the dashed lines. The key term appears on one side of the flash card, and its definition appears on the opposite side.

adenosine triphosphate (ATP)	dehydration synthesis
amino acid	denatured
base	deoxyribonucleic acid (DNA)
carbohydrate	disaccharide
cellulose	disulfide bond
chitin	enzyme

A chemical reaction in which two molecules are joined by a covalent bond with the simultaneous removal of a hydrogen from one molecule and a hydroxyl group from the other, forming water; the reverse of hydrolysis.

A molecule composed of the sugar ribose, the base adenine, and three phosphate groups; the major energy carrier in cells. The last two phosphate groups are attached by "high-energy" bonds.

Having the secondary and/or tertiary structure of a protein disrupted, while leaving the amino acid sequence unchanged. Denatured proteins can no longer perform their biological functions.

The individual subunit of which proteins are made, composed of a central carbon atom bonded to an amino group ($-NH_2$), a carboxyl group ($-COOH$), a hydrogen atom, and a variable group of atoms denoted by the letter R.

A molecule composed of deoxyribose nucleotides; contains the genetic information of all living cells.

One of the nitrogen-containing, single- or double-ringed structures that distinguishes one nucleotide from another. In DNA, the bases are adenine, guanine, cytosine, and thymine.

A carbohydrate formed by the covalent bonding of two monosaccharides.

A compound composed of carbon, hydrogen, and oxygen, with the approximate chemical formula $(CH_2O)_n$; includes sugars, starches, and cellulose.

The covalent bond formed between the sulfur atoms of two cysteines in a protein; typically causes the protein to fold by bringing otherwise distant parts of the protein close together.

An insoluble carbohydrate composed of glucose subunits; forms the cell wall of plants.

A protein catalyst that speeds up the rate of specific biological reactions.

A compound found in the cell walls of fungi and the exoskeletons of insects and some other arthropods; composed of chains of nitrogen-containing, modified glucose molecules.

fat	helix
fatty acid	hydrolysis
functional group	inorganic
glucose	lactose
glycerol	lipid
glycogen	maltose

A coiled, springlike secondary structure of a protein.

A lipid composed of three saturated fatty acids covalently bonded to glycerol; fats are solid at room temperature.

The chemical reaction that breaks a covalent bond by means of the addition of hydrogen to the atom on one side of the original bond and a hydroxyl group to the atom on the other side; the reverse of dehydration synthesis.

An organic molecule composed of a long chain of carbon atoms, with a carboxylic acid (COOH) group at one end; may be saturated (all single bonds between the carbon atoms) or unsaturated (one or more double bonds between the carbon atoms).

Describing any molecule that does not contain both carbon and hydrogen.

One of several groups of atoms commonly found in an organic molecule, including hydrogen, hydroxyl, amino, carboxyl, and phosphate groups, that determine the characteristics and chemical reactivity of the molecule.

A disaccharide composed of glucose and galactose; found in mammalian milk.

The most common monosaccharide, with the molecular formula $C_6H_{12}O_6$; most polysaccharides, including cellulose, starch, and glycogen, are made of glucose subunits covalently bonded together.

One of a number of organic molecules containing large nonpolar regions composed solely of carbon and hydrogen, which make lipids hydrophobic and insoluble in water; includes oils, fats, waxes, phospholipids, and steroids.

A three-carbon alcohol to which fatty acids are covalently bonded to make fats and oils.

A disaccharide composed of two glucose molecules.

A highly branched polymer of glucose that is stored by animals in the muscles and liver and metabolized as a source of energy.

monomer	peptide
monosaccharide	peptide bond
nucleic acid	phospholipid
nucleotide	pleated sheet
oil	polymer
organic	polysaccharide

A chain composed of two or more amino acids linked together by peptide bonds.

A small organic molecule, several of which may be bonded together to form a chain called a polymer.

The covalent bond between the nitrogen of the amino group of one amino acid and the carbon of the carboxyl group of a second amino acid, joining the two amino acids together in a peptide or protein.

The basic molecular unit of all carbohydrates, normally composed of a chain of carbon atoms bonded to hydrogen and hydroxyl groups.

A lipid consisting of glycerol bonded to two fatty acids and one phosphate group, which bears another group of atoms, typically charged and containing nitrogen. A double layer of phospholipids is a component of all cellular membranes.

An organic molecule composed of nucleotide subunits; the two common types of nucleic acids are ribonucleic acid (RNA) and deoxyribonucleic acid (DNA).

A form of secondary structure exhibited by certain proteins, such as silk, in which many protein chains lie side by side, with hydrogen bonds holding adjacent chains together.

A subunit of which nucleic acids are composed; a phosphate group bonded to a sugar (deoxyribose in DNA), which is in turn bonded to a nitrogen-containing base (adenine, guanine, cytosine, or thymine in DNA).

A molecule composed of three or more (perhaps thousands) smaller subunits called monomers, which may be identical (for example, the glucose monomers of starch) or different (for example, the amino acids of a protein).

A lipid composed of three fatty acids, some of which are unsaturated, covalently bonded to a molecule of glycerol; oils are liquid at room temperature.

A large carbohydrate molecule composed of branched or unbranched chains of repeating monosaccharide subunits, normally glucose or modified glucose molecules; includes starches, cellulose, and glycogen.

Describing a molecule that contains both carbon and hydrogen.

primary structure	starch
protein	steroid
quaternary structure	sucrose
ribonucleic acid (RNA)	sugar
saturated	tertiary structure
secondary structure	trans fat

A polysaccharide that is composed of branched or unbranched chains of glucose molecules; used by plants as a carbohydrate-storage molecule.

The amino acid sequence of a protein.

A lipid consisting of four fused carbon rings, with various functional groups attached.

A polymer composed of amino acids joined by peptide bonds.

A disaccharide composed of glucose and fructose.

The complex three-dimensional structure of a protein consisting of more than one peptide chain.

A simple carbohydrate molecule, either a monosaccharide or a disaccharide.

A molecule composed of ribose nucleotides, each of which consists of a phosphate group, the sugar ribose, and one of the bases adenine, cytosine, guanine, or uracil; involved in converting the information in DNA into protein; also the genetic material of some viruses.

The complex three-dimensional structure of a single peptide chain; held in place by disulfide bonds between cysteines.

Referring to a fatty acid with as many hydrogen atoms as possible bonded to the carbon backbone (therefore, a saturated fatty acid has no double bonds in its carbon backbone).

A type of fat produced during the process of hydrogenating oils that may increase the risk of heart disease. The fatty acids of trans fats include an unusual configuration of double bonds that is not normally found in fats of biological origin.

A repeated, regular structure assumed by a protein chain, held together by hydrogen bonds; for example, a helix.

triglyceride

wax

unsaturated

A lipid composed of fatty acids covalently bonded to long-chain alcohols.

A lipid composed of three fatty-acid molecules bonded to a single glycerol molecule.

Referring to a fatty acid with fewer than the maximum number of hydrogen atoms bonded to its carbon backbone (therefore, an unsaturated fatty acid has one or more double bonds in its carbon backbone).

SELF TEST

1. Characteristics of carbon that contribute to its ability to form an immense diversity of organic molecules include all of the following, except for its
   a. tendency to form covalent bonds.
   b. behavior of always forming bonds with four other atoms.
   c. capacity to form single and double bonds.
   d. ability to bond to form extensive, branched, or unbranched carbon skeletons.

2. *Organic* is a term we often see in common usage. To a chemist, organic compounds
   a. are compounds of carbon and hydrogen.
   b. are found only in health food stores.
   c. can be made only by living organisms.
   d. are based on a skeleton of oxygen.

3. Imagine a molecule that includes two carbon atoms and an unknown number of hydrogen atoms. If the carbon atoms are covalently bonded to each other, sharing six electrons between them, then how many hydrogen atoms should this molecule contain?
   a. 1   c. 3
   b. 2   d. 4

4. People who are not lactose intolerant are able to make the enzyme lactase, which cleaves the disaccharide milk sugar lactose into the two smaller monosaccharides, glucose and galactose. Because this process _____ water, glucose and galactose together weigh _____ lactose.
   a. uses up; more than
   b. uses up; less than
   c. produces; more than
   d. produces; less than

5. How are large organic molecules (macromolecules) synthesized?
   a. hydrolysis of monomers
   b. hydrolysis of polymers
   c. dehydration reaction utilizing monomers
   d. dehydration reaction utilizing polymers

6. Carbohydrates do not
   a. exist as monomers (monosaccharides), dimers (disaccharides), and polymers (polysaccharides).
   b. have the general chemical formula of $(CH_2O)_n$.
   c. serve as the major reservoir of long-term, stored energy in animals.
   d. function as a source of energy or as an extremely durable structural material, depending on the specific nature of the chemical bonds between subunits (monomers).

7. What is the difference between a simple sugar and a complex carbohydrate?
   a. Complex carbohydrates are polymers of simple sugars.
   b. Sugars are found in proteins, and carbohydrates are found in nucleic acids.
   c. Sugars are made by plants, and carbohydrates are made by animals.
   d. Carbohydrates are liquid at room temperature, and sugars are solid.

8. Complex carbohydrates have numerous roles in plant cells, including providing
   a. short-term energy storage and acting as enzymes.
   b. long-term energy storage and giving structure to plant cell walls.
   c. short-term energy storage and giving structure to plant cell walls.
   d. a storage medium for genetic information.

9. The cellulose molecule is much stronger than starch because
   a. the starch molecule is much smaller than cellulose.
   b. cellulose uses nitrogen-containing functional groups to form strong cross-links.
   c. cellulose is constructed from sucrose disaccharides, which make it twice as strong as starch.
   d. individual cellulose molecules are cross-linked with each other by hydrogen bonds, whereas starch molecules do not form such cross-links.
   e. the glucose monomers in cellulose are all oriented in the same direction, whereas in starch, the monomers alternate between "up" and "down" orientations.

10. The general class of biological molecules that contains large, nonpolar regions that make these molecules insoluble in water is called
    a. phospholipids.   c. lipids.
    b. fats.   d. waxes.

11. Phospholipids contain a "head group" that is _____ and two fatty acid "tails" that are _____.
    a. hydrophobic; hydrophilic
    b. hydrophilic; hydrophobic
    c. hydrolyzed; nonhydrolyzed
    d. hydrophilic; hydrophilic

12. Lipids are made of fatty acids, which are made of mostly nonpolar bonds. Will these molecules be soluble or insoluble in water?
    a. soluble in water
    b. insoluble in water
    c. hydrophilic
    d. both b and c

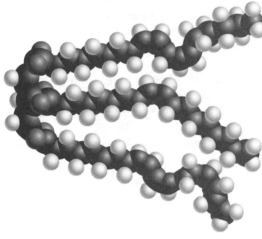

(a) A fat

(b) An oil

13. Examine the figure above. Oleic acid is a monounsaturated fatty acid, common in olive oil. That is, it contains one double covalent bond between carbon atoms. On the other hand, stearic acid is a saturated fatty acid, common in beef fat. Comparing a triglyceride containing entirely oleic acid (an oil, similar to that in the figure) with another triglyceride containing entirely stearic acid (a fat, similar to that in the figure), how many more hydrogen atoms does the fat triglyceride contain?
    a. 2   c. 6
    b. 4   d. 8

14. Fat accumulates in our bodies when we overeat because lipids
    a. are important for cell membranes.
    b. have enzymatic activity.
    c. are found only in plants.
    d. are a form of long-term energy storage.

15. Testosterone, cholesterol, and estrogen
    a. are composed of essentially the same kinds of fatty acids.
    b. have the same biological functions in the human body, although their structures are different.
    c. all contain nitrogen, unlike other lipids.
    d. have the same core structure of four interlocked rings, but have different functional groups attached to the rings.

16. Proteins are polymers of
    a. peptides.   c. nucleotides.
    b. amino acids.   d. sugars.

17. Proteins are macromolecules that can perform many different functions within an organism. This diversity of function for proteins is due to the unique nature of one of its four functional groups, which is bound to a central carbon atom in each amino acid. Which functional group is responsible for the wide diversity of function attributed to proteins?
    a. carboxyl group
    b. amino group
    c. hydrogen atom group
    d. R group

18. You have identified a protein that is unable to form disulfide bridges. Which of the following would this affect?
    a. primary structure of the protein
    b. secondary structure of the protein
    c. tertiary structure of the protein
    d. dehydration synthesis

19. A scientist is studying the metabolism of proteins in yeast and wants to follow the formation of proteins from the earliest possible point. In her experiment, she will feed the yeast radioactive nutrients and follow the fate of the radioactivity in the cells. Which of the following atoms will allow her to exclusively follow proteins in the cell?
    a. radioactive carbon
    b. radioactive nitrogen
    c. radioactive oxygen
    d. radioactive sulfur

20. A particular enzyme, originally found in bacteria that thrive in extremely hot pools in Yellowstone National Park, is now being used to rapidly make many copies of DNA recovered from crime scenes. The enzyme is valuable because it remains functional at high temperatures, which greatly speeds up the DNA copying process. Thus, the enzyme is resistant to
    a. polymerization.
    b. dehydration synthesis.
    c. being denatured.
    d. becoming unsaturated.
    e. functional groups.

21. What is the most correct statement comparing DNA and RNA?
    a. Single-stranded DNA, containing ribose sugars, contains information that is copied from the double-stranded RNA.
    b. Single-stranded RNA, containing ribose sugars, contains information that is copied from the double-stranded DNA.

c. Single-stranded DNA, containing deoxyribose sugars, contains information that is copied from the double-stranded RNA.

d. Single-stranded RNA, containing deoxyribose sugars, contains information that is copied from the double-stranded DNA.

22. ATP stores energy in the
    a. bonds between its three phosphate groups.
    b. double bonds between the phosphorus atoms and certain oxygen atoms.
    c. single bond between its nitrogenous base and its deoxyribose sugar.
    d. single bond between its deoxyribose sugar and a phosphate group.
    e. bonds between the different nitrogen atoms in its nitrogenous base.

23. A nucleotide is composed of
    a. a sugar and a phosphate group.
    b. a phosphate group and a nitrogen-containing base.
    c. a sugar and a nitrogen-containing base.
    d. a sugar, a phosphate group, and a nitrogen-containing base.

24. Study the figure at right. What would happen if you broke all of the hydrogen bonds in a DNA molecule?
    a. The phosphates would separate from the deoxyribose sugars in each strand.
    b. The two strands of the double helix would separate from each other.
    c. The DNA molecule, originally a single molecule millions of nucleotides long, would break into millions of very short segments.
    d. Vast amounts of chemical energy would be released as the bonds between adjacent phosphate groups are broken.

25. The liver converts excess nitrogen wastes from our food into urea. Which of the following types of foods would you expect to increase the levels of urea in your blood?
    a. vegetables that are high in carbohydrates
    b. vegetable oils that are high in lipids
    c. meats that contain a lot of protein
    d. legumes that contain a lot of nucleic acid
    e. both a and b
    f. both c and d

26. If a biological polymer includes both nitrogen and phosphorus, it is most likely to be
    a. DNA or RNA.
    b. a polypeptide.
    c. a polysaccharide.
    d. a triglyceride.

27. At the gym one day, you notice a new "Energy Bar" being sold that advertises quick energy for your workout. To impress you further, the packaging

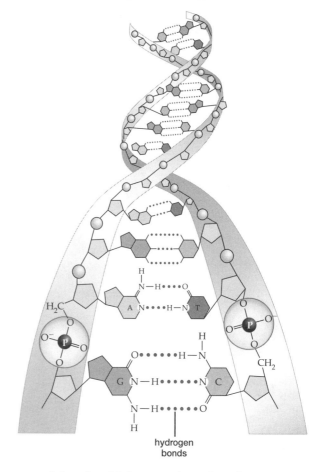

hydrogen bonds

claims that this bar contains only carbon, oxygen, and hydrogen atoms. What kind of biological molecule(s) would you be eating if you ate this "Energy Bar"?
    a. carbohydrates
    b. proteins
    c. DNA and RNA
    d. lipids
    e. carbohydrates and lipids
    f. carbohydrates and proteins

28. Which of the following organic molecules would you expect to be insoluble in water?
    a. a carbon backbone with many hydroxyl groups, OH
    b. the carboxylic acid portion of an amino acid, which may donate a hydrogen ion (H^+) to the surrounding solution
    c. a carbon backbone with methyl groups, CH_3
    d. the "head" region of a phospholipid molecule, which contains both positively and negatively charged regions

29. The earliest ancient Egyptians buried their dead in pits in the desert. The heat and dryness of the sand dehydrated the bodies relatively quickly, leaving lifelike mummies. The trunk region of the body dehydrates much more slowly than peripheral regions such as the hands or feet. Based

on what you know about how biological molecules are broken apart, which regions of a mummy's body would give the most intact pieces of DNA?

a. liver
b. fingers
c. stomach
d. thigh muscles

30. Rearranging the order of the subunits in which of the following molecules would have the greatest effect on its function?

a. starch
b. protein
c. DNA
d. both a and b
e. both b and c

ANSWER KEY

1. b
2. a
3. b
4. a
5. c
6. c
7. a
8. b
9. d
10. c
11. b
12. b
13. c
14. d
15. d

16. b
17. d
18. c
19. d
20. c
21. b
22. a
23. d
24. b
25. f
26. a
27. e
28. c
29. b
30. e

CHAPTER 4 CELL STRUCTURE AND FUNCTION

OUTLINE

Section 4.1 What Is the Cell Theory?

Cell theory states that (1) all living things are made of one or more cells; (2) the smallest organisms are single cells, and cells are the functional units of multicellular organisms; and (3) all cells arise from preexisting cells.

Section 4.2 What Are the Basic Attributes of Cells?

Although cells range in size (**Figure 4-1**), the majority are very small so as to optimize the movements of nutrients and wastes in or out of them.

Cells are enclosed by a thin, double layer of phospholipids with embedded proteins called the **plasma membrane** (**Figure 4-2**). This membrane mediates the movements of substances in and out of the cell.

All cells contain **cytoplasm**, which consists of the material between the plasma membrane and DNA (**Figures 4-3 and 4-4**). The fluid portion of the cytoplasm is called the **cytoplasmic fluid**.

All cells use **deoxyribonucleic acid (DNA)** as genetic material, and **ribonucleic acid (RNA)** is used to copy genes on DNA and to help make proteins.

All cells obtain energy and nutrients from their environment.

There are two basic types of cells, *eukaryotic* and *prokaryotic*.

Section 4.3 What Are the Major Features of Eukaryotic Cells?

Eukaryotic cells contain their DNA within a **nucleus** and possess membrane-bound **organelles** within their cytoplasm (**Figures 4-3 and 4-4**).

The outer surfaces of plant, fungi, and some protist cells are covered by a supportive and protective **cell wall** (**Figure 4-4**). Plant cell walls are made of cellulose and polysaccharides; fungal cells are made of polysaccharides and chitin; and protist cell walls can be made of cellulose, protein, or silica. Cell walls are typically porous, allowing the passage of chemicals through the plasma membrane.

Organelles are attached to a network of protein fibers called the **cytoskeleton** (**Figure 4-5**). The cytoskeleton (**Table 4-2**) helps maintain cell shape, causes cell and organelle movement, and assists in cell division.

Some cells have bristle-like **cilia** or whip-like **flagella** (**Figures 4-6 and 4-7**) that can propel individual cells or move fluids along their surfaces.

The nucleus of the cell houses the DNA of a eukaryotic cell (**Figure 4-8a**) by enclosing it in a porous, double-layered membrane (the **nuclear envelope**). **Chromatin** inside the nucleus forms long **chromosome** strands of DNA (**Figure 4-9**). The **nucleoli** inside the nucleus make **ribosomes** that act as a workbench during protein synthesis (**Figure 4-10**). Ribosomes are distributed within the cytoplasm or along the membranes of the nucleus and **endoplasmic reticulum (ER)**.

The endoplasmic reticulum is a folded series of membranes that are composed of a smooth and rough portion, named for the presence or absence of ribosomes embedded in the membrane (**Figure 4-11a**). The smooth ER can manufacture lipids, break down carbohydrates, or metabolize substances depending on the type of cell. The rough ER manufactures proteins.

The **Golgi apparatus** (**Figure 4-12**) modifies, sorts, and packages the products of the rough ER (as **vesicles**), which then are then exported from the cell (**Figure 4-13**).

In some cells, the Golgi apparatus can form **lysosomes**, whose enzymes digest food particles that have been internalized by **food vacuoles** (**Figure 4-14**). Other cells form **contractile vacuoles** for water regulation (such as in freshwater protists, **Figure 4-15**) or **central vacuoles** for support and waste storage (such as in plants, **Figure 5-10**).

Mitochondria are specialized to metabolize food molecules through **aerobic** means to make large amounts of ATP as an energy source (**Figure 4-16**). These reactions are associated with a pair of mitochondrial membranes that enclose a fluid matrix.

Chloroplasts are found in photosynthetic eukaryotic cells. They are surrounded by a double membrane inside of which is the green pigment **chlorophyll** (**Figure 4-17**). Chlorophyll captures the energy in sunlight and uses it to make sugars from carbon dioxide and water.

Plastids are used by plants and photosynthetic protists to store sugars that were made during photosynthesis in the form of starches (**Figure 4-18**).

Section 4.4 What Are the Major Features of Prokaryotic Cells?

Prokaryotic cells are small; have a cell wall; and have rodlike, spherical, or helix shapes. Some move by means of flagella. Infectious bacteria possess capsules, slime layers, and/or pili that help them adhere to host tissues (**Figure 4-19a**).

Prokaryotes lack nuclei and membranous organelles.

Prokaryotes possess a single, circular chromosome, which is coiled in the central region of the cell (the **nucleoid**). Most prokaryotes have small DNA rings (plasmids) that carry genes that impart special properties to the cell.

Prokaryotic cytoplasm contains ribosomes that function during protein synthesis.

FLASH CARDS

To use the flash cards, tear the page from the book and cut along the dashed lines. The key term appears on one side of the flash card, and its definition appears on the opposite side.

aerobic	cell wall
anaerobic	central vacuole
archaea	centriole
bacteria (sing., bacterium)	chlorophyll
basal body	chloroplast
cell theory	chromatin

A layer of material, normally made up of cellulose or cellulose-like materials, that is outside the plasma membrane of plants, fungi, bacteria, and some protists.

Using oxygen.

A large, fluid-filled vacuole occupying most of the volume of many plant cells; performs several functions, including maintaining turgor pressure.

Not using oxygen.

In animal cells, a short, barrel-shaped ring consisting of nine microtubule triplets; a pair of centrioles is found near the nucleus and may play a role in the organization of the spindle; centrioles also give rise to the basal bodies at the base of each cilium and flagellum that give rise to the microtubules of cilia and flagella.

Prokaryotes that are members of the domain Archaea, one of the three domains of living organisms; only distantly related to members of the domain Bacteria.

A pigment found in chloroplasts that captures light energy during photosynthesis; chlorophyll absorbs violet, blue, and red light but reflects green light.

Prokaryotes that are members of the domain Bacteria, one of the three domains of living organisms; only distantly related to members of the domain Archaea.

The organelle in plants and plantlike protists that is the site of photosynthesis; surrounded by a double membrane and containing an extensive internal membrane system that bears chlorophyll.

A structure derived from a centriole that produces a cilium or flagellum and anchors this structure within the plasma membrane.

The complex of DNA and proteins that makes up eukaryotic chromosomes.

The scientific theory stating that every living organism is made up of one or more cells; cells are the functional units of all organisms; and all cells arise from preexisting cells.

chromosome

deoxyribonucleic acid (DNA)

cilium (plural, cilia)

endoplasmic reticulum (ER)

contractile vacuole

endosymbiont hypothesis

cytoplasm

eukaryotic

cytoplasmic fluid

flagellum (plural, flagella)

cytoskeleton

food vacuole

A molecule composed of deoxyribose nucleotides; contains the genetic information of all living cells.

A DNA double helix together with proteins that help to organize and regulate the use of the DNA.

A system of membranous tubes and channels in eukaryotic cells; the site of most protein and lipid synthesis.

A short, hairlike projection from the surface of certain eukaryotic cells that contains microtubules in a 9 + 2 arrangement. The movement of cilia may propel cells through a fluid medium or move fluids over a stationary surface layer of cells.

The hypothesis that certain organelles, especially chloroplasts and mitochondria, arose as mutually beneficial associations between the ancestors of eukaryotic cells and captured bacteria that lived within the cytoplasm of the pre-eukaryotic cell.

A fluid-filled vacuole in certain protists that takes up water from the cytoplasm, contracts, and expels the water outside the cell through a pore in the plasma membrane.

Referring to cells of organisms of the domain Eukarya (plants, animals, fungi, and protists). Eukaryotic cells have genetic material enclosed within a membrane-bound nucleus and contain other organelles.

The material contained within the plasma membrane of a cell, exclusive of the nucleus.

A long, hairlike extension of the plasma membrane; in eukaryotic cells, it contains microtubules arranged in a 9 + 2 pattern. The movement of flagella propels some cells through fluids.

The fluid portion of the cytoplasm; the substance within the plasma membrane exclusive of the nucleus and organelles; also called cytosol.

A membranous sac, within a single cell, in which food is enclosed. Digestive enzymes are released into the vacuole, where intracellular digestion occurs.

A network of protein fibers in the cytoplasm that gives shape to a cell, holds and moves organelles, and is typically involved in cell movement.

Golgi apparatus

nuclear envelope

intermediate filament

nuclear pore complex

lysosome

nucleoid

microfilament

nucleolus (plural, nucleoli)

microtubule

nucleus

mitochondrion (plural, mitochondria)

organelle

The double-membrane system surrounding the nucleus of eukaryotic cells; the outer membrane is typically continuous with the endoplasmic reticulum.

A stack of membranous sacs, found in most eukaryotic cells, that is the site of processing and separation of membrane components and secretory materials.

An array of proteins that line pores in the nuclear membrane and control which substances enter and leave the nucleus.

Part of the cytoskeleton of eukaryotic cells that is composed of several types of proteins and probably functions mainly for support.

The location of the genetic material in prokaryotic cells; not membrane enclosed.

A membrane-bound organelle containing intracellular digestive enzymes.

The region of the eukaryotic nucleus that is engaged in ribosome synthesis; consists of the genes encoding ribosomal RNA, newly synthesized ribosomal RNA, and ribosomal proteins.

Part of the cytoskeleton of eukaryotic cells that is composed of the proteins actin and (in some cases) myosin; functions in the movement of cell organelles and in locomotion by extension of the plasma membrane.

The membrane-bound organelle of eukaryotic cells that contains the cell's genetic material.

A hollow, cylindrical strand, found in eukaryotic cells, that is composed of the protein tubulin; part of the cytoskeleton used in the movement of organelles, cell growth, and the construction of cilia and flagella.

A membrane-bound structure, found in the cytoplasm of eukaryotic cells, that performs a specific function.

An organelle, bounded by two membranes, that is the site of the reactions of aerobic metabolism.

plasma membrane

ribosome

plastid

vacuole

prokaryotic

vesicle

ribonucleic acid (RNA)

A complex consisting of two subunits, each composed of ribosomal RNA and protein, found in the cytoplasm of cells or attached to the endoplasmic reticulum, that is the site of protein synthesis.

A vesicle that is typically large and consists of a single membrane enclosing a fluid-filled space.

A small, membrane-bound sac within the cytoplasm.

The outer membrane of a cell, composed of a bilayer of phospholipids in which proteins are embedded.

In plant cells, an organelle bounded by two membranes that may be involved in photosynthesis (chloroplasts), pigment storage, or food storage.

Referring to cells of the domains Bacteria or Archaea. Prokaryotic cells have genetic material that is not enclosed in a membrane-bound nucleus; they also lack other membrane-bound organelles.

A molecule composed of ribose nucleotides, each of which consists of a phosphate group, the sugar ribose, and one of the bases adenine, cytosine, guanine, or uracil; involved in converting the information in DNA into protein; also the genetic material of some viruses.

SELF TEST

1. Suppose we discover a form of alien life that does not conform to the cell theory. What could be one feature of this new life-form that violates the cell theory?
   a. It consists of one or more small compartmentalized structures, each of which is separated from the outside environment by a plasma membrane.
   b. Much of the life-form consists of a water-based fluid, within which many organic molecules are dissolved.
   c. There is a critical lower limit to the size of living material, and if such a unit is cut into smaller pieces, the individual pieces do not, on their own, behave as life-forms.
   d. The life-forms usually originate directly from interactions among the nonliving chemicals on the planet, rather than from other, similar, living organisms.

2. The cell theory states that
   a. cells are generally small to allow for diffusion.
   b. all cells contain cytoplasm.
   c. cells are either prokaryotes or eukaryotes.
   d. all living things are composed of cells.
   e. all cells arise from organic molecules such as DNA.

3. Which of the following BEST describes the structures found in all cells?
   a. cell wall
   b. plasma membrane
   c. nucleus
   d. ribosomes
   e. both a and c
   f. both b and d

4. The smallest living cells are mycoplasmas, which range in size from about 0.1 micrometer to 1 micrometer. What could be one reason that no cells are smaller than about 1/10 of a micrometer in size?
   a. A plasma membrane could not effectively isolate the contents of a much smaller cell from the outside environment.
   b. Cells that are any smaller could not obtain oxygen rapidly enough by diffusion.
   c. A much smaller cell would not have space for essential structures such as ribosomes.
   d. Chemical waste products could not possibly diffuse rapidly enough out of a much smaller cell.

5. Why must living cells remain microscopic in size?
   a. Cells produce a limited number of enzymes.
   b. The energy needs of giant cells would outstrip the available supply of energy from the environment.

   c. Exchanges of substances at the membrane surface would take too long to diffuse throughout the interior of the cell.
   d. both a and b

6. Examine the figure on the following page. Suppose that light microscopes had never been invented, but that we had still invented the various types of electron microscopes. Which microscopic structures would be the most difficult to see in their entirety?
   a. organelles such as mitochondria
   b. large plant cells
   c. bacteria
   d. viruses

7. In what manner do molecules such as proteins and RNA enter into or exit from the nucleus?
   a. diffusion through the lipid bilayers of the nuclear envelope
   b. movement through pores in the nuclear envelope
   c. osmosis through the lipid bilayers of the nuclear envelope
   d. breakdown of the nuclear envelope

8. Among other things, the endoplasmic reticulum is needed for
   a. storage of a cell's genetic material.
   b. oxidation of organic food molecules, to release the chemical energy that powers a cell.
   c. transport of proteins that are moving from the Golgi apparatus to the outside of a cell.
   d. protein synthesis.

9. Chromosomes consist of
   a. DNA.
   b. proteins.
   c. RNA.
   d. proteins and RNA.
   e. proteins and DNA.

10. Sorting and modification of proteins is an important function of
    a. mitochondria.
    b. chloroplasts.
    c. lysosomes.
    d. the Golgi complex.
    e. the plasma membrane.

11. Which of the following correctly lists organelles that are part of the internal membrane system of eukaryotic cells?
    a. endoplasmic reticulum, Golgi apparatus, and lysosomes
    b. endoplasmic reticulum and mitochondria
    c. Golgi apparatus and nucleus
    d. endoplasmic reticulum, Golgi apparatus, and cell wall

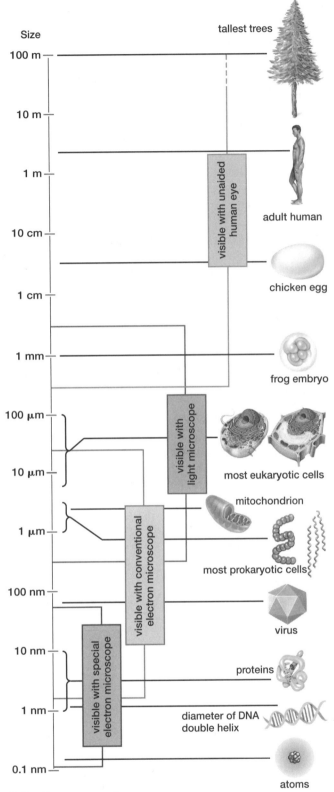

Size

100 m — tallest trees

10 m —

1 m — adult human

10 cm — chicken egg

1 cm —

1 mm — frog embryo

100 μm — most eukaryotic cells

10 μm —

mitochondrion

1 μm —

most prokaryotic cells

100 nm — virus

10 nm —

proteins

1 nm — diameter of DNA double helix

0.1 nm — atoms

visible with unaided human eye

visible with light microscope

visible with conventional electron microscope

visible with special electron microscope

Units of measurement:
1 meter (m) = 39.37 inches
1 centimeter (cm) = 1/100 m
1 millimeter (mm) = 1/1,000 m
1 micrometer (μm) = 1/1,000,000 m
1 nanometer (nm) = 1/1,000,000,000 m

12. Which of the following lists the correct order in which newly synthesized proteins are delivered to the plasma membrane?
    a. endoplasmic reticulum to lysosomes to the Golgi apparatus to the plasma membrane
    b. endoplasmic reticulum to the Golgi apparatus to the plasma membrane
    c. Golgi apparatus to the endoplasmic reticulum to the plasma membrane
    d. endoplasmic reticulum to the plasma membrane
    e. endoplasmic reticulum to the Golgi apparatus to lysosomes to the plasma membrane

13. Imagine that you are late for a date and you reach your friend's door out of breath because you just ran the last three blocks from the bus stop. In a lame effort to impress and to try to make your date forget that you are half an hour late, you describe what oxygen is used for in your cells. Which of the following is correct?
    a. The lysosomes in my muscle cells need this extra oxygen to digest sugars and provide me with energy for running.
    b. The cellular enzymes in my leg muscles need this extra oxygen to repair the damage that occurs to my muscle cells as I run.
    c. The mitochondria in my muscle cells need the extra oxygen to produce sugars that, in turn, provide the energy I need to run.
    d. The mitochondria in my muscle cells need this extra oxygen to break down sugars and produce the energy I need to run.

14. Which organelle or other cell structure is found in both plant and animal cells?
    a. smooth ER
    b. cell wall
    c. central vacuole
    d. plastid
    e. chloroplast

15. Which organelle would you expect to be in abundance in the liver of a drug addict?
    a. Golgi complex
    b. smooth endoplasmic reticulum
    c. rough endoplasmic reticulum
    d. nucleolus
    e. ribosome

16. In certain types of genetic engineering, DNA is injected into the nucleus of a recipient animal cell. What is the fewest number of membranes that must be pierced by the microscopic needle in order to inject the DNA? (Note that the needles used are not small enough to pass through a nuclear pore.)
    a. one
    b. two

c. three
d. four

17. Which of the following originated by endosymbiosis?
   a. chloroplasts
   b. mitochondria
   c. flagella
   d. lysosomes
   e. both a and b

18. A researcher has discovered an unusual organism deep in the crust of Earth. She wants to know whether it is prokaryotic or eukaryotic. Imagine that she has rapid tests to determine if the following molecules are present: DNA, RNA, and the proteins that form microtubules. You would advise her to test for
   a. DNA, because only eukaryotes have a nucleus.
   b. RNA, because only eukaryotes have ribosomes.
   c. microtubule proteins, because only eukaryotes have microtubules.
   d. DNA and RNA, because only eukaryotes contain both of these at the same time.

19. Ribosomes are
   a. essential in all living cells.
   b. essential in animal cells but optional in plant cells, which also have an alternative means of constructing proteins.
   c. smaller than animal cells, but somewhat larger than mitochondria.
   d. small structures composed of DNA and carbohydrates.

20. Lysosomes contain very powerful digestive enzymes that can break down proteins, carbohydrates, and other molecules. Why don't these enzymes digest the cell itself?
   a. The enzymes will digest only foreign material.
   b. The enzymes are separated from the cytoplasm by the lysosomal membrane.
   c. The enzymes are inactive until secreted from the cell.
   d. Different species produce slightly different versions of digestive enzymes, and the cells of each species are immune to the particular enzyme versions that they produce.

21. Which plant part should contain the greatest number of plastids of the type shown in the figure?
   b. a leaf of green grass
   b. a tuber of a potato plant
   c. the juicy flesh of a peach
   d. the wood of a pine tree

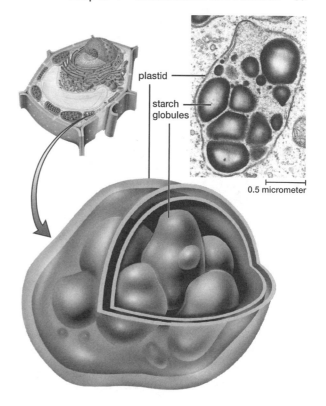

plastid
starch globules
0.5 micrometer

22. If the nucleus is the control center of the cell, how is information encoded and shipped to the cytoplasm?
   a. by RNA
   b. by chromosomes
   c. by nuclear pores
   d. by the nucleolus

23. Why do cells have so many ribosomes?
   a. Proteins are essential to the life of a cell, and an active cell must constantly create many new ones, through the activity of many ribosomes.
   b. Cells need to have thousands of different kinds of ribosomes, because each different kind of protein is constructed by a different kind of ribosome.
   c. Ribosomes are used up every time they create proteins, so they must constantly be replaced.
   d. After they create proteins, ribosomes must return to the nucleolus for repairs, so many spare ribosomes must circulate in the cytoplasm at all times.

24. Which of the following characteristics of mitochondria are true?
   a. They are able to take energy from food molecules and store it in high-energy bonds of ATP.
   b. All metabolic conversion of high-energy molecules (e.g., glucose) to ATP occurs within the mitochondria.

c. They are able to convert solar energy directly into high-energy sugar molecules.

d. They release oxygen during the process of aerobic metabolism.

25. Proteins that are going to be exported from the cell are synthesized
   a. in the Golgi complex.
   b. on the rough endoplasmic reticulum.
   c. in mitochondria.
   d. on the smooth endoplasmic reticulum.

26. Which of the following generalizations can you make about the cytoskeleton?
   a. The name implies a fixed structure like the bones of a vertebrate or 2×4 boards in the wall of a building.
   b. A variety of cytoskeletal elements are integral in the performance of numerous essential cellular functions.
   c. It provides a type of cellular armor on the outside of cells that serves a protective function.
   d. Although it provides physical support, the cytoskeleton does not play any role in the movement either of cells or their internal structures.

27. Suppose that a plant cell is placed into distilled water. Water will tend to flow into the cell from the outside. What will happen next?
   a. The plant cell will rapidly grow a thicker cell wall, to block the inflow of excessive water.
   b. The central vacuole will take up the water and increase in size, but the rigid cell wall will prevent the cell from bursting open.
   c. The plant cell will collect the water into a contractile vacuole, which it will then squeeze to expel the excessive water.

d. Excessive water will cause the cellulose in the cell wall to rapidly break down, by hydrolysis, into glucose monomers, thus causing the cell to disintegrate.

28. Prokaryotic cells
   a. are large cells, typically greater than 10 mm in diameter.
   b. include numerous membrane-enclosed structures known as organelles.
   c. possess a single strand of DNA but no definable membrane-enclosed nucleus.
   d. never contain ribosomes, which are too large to fit inside a prokaryotic cell.

29. Plasmids are located
   a. in the nucleus.
   b. in the nucleolus.
   c. in the cytoplasm.
   d. continuous with the nuclear envelope.

30. A difference between prokaryotic cells, on the one hand, and the eukaryotic organelles, mitochondria and chloroplasts, on the other, is that
   a. prokaryotic cells contain DNA, whereas mitochondria and chloroplasts do not.
   b. prokaryotic cells are free-living organisms, whereas mitochondria and chloroplasts are not.
   c. prokaryotic cells contain ribosomes, whereas mitochondria and chloroplasts do not.
   d. typical prokaryotic cells are much smaller than mitochondria and chloroplasts.
   e. prokaryotic cells can synthesize ATP, whereas mitochondria and chloroplasts cannot.

ANSWER KEY

1. d		16. c	
2. d		17. e	
3. f		18. c	
4. c		19. a	
5. c		20. b	
6. b		21. b	
7. b		22. a	
8. d		23. a	
9. e		24. a	
10. d		25. b	
11. a		26. b	
12. b		27. b	
13. d		28. c	
14. a		29. c	
15. b		30. b	

Chapter 5 Cell Membrane Structure and Function

OUTLINE

Section 5.1 How Is the Structure of a Membrane Related to Its Function?

Cell membranes perform five crucial functions: (1) isolation of cell contents from the environment, (2) regulation of the exchange of substances between the cell and extracellular **fluid**, (3) cell-to-cell communication, (4) attachment within and between cells, and (5) regulation of biochemical reactions.

Cell membranes are composed of a flexible double layer of phospholipids that contains a shifting patchwork of proteins (the **fluid mosaic model, Figure 5-1**).

Phospholipids are composed of a hydrophilic (polar) phosphate head and two hydrophobic (nonpolar) fatty acid tails (**Figure 5-2**). Phospholipids form the cell membrane **phospholipid bilayer**, with hydrophobic tails facing each other and hydrophilic heads facing the watery environment outside and inside the cell (**Figure 5-3**).

Most animal cell membranes also contain cholesterol, which adds strength and flexibility, as well as reducing permeability to water-soluble substances.

Many types of proteins are found in the cell membrane. **Receptor proteins** bind to specific molecules and trigger reactions within the cell (**Figure 5-5**). **Recognition proteins** are found on cell surfaces and act as identification tags. **Enzymes** catalyze chemical reactions. **Attachment proteins** anchor a cell membrane to another cell or to internal cytoskeleton protein filaments. **Transport proteins** move hydrophilic molecules through the cell membrane via pores (**channel proteins**) or by binding to the molecule, changing its shape, and depositing it on the other side of the cell membrane (**carrier proteins**).

Section 5.2 How Do Substances Move Across Membranes?

Molecules in fluids move in response to their **concentration gradient**, from regions of high to regions of low **concentration** (a process called **diffusion**), until the concentrations are equalized (**Figure 5-6**). Diffusion is more rapid at higher temperatures and greater concentration gradients. Diffusion cannot move molecules rapidly over long distances.

Transport of molecules across the cell membrane occurs by *passive transport* (does not require energy) or *energy-requiring transport*.

Passive transport occurs in a number of ways. Lipid-soluble molecules and dissolved gases move directly through the cell membrane by simple diffusion (**Figure 5-7a**). Ions and water-soluble molecules require channel or carrier proteins to diffuse through the cell membrane, a process called **facilitated diffusion** (**Figures 5-7b, d**). Water diffuses across the cell membrane by **osmosis**, which is influenced by the tonicity differences across a **selectively permeable** membrane (**Figures 5-8 and 5-9**).

Energy-requiring transport occurs by three mechanisms: (1) **Active transport** uses energy (ATP) and carrier proteins (pumps) to move molecules against their concentration gradients (**Figure 5-11**). (2) **Endocytosis** allows a cell to engulf particles of fluids by surrounding them with its cell membrane. Endocytosis of a fluid is called **pinocytosis** (**Figure 5-12**) and endocytosis of a large particle is called **phagocytosis** (**Figure 5-14**). **Receptor-mediated endocytosis** allows a cell to engulf a specific molecule that binds to cell membrane receptor proteins (**Figure 5-13**). (3) Unwanted materials can be expelled from a cell by **exocytosis**, which is the reverse of endocytosis (**Figure 5-15**).

Generally, most cells are small to optimize their surface area/volume ratios (**Figure 5-16**). This allows them to transport adequate amounts of materials across their cell membranes to sustain themselves.

Section 5.3 How Do Specialized Junctions Allow Cells to Connect and Communicate?

> **Desmosomes** attach adjacent cells together even when stretched (**Figure 5-17a**).
>
> **Tight junctions** attach cells to each other and prevent substances from leaking between them, forming a waterproof barrier (**Figure 5-17b**).
>
> **Gap junctions** form channels that join the cytosol between adjacent animal cells (**Figure 5-18a**). These junctions allow cells to communicate easily with each other by allowing specific substances to pass from cell interior to cell interior.
>
> **Plasmodesmata**, located in the walls of plant cells, are the functional equivalents of gap junctions (**Figure 5-18b**).

FLASH CARDS

To use the flash cards, tear the page from the book and cut along the dashed lines. The key term appears on one side of the flash card, and its definition appears on the opposite side.

active transport	concentration gradient
aquaporin	desmosome
attachment protein	diffusion
carrier protein	endocytosis
channel protein	energy-requiring transport
concentration	enzyme

The difference in concentration of a substance between two parts of a fluid or across a barrier such as a membrane.

The movement of materials across a membrane through the use of cellular energy, normally against a concentration gradient.

A strong cell-to-cell junction that attaches adjacent cells to one another.

A channel protein in the plasma membrane of a cell that is selectively permeable to water.

The net movement of particles from a region of high concentration of that particle to a region of low concentration, driven by the concentration gradient; may occur entirely within a fluid or across a barrier such as a membrane.

A protein in the plasma membrane of a cell that attaches either to the cytoskeleton inside the cell, to other cells, or to the extracellular matrix.

The process in which the plasma membrane engulfs extracellular material, forming membrane-bound sacs that enter the cytoplasm and thereby move material into the cell.

A membrane protein that facilitates the diffusion of specific substances across the membrane. The molecule to be transported binds to the outer surface of the carrier protein; the protein then changes shape, allowing the molecule to move across the membrane.

The transfer of substances across a cell membrane using energy from ATP; includes active transport, endocytosis, and exocytosis.

A membrane protein that forms a channel or pore completely through the membrane and that is usually permeable to one or to a few water-soluble molecules, especially ions.

A protein catalyst that speeds up the rate of specific biological reactions.

The number of particles of a dissolved substance in a given unit of volume.

exocytosis	glycoprotein
facilitated diffusion	gradient
fluid	hypertonic
fluid mosaic model	hypotonic
food vacuole	isotonic
gap junction	microvilli

A protein to which a carbohydrate is attached.

The process in which intracellular material is enclosed within a membrane-bound sac that moves to the plasma membrane and fuses with it, releasing the material outside the cell.

A difference in concentration, pressure, or electrical charge between two regions.

The diffusion of molecules across a membrane, assisted by protein pores or carriers embedded in the membrane.

Referring to a solution that has a higher concentration of dissolved particles (and therefore a lower concentration of free water) than has the cytoplasm of a cell.

A liquid or gas.

Referring to a solution that has a lower concentration of dissolved particles (and therefore a higher concentration of free water) than has the cytoplasm of a cell.

A model of cell membrane structure; according to this model, membranes are composed of a double layer of phospholipids in which various proteins are embedded. The phospholipid bilayer is a somewhat fluid matrix that allows the movement of proteins within it.

Referring to a solution that has the same concentration of dissolved particles (and therefore the same concentration of free water) as has the cytoplasm of a cell.

A membranous sac within a cell, in which food is enclosed. Digestive enzymes are released into the vacuole, where intracellular digestion occurs.

A microscopic projection of the plasma membrane, which increases the surface area of a cell.

A type of cell-to-cell junction in animals in which channels connect the cytoplasm of adjacent cells.

osmosis	receptor-mediated endocytosis
passive transport	receptor protein
phagocytosis	recognition protein
phospholipid bilayer	selectively permeable
pinocytosis	simple diffusion
plasmodesmata	solute

The selective uptake of molecules from the extracellular fluid by binding to a receptor located at a coated pit on the plasma membrane and pinching off the coated pit into a vesicle that moves into the cytoplasm.

The diffusion of water across a differentially permeable membrane, normally down a concentration gradient of free water molecules. Water moves into the solution that has a lower concentration of free water from a solution that has a higher concentration of free water.

A protein, located on a membrane or in the cytoplasm of a cell, that recognizes and binds to specific molecules. Binding by receptor proteins typically triggers a response by a cell, such as endocytosis, increased metabolic rate, or cell division.

The movement of materials across a membrane down a gradient of concentration, pressure, or electrical charge without using cellular energy.

A protein or glycoprotein protruding from the outside surface of a plasma membrane that identifies a cell as belonging to a particular species, to a specific individual of that species, and in many cases to one specific organ within the individual.

A type of endocytosis in which extensions of a plasma membrane engulf extracellular particles, enclose them in a membrane-bound sac, and transport them into the interior of the cell.

The quality of a membrane that allows certain molecules or ions to move through it more readily than others.

A double layer of phospholipids that forms the basis of all cellular membranes. The phospholipid heads, which are hydrophilic, face the water of extracellular fluid or the cytoplasm; the tails, which are hydrophobic, are buried in the middle of the bilayer.

The diffusion of water, dissolved gases, or lipid-soluble molecules through the phospholipid bilayer of a cellular membrane.

The nonselective movement of extracellular fluid, enclosed within a vesicle formed from the plasma membrane, into a cell.

A substance dissolved in a solvent (usually a liquid).

A cell-to-cell junction in plants that connects the cytoplasm of adjacent cells.

solvent

transport protein

tight junction

turgor pressure

A protein that regulates the movement of water-soluble molecules through the plasma membrane.

A liquid capable of dissolving (uniformly dispersing) other substances in itself.

Pressure developed within a cell (especially the central vacuole of plant cells) as a result of osmotic water entry.

A type of cell-to-cell junction in animals that prevents the movement of materials through the spaces between cells.

SELF TEST

1. The fluid mosaic model describes membranes as fluid because the
   a. phospholipids of membranes are constantly switching from one layer to the other layer.
   b. membrane is composed mainly of water.
   c. phospholipid molecules are bonded to one another, making them more movable.
   d. phospholipids move from place to place around the motionless membrane proteins.
   e. phospholipids and proteins move from place to place within the bilayer.

2. Membrane fluidity within a phospholipid bilayer is based on
   a. interactions among nonpolar (hydrophobic) lipid tails.
   b. hydrophilic interactions among polar phospholipid heads.
   c. the presence of transport proteins in the lipid bilayer.
   d. the presence of water in the lipid bilayer.

3. Recognition proteins function to
   a. regulate the movement of ions across the cell membrane.
   b. bind hormones and alter the intracellular physiology of a cell.
   c. permit the cells of the immune system to distinguish between pathogens such as bacteria and the cells of your own body.
   d. allow a cell to recognize the particular nutrient molecules that it needs.

4. A hormone circulating in the bloodstream would most likely bind to
   a. a recognition protein.
   b. a receptor protein.
   c. a channel protein.
   d. protein filaments in the cytoplasm.

5. Phospholipids containing more unsaturated fatty acids tend to pack together more _____, and to thus form membranes remaining relatively _____ at higher temperatures, than do phospholipids containing more saturated fatty acids.
   a. tightly; solid
   b. tightly; liquid
   c. loosely; solid
   d. loosely; liquid

6. A membrane transport protein, such as the one toward the right side of the figure, probably has a _____ region that is in contact with the fatty acid tails, as well as two _____ regions that are in contact with the head groups of the phospholipid molecules.
   a. polar; polar
   b. polar; nonpolar
   c. nonpolar; polar
   d. nonpolar; nonpolar

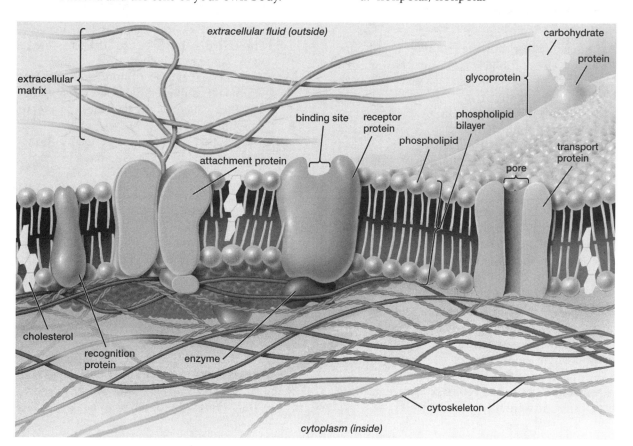

7. Insulin is a protein that is released by the pancreas in response to high concentrations of glucose in the blood. It binds with certain cells in the body, and signals them to allow glucose to enter, from the blood. Insulin thus binds to _____ proteins on these cells.
   a. recognition
   b. transport
   c. enzymatic
   d. attachment
   e. receptor

8. The phospholipid bilayer of the plasma membrane
   a. accelerates the rate at which substances of most types can move into and out of a cell.
   b. allows large, water-soluble molecules such as starch to freely diffuse through, but blocks diffusion of small, nonpolar molecules such as O_2 and CO_2.
   c. allows small nonpolar molecules such as O_2 and CO_2 to freely diffuse through, but blocks the diffusion of most larger, polar molecules into or out of a cell.
   d. forces all water that enters or leaves the cell to pass through aquaporin channels.

9. Which of the following types of molecules must pass through membranes via the aqueous pores formed by membrane proteins?
   a. gases such as carbon dioxide and oxygen
   b. water
   c. large particles such as bacteria
   d. small charged ions, such as Na^+ and Ca^{2+}

10. Diffusion is the movement of molecules from
    a. an area of higher concentration of that type of molecule to an area of lower concentration.
    b. regions that are hydrophobic to regions that are hydrophilic.
    c. outside the cell to inside the cell.
    d. an area of lower concentration of that type of molecule to an area of higher concentration.
    e. the cytoplasm to the interior of an organelle, such as a mitochondrion.

11. A list of substances follows. Each substance needs a different type of transport to get INTO a cell. Pick the choice below that accurately matches the type of transport needed to convey each substance INTO a cell.

    Substance list: oxygen, water, sodium ions, potassium ions, bacterium. (Note that sodium ions are more concentrated outside cells than inside; potassium ions are more concentrated inside cells than outside.)
    a. simple diffusion, osmosis, facilitated diffusion, active transport, exocytosis

    b. osmosis, simple diffusion, facilitated diffusion, active transport, endocytosis
    c. passive transport, osmosis, simple diffusion, facilitated diffusion, active transport
    d. simple diffusion, osmosis, facilitated diffusion, active transport, endocytosis
    e. endocytosis, active transport, active transport, facilitated diffusion, exocytosis

12. In osmosis, water diffuses from the side of the membrane with a higher concentration of water to the side with a lower concentration of water. What is the main determinant of the concentration of water in a solution?
    a. the volume of the solution
    b. the amount of molecules other than water dissolved in the solution
    c. the size of the container
    d. the temperature of the solution

13. Imagine that you are studying cell structure in various organisms in your biology lab. Your instructor gives you a microscope slide showing two types of cells that have been suspended in pure water. One type of cell swells up until it bursts. The other cell maintains its shape throughout the experiment. Suggest an explanation for these observations. Assume that both cells were alive at the start of the experiment and that the concentration of water inside both types of cells is similar.
    a. The cell that burst lacked gap junctions, so water that entered the cell via osmosis could not leak back out through the junctions.
    b. The cell that burst lacked a plasma membrane for regulating osmosis.
    c. The cell that remained intact had plasmodesmata that allowed the excess water to leak out, thus balancing the tendency of water to enter the cell via osmosis.
    d. The cell that remained intact had a contractile vacuole for pumping out the excess water that entered the cell via osmosis.

14. Which of the following processes does a cell use to take up molecules against their concentration gradient?
    a. simple diffusion
    b. facilitated diffusion
    c. active transport
    d. endocytosis
    e. both c and d

15. Imagine that we carefully place a single drop of a particular colored dye into water in beaker A, and place another drop of a different colored dye into water in beaker B. We examine the beakers 1 hour later, and are surprised to see that although the dye in beaker A has

spread out to become uniformly dispersed, the dye in beaker B has not, at least not as completely. What could account for these different behaviors?

   a. The water in beaker A is hotter than the water in beaker B.

   b. The dye in beaker A has a much larger molecular size than the dye in beaker B.

   c. The dye in beaker A is hydrophobic, whereas the dye in beaker B is hydrophilic.

   d. The water in beaker A has remained stationary, whereas the water in beaker B has been agitated.

16. During endocytosis, the contents of the endocytic vesicle

   a. enter the cell.

   b. exit the cell.

   c. enter or exit the cell, always moving down a concentration gradient.

   d. cannot ever be larger than large molecules, such as proteins.

17. Imagine that a sailor is adrift on the ocean in a life raft. Dying of thirst, he gives in to the temptation to drink salty seawater. After this salt is absorbed into the plasma (liquid portion) of his blood, what happens to his red blood cells?

   a. They enlarge, because salt diffuses into the cells, where it then occupies space.

   b. They enlarge and burst, because water moves into the hypotonic cells from the plasma.

   c. They shrink, because the contents of the central vacuoles are drawn out into the surrounding hypotonic plasma.

   d. They shrink, because water passes out of them into the hypertonic plasma.

18. Which of the following transport processes require(s) energy?

   a. facilitated diffusion

   b. osmosis

   c. endocytosis

   d. facilitated diffusion and osmosis

   e. facilitated diffusion, osmosis, and endocytosis

19. Which of the following substances will diffuse most rapidly across the plasma membrane?

   a. amino acid

   b. sodium ion

   c. water

   d. oxygen

   e. Na^+ and Ca^{2+} ions

20. Imagine that an active muscle cell is rapidly using oxygen. Oxygen is being provided by the circulating, nearby blood. We expect oxygen levels inside the muscle cell to be _____ than they are outside the cell, even though the net flow of oxygen is moving _____ the cell.

   a. higher; into

   b. higher; out of

   c. lower; into

   d. lower; out of

21. What will happen if there is a high concentration of oxygen on one side of a biological membrane?

   a. The oxygen will stay on the one side because it cannot cross the membrane.

   b. The oxygen will diffuse across the membrane until the concentration of each side is the same.

   c. The oxygen will be actively transported across the membrane.

   d. Many oxygen molecules will be transported at once, via phagocytosis.

22. Substances are able to cross the lipid bilayer of a cell at different rates that are unique for each substance. Which of the following characteristics would favor the simple diffusion of a substance across a cell membrane?

   a. low lipid solubility

   b. small molecule size

   c. small concentration gradients

   d. the number of membrane transport proteins

23. A chloride ion is in a cell that has no channel proteins in its membrane. Will the chloride be able to get out of the cell?

   a. Yes, the chloride ion has a concentration gradient from inside to outside, so it can diffuse out of the cell.

   b. No, there is no channel to diffuse through, and a charged ion or molecule cannot diffuse down its concentration gradient without a channel to pass it through the hydrophobic portion of the membrane.

   c. No, the chloride ion can only follow a concentration gradient to get into the cell.

   d. No, the chloride already entered the cell because of a concentration gradient.

24. Many metabolic poisons work by inhibiting ATP production. Which type of transport would be most affected?

   a. osmosis

   b. facilitated diffusion

   c. active transport

   d. simple diffusion

25. While the long fiber (axon) of a nerve cell is in a resting state, the concentration of sodium ions (Na^+) is much higher outside the cell than inside. Later, as an impulse travels down the axon, Na^+ ions briefly move into the cell, and the

concentration of Na$^+$ ions inside the cell becomes about equal to the concentration outside. These two different conditions could be caused, respectively, by _____ and _____ of the Na$^+$ ions.

a. passive transport; active transport
b. endocytosis; passive transport
c. active transport; exocytosis
d. active transport; passive transport

26. Imagine we have a small bag, made from an artificial material that behaves like a phospholipid bilayer. We fill the bag with a sucrose-water solution containing 10% sucrose (by weight) and seal it. We place it into a beaker, in which we have already dissolved 150 grams of sucrose into 850 grams of water. What is the best description of what happens next?

a. Water enters the bag, because the solution inside the bag is hypotonic to its surroundings.
b. Water leaves the bag, because the solution in the beaker is hypertonic to the solution inside the bag.
c. Sucrose rapidly leaves the bag, until the concentration of sucrose inside the bag is equal to the concentration outside.
d. Sucrose enters the bag and water leaves the bag, until the contents of the bag become isotonic to their surroundings.

27. One possible drawback of having plasmodesmata is that

a. the cell walls may be structurally weakened.
b. they make it impossible for water to flow from one cell to another.
c. they make it much harder for cells to communicate.
d. they can only exist in roots, and not in any other parts of the plant.

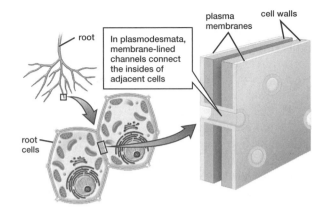

28. Which of the following characteristics does NOT describe cell walls?

a. porous
b. incapable of resisting the expansion of a plant cell placed into a hypotonic solution
c. composed of cellulose or polysaccharides
d. hydrophilic
e. strong

29. Which cell junction functions to make the connections between cells leakproof?

a. desmosomes
b. tight junctions
c. gap junctions
d. plasmodesmata

30. Which of the following associations is NOT correct?

a. gap junctions—allow communication between animal cells
b. plasmodesmata—allow communication between plant cells
c. desmosomes—allow inflexible attachments among cells
d. tight junctions—create watertight junctions between cells

ANSWER KEY

1. e
2. a
3. c
4. b
5. d
6. c
7. e
8. c
9. d
10. a
11. d
12. b
13. d
14. e
15. a

16. a
17. d
18. c
19. d
20. c
21. b
22. b
23. b
24. c
25. d
26. b
27. a
28. b
29. b
30. c

CHAPTER 6 ENERGY FLOW IN THE LIFE OF A CELL

OUTLINE

Section 6.1 What Is Energy?

Energy is defined as the capacity to do **work**, which is the force acting on an object that causes it to move. **Kinetic energy** is the energy of movement, whereas **potential energy** is stored energy. Energy can be changed from potential to kinetic energy (and vice versa) under the right conditions (**Figure 6-1**).

The basic properties of energy are described by the **laws of thermodynamics**. The **first law of thermodynamics** (i.e., the **law of conservation of energy**) states that energy can be neither created nor destroyed, but it can change forms. The **second law of thermodynamics** states that when energy changes forms, the amount of useful energy decreases (**Figure 6-2**).

Section 6.2 How Does Energy Flow in Chemical Reactions?

Chemical reactions form or break chemical bonds between atoms by converting **reactants** into **products**.

All chemical reactions either release energy (are **exergonic**, **Figure 6-3**) or require energy (are **endergonic**, **Figure 6-4**). The synthesis of large biological molecules from small subunits is an endergonic process.

All chemical reactions require **activation energy** to begin, usually in the form of kinetic energy from moving molecules (**Figure 6-6**).

Section 6.3 How Is Energy Transported Within Cells?

Adenosine triphosphate (ATP) is the most common **energy-carrier molecule** used by cells. ATP releases energy from its high-energy phosphate bonds when it is broken down into **adenosine diphosphate (ADP)** and phosphate (**Figure 6-8b**). Energy is stored when phosphate forms chemical bonds with ADP (**Figure 6-8a**).

Electron carriers can also transport energy by transferring energetic electrons within cells.

Coupled reactions occur when an exergonic reaction provides the energy needed for an endergonic reaction to occur. In cells, this often occurs through the use of energy-carrier molecules. During energy transfers between exergonic and endergonic reactions (**Figure 6-9**), some energy is lost as waste heat.

Section 6.4 How Do Enzymes Promote Biochemical Reactions?

Spontaneous reactions that sustain life proceed too slowly at body temperatures due to the high activation energies required. **Enzymes** are proteins that act as biological **catalysts** that reduce the activation energy and increase the speed of these reactions, thus allowing life to exist (**Figure 6-10**).

An enzyme functions when one or more **substrate** molecules fit into its **active site**. The substrates and active site change shape, promoting a reaction that forms a new molecular product (**Figure 6-11**).

Section 6.5 How Do Cells Regulate Their Metabolic Reactions?

The metabolic reactions of a cell are linked by **metabolic pathways** (**Figure 6-12**) that synthesize or break down molecules.

The rates of reactions tend to increase as the concentrations of substrates or enzymes increase.

Cells regulate metabolic reactions by controlling the rates of enzyme synthesis, by synthesizing enzymes in inactive forms, or by producing regulatory molecules that change enzyme activity (called **allosteric regulation**, **Figure 6-13**). **Feedback inhibition** is a form of allosteric regulation in which a metabolic pathway stops producing a product when it reaches a specific level (**Figure 6-14**).

Some poisons and drugs compete with substrates for the active site of an enzyme, disrupting that specific metabolic pathway (called **competitive inhibition**, **Figure 6-13b**).

Enzymes function optimally within narrow temperature and pH ranges (**Figure 6-15**) because suboptimal conditions disrupt their three-dimensional protein structure.

FLASH CARDS

To use the flash cards, tear the page from the book and cut along the dashed lines. The key term appears on one side of the flash card, and its definition appears on the opposite side.

activation energy	chemical reaction
active site	closed system
adenosine diphosphate (ADP)	coenzyme
adenosine triphosphate (ATP)	competitive inhibition
allosteric regulation	coupled reaction
catalyst	denature
chemical energy	electron carrier

The process that forms and breaks chemical bonds that hold atoms together in molecules.

In a chemical reaction, the energy needed to force the electron shells of reactants together, prior to the formation of products.

A hypothetical space where neither energy nor matter can enter or leave.

The region of an enzyme molecule that binds substrates and performs the catalytic function of the enzyme.

An organic molecule that is bound to certain enzymes and is required for the enzymes' proper functioning; typically, a nucleotide bound to a water-soluble vitamin.

A molecule composed of the sugar ribose, the base adenine, and two phosphate groups; a component of ATP.

The process by which two or more molecules that are somewhat similar in structure compete for the active site of an enzyme.

A molecule composed of the sugar ribose, the base adenine, and three phosphate groups; the major energy carrier in cells. The last two phosphate groups are attached by "high-energy" bonds.

A pair of reactions, one exergonic and one endergonic, that are linked together such that the energy produced by the exergonic reaction provides the energy needed to drive the endergonic reaction.

The process by which enzyme action is enhanced or inhibited by small organic molecules that act as regulators by binding to the enzyme at a regulatory site distinct from the active site, and altering the shape and/or function of the active site.

To disrupt the secondary and/or tertiary structure of a protein while leaving its amino acid sequence intact. Denatured proteins can no longer perform their biological functions.

A substance that speeds up a chemical reaction without itself being permanently changed in the process; a catalyst lowers the activation energy of a reaction.

A molecule that can reversibly gain or lose electrons. Electron carriers generally accept high-energy electrons produced during an exergonic reaction and donate the electrons to acceptor molecules that use the energy to drive endergonic reactions.

A form of potential energy that is stored in molecules and may be released during chemical reactions.

endergonic	first law of thermodynamics
energy	kinetic energy
energy-carrier molecule	lactose intolerance
entropy	law of conservation of energy
enzyme	laws of thermodynamics
exergonic	metabolic pathway
feedback inhibition	metabolism

The principle of physics that states that within any closed system, energy can be neither created nor destroyed, but can be converted from one form to another; also called the law of conservation of energy.

Pertaining to a chemical reaction that requires an input of energy to proceed; an "uphill" reaction.

The energy of movement; includes light, heat, mechanical movement, and electricity.

The capacity to do work.

The inability to break down lactose (milk sugar) because of the inability to produce the enzyme lactase.

A molecule that stores energy in "high-energy" chemical bonds and releases the energy to drive coupled endothermic reactions. In cells, ATP is the most common energy-carrier molecule.

The principle of physics that states that within any closed system, energy can be neither created nor destroyed, but can be converted from one form to another; also called the first law of thermodynamics.

A measure of the amount of randomness and disorder in a system.

The physical laws that define the basic properties and behavior of energy.

A protein catalyst that speeds up the rate of specific biological reactions.

A sequence of chemical reactions within a cell, in which the products of one reaction are the reactants for the next reaction.

Pertaining to a chemical reaction that releases energy (either as heat or in the form of increased entropy); a "downhill" reaction.

The sum of all chemical reactions that occur within a single cell or within all the cells of a multicellular organism.

In enzyme-mediated chemical reactions, the condition in which the product of a reaction inhibits one or more of the enzymes involved in synthesizing the product.

noncompetitive inhibition	reactant
phenylketonuria (PKU)	second law of thermodynamics
potential energy	substrate
product	work

An atom or molecule that is used up in a chemical reaction to form a product.

The process by which an inhibitory molecule binds to a site on an enzyme that is distinct from the active site. As a result, the enzyme's active site is distorted, making it less able to catalyze the reaction involving its normal substrate.

The principle of physics that states that any change in a closed system causes the quantity of concentrated, useful energy to decrease and the amount of randomness and disorder (entropy) to increase.

A disorder caused by a defect in the enzyme that normally oxidizes the amino acid phenylalanine; if untreated, causes severe mental retardation.

The atoms or molecules that are the reactants for an enzyme-catalyzed chemical reaction.

"Stored" energy; normally chemical energy or energy of position within a gravitational field.

Energy transferred to an object, usually causing the object to move.

An atom or molecule that is formed from reactants in a chemical reaction.

SELF TEST

1. The laws of thermodynamics define the properties and behavior of energy. The first law states that energy
   a. equals mass times the speed of light, squared (i.e., $E = mc^2$).
   b. can be created by thermonuclear explosions.
   c. cannot be created or destroyed but can be changed from one form into another.
   d. is the basic structure of the universe.

2. If all matter tends toward increasing randomness and disorder, how can life exist?
   a. Living things do not obey the second law of thermodynamics.
   b. There is a continuous input of energy from the sun.
   c. Living things do not require energy.
   d. Taken as a whole, the living biosphere of the planet constantly discards disorderly matter and replaces it with more orderly matter.

3. As a space shuttle leaves orbit, and enters the atmosphere, it begins to lose velocity, even though it also loses gravitational potential energy as it descends. What is happening to this potential energy?
   a. Objects are weightless in space, so the question is irrelevant.
   b. It is heating the shuttle and the atmosphere.
   c. It is being converted into the kinetic energy possessed by the moving shuttle.
   d. All of it is being captured in the fuel tanks of the reusable shuttle, to be used again during the next space launch.

4. Which of the following does not represent a form of potential energy?
   a. the load of fuel inside the storage tanks of a commercial jetliner.
   b. nutrition labels on packages of snack pretzels, which show that they each contain 50 calories.
   c. a plane traveling from New York to London at 0.8 times the speed of sound.
   d. a plane at an altitude of 10,000 meters.

5. When electrical energy is used to turn on a light bulb, the conversion from electrical energy to light energy is not 100% efficient. This loss of usable energy can be explained by
   a. the first law of thermodynamics.
   b. the second law of thermodynamics.
   c. a destruction of energy.
   d. a conversion to potential energy.

6. The second law of thermodynamics relates the organization of matter to energy. It states that unless additional energy is used, the orderliness of a system tends to _____, whereas entropy _____.
   a. increase; decreases
   b. decrease; increases
   c. stay the same; increases
   d. decrease; stays the same

7. In a chemical reaction, the _____ are the atoms or molecules that enter into the reaction; the _____ are the chemicals or atoms produced by the reaction.
   a. products; reactants
   b. reactants; products
   c. receptors; products
   d. catalysts; reactants

8. Cellulose (a major constituent of paper) is made from glucose. In the exergonic reaction in which cellulose reacts with oxygen (O_2) to produce carbon dioxide (CO_2) and water (H_2O), a common nonbiological source of the activation energy that triggers this reaction is
   a. a lit match.
   b. human body heat.
   c. atmospheric oxygen.
   d. decreasing entropy.
   e. ATP.

9. What is the ultimate source of energy that powers the endergonic reactions occurring in animals?
   a. the oxygen that plants release into the atmosphere as a waste product of photosynthesis
   b. the green wavelengths of light that are reflected by plants
   c. the sunlight absorbed by plants
   d. the carbohydrates produced by plants during photosynthesis

10. Activation energy is
    a. required for endergonic reactions.
    b. produced by exergonic reactions.
    c. required for all chemical reactions.
    d. produced by chemical reactions.

11. The importance of coupled reactions is that they
    a. allow an endergonic reaction to drive an exergonic reaction.
    b. allow an exergonic reaction to drive an endergonic reaction.
    c. turn potential energy into kinetic energy.
    d. violate the second law of thermodynamics.

12. What is the best interpretation of the figure, if we are considering the sizes of the reactant and product molecules?
    a. Exergonic processes create large products from small reactant molecules, and absorb the energy that they require from the endergonic processes that are forming small products from large reactant molecules.

b. Exergonic processes create small products from large reactant molecules, and absorb the energy that they require from the endergonic processes that are forming large products from small reactant molecules.

c. Exergonic processes create large products from small reactant molecules, and release the energy that is then required by the endergonic processes that are forming small products from large reactant molecules.

d. Exergonic processes create small products from large reactant molecules, and release the energy that is then required by the endergonic processes that are forming large products from small reactant molecules.

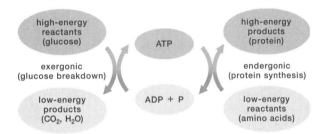

13. ATP is
   a. formed in an exergonic process that draws its energy from the endergonic breakdown of glucose into CO_2 and H_2O.
   b. used in the body almost as soon as it is generated.
   c. stored in the liver and muscles as a long-term energy reserve.
   d. used in very small amounts, so that our bodies only need the equivalent of micrograms of ATP per day.

14. ATP is well suited to its role as an energy-carrier molecule in cells because
   a. the covalent bond between the last two phosphates can be broken to release substantial amounts of energy.
   b. it contains covalent bonds.
   c. it is small and can fit into a lot of places in the cell.
   d. the covalent bonds between the last two phosphates are high-energy bonds that can absorb a substantial amount of energy when the bonds are broken.

15. Electron carrier molecules that transport energy include
   a. NAD^+.   d. CO_2.
   b. FAD.   e. both a and b.
   c. ADP.

16. How is ATP used in energy metabolism?
   a. ATP synthesis is coupled to the endergonic reactions of photosynthesis.
   b. ATP synthesis results from the reactions of protein synthesis.
   c. ATP synthesis is coupled to an exergonic reaction, the breakdown of glucose.
   d. ATP hydrolysis increases entropy.

17. Which of the following statements correctly describes how enzymes work?
   a. Enzymes catalyze specific chemical reactions because the shapes of their active sites allow only certain substrate molecules to enter.
   b. Each enzyme may catalyze a wide range of chemical reactions.
   c. Although substrates will change shape during their reactions, enzymes have fixed and unchangeable shapes.
   d. Both a and b are correct.
   e. Both b and c are correct.

18. Biological catalysts
   a. are organic substances that lower the activation energy required to initiate and speed up a reaction.
   b. are broken down and destroyed as part of the chemical process they help initiate.
   c. are seldom regulated by the molecules that participate in the catalyzed reactions.
   d. can speed up any reaction, even those that would not occur naturally.

19. Essentially all humans make the enzyme sucrase, which breaks down the disaccharide sucrose into its two monomers. However, only some adult humans make the enzyme lactase, which breaks down the disaccharide lactose into its two monomers. Given that both lactose and sucrose have the same numbers of oxygen, carbon, and hydrogen atoms, what is the best explanation for the fact that adults who are lactose-intolerant cannot use sucrase to break down lactose?
   a. The two monomers in sucrose are linked by weak hydrogen bonds, whereas the two monomers in lactose are linked by stronger covalent bonds, which the sucrase enzyme cannot break.
   b. Sucrase can use hydrolysis to break apart a wide variety of small polymers into their monomers.
   c. The sucrose and lactose molecules are shaped differently, and only sucrose fits properly into the active site of sucrase.
   d. Sucrose is soluble in water, whereas lactose is nonpolar and is thus always surrounded by

lipid molecules (the butterfat in milk) that block the water-soluble sucrase from reaching it.
  e. The activation energy for the breakdown of lactose is too low.

20. Alcohol dehydrogenase facilitates the transformation of ethanol (alcohol) into a nontoxic chemical form. In this situation, alcohol dehydrogenase is
  a. used up during the reaction, and must be continually regenerated every time that a person consumes an alcoholic beverage.
  b. a chemical product of the detoxification of ethanol.
  c. so specific that no other molecule, besides ethanol, can enter its active site.
  d. a catalyst that may facilitate the same chemical reaction many times.

21. The speed of a reaction is determined by its
  a. reactants.   c. activation energy.
  b. products.   d. potential energy.

22. The most important reason a particular enzyme can function only within certain limits of temperature, salt conditions, and pH is that changes in temperature, salt, and pH
  a. change the shape of an enzyme.
  b. alter the amount of substrates.

23. Pepsin is an enzyme that breaks down protein but will not act on starch. This fact is an indication that enzymes are
  a. catalytic.   c. specific.
  b. hydrolytic.   d. temperature sensitive.

24. A substance that is acted on by an enzyme to produce a product is called a(n)
  a. allosteric inhibitor.   c. substrate.
  b. coenzyme.   d. electron carrier.

25. Examine the figure showing the production of isoleucine from threonine, and how it is regulated. Each enzyme performs a crucial chemical transformation of its substrate to create a particular product. Suppose that a cell needs isoleucine, but that its enzyme #2 is faulty and is not able to convert intermediate A into B. How do you expect the concentrations of isoleucine, threonine, and intermediates A, B, C, and D to then differ, in comparison to the ideal situation where enzyme #2 is normal? Label these molecules, using greater than (>), less than (<), or equal (=) signs to indicate how you think their concentrations will differ (or not), in comparison to the ideal situation.

c. alter the amount of products.
d. lower the activation energy of a reaction.

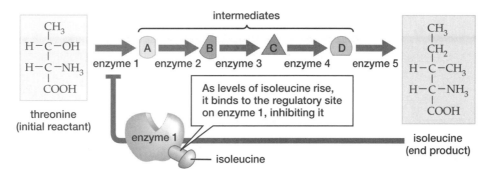

threonine (initial reactant) — enzyme 1 — A — enzyme 2 — B — enzyme 3 — C — enzyme 4 — D — enzyme 5 — isoleucine (end product)

intermediates

As levels of isoleucine rise, it binds to the regulatory site on enzyme 1, inhibiting it

enzyme 1 — isoleucine

26. Suppose that a person eats several entire packages of antacid tablets shortly after eating a large meal. What do we expect to happen?
  a. Digestion in the person's stomach will be slowed, because the resulting rise in pH will lower the activity of pepsin.
  b. Digestion in the person's stomach will happen more quickly, because the lowered pH will accelerate the activity of pepsin.
  c. With less stomach acid present, pepsin is much more likely to digest and damage the person's own stomach lining.
  d. Stomach acid will no longer be able to denature and destroy pepsin, thus the speed of digestion will increase.

27. Enzyme regulation can be precisely controlled through all of the following mechanisms, except by
  a. producing an inactive form of an enzyme that is activated only when needed.
  b. using a coenzyme that is necessary for function.
  c. binding a competitive inhibitor to the active site.
  d. having enzymes automatically stop functioning, after they have been used for a specified number of times.

28. The amino acid threonine is converted to isoleucine by a sequence of five enzymatic reactions. When isoleucine levels are high, the first

reaction in this sequence is "turned off." This is an example of

a. substrate activation.
b. feedback inhibition.
c. competitive inhibition.
d. coenzyme activation.

29. Lysosomes have an acidic interior (pH = 5), unlike the rest of the cell (pH = 7). Lysosomal enzymes are most active at

a. pH = 4.   c. pH = 7.
b. pH = 5.   d. pH = 9.

30. Why do most reactions occur more rapidly at high temperature?

a. The lower a temperature becomes, the faster molecules move.
b. Collisions between molecules are less frequent because the atoms in cells are farther apart at higher temperatures.
c. Collisions are hard enough to force electron shells to interact.
d. Higher temperatures cause substrate concentrations to increase.

ANSWER KEY

1. c
2. b
3. b
4. c
5. b
6. b
7. b
8. a
9. c
10. c
11. b
12. d
13. b
14. a
15. e
16. c
17. a
18. a
19. c
20. d
21. c
22. a
23. c
24. c
25. See figure below

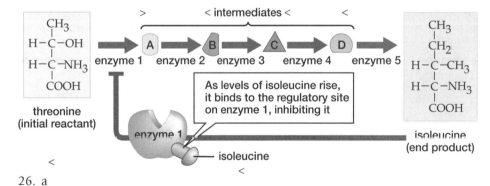

26. a
27. d
28. b
29. b
30. c

CHAPTER 7 CAPTURING SOLAR ENERGY: PHOTOSYNTHESIS

OUTLINE

Section 7.1 What Is Photosynthesis?

In **photosynthesis**, autotrophic cells capture light energy (in the presence of CO_2 and H_2O) and store it in the chemical bonds of glucose, releasing O_2 in the process.

Leaves are flattened structures that expose a large surface area to the sun. This allows **chloroplasts** to collect sunlight for use in photosynthesis (**Figure 7-1**).

Two chemical **light reactions** occur during photosynthesis: (1) *light-dependent reactions* that convert sunlight energy to energy-carrier molecules and (2) *light-independent reactions* that convert the energy in carrier molecules into glucose (**Figure 7-3**).

Section 7.2 Light Reactions: How Is Light Energy Converted to Chemical Energy?

Sunlight **photons** are captured by **chlorophyll** and **accessory pigment** molecules that allow chloroplasts to absorb light at different wavelengths (**Figure 7-4**).

Light-dependent reactions occur within the **thylakoid** membranes of chloroplasts in specialized **photosystems** (**Figure 7-6**). Photosystem II generates ATP by **chemiosmosis**, that is, by passing an energized electron along an **electron transport chain**. This electron becomes reenergized by photosystem I, which passes it to an electron transport chain that forms NADPH (**Figure 7-7**).

Some of the energy formed from these reactions is used to split H_2O, which provides O_2 as a waste product, and H^+ ions that facilitate ATP production.

Section 7.3 The Calvin Cycle: How Is Chemical Energy Stored in Sugar Molecules?

The **Calvin cycle** captures CO_2 and stores it in glucose molecules.

ATP and NADPH formed during the light-dependent reactions dissolve in the **stroma** fluid and are used by light-independent reactions to synthesize glucose from CO_2 and H_2O.

This process begins with the Calvin cycle (C_3 **pathway**), which has three major parts: (1) **carbon fixation**, (2) synthesis of G3P, and (3) regeneration of RuBP (**Figure 7-10**). During carbon fixation, plants capture CO_2 and fix its atoms in a larger molecule (RuBP), which then splits into two PGA molecules. During synthesis of G3P, energy in ATP and NADPH is used to convert PGA into G3P. During regeneration of RuBP, ATP is used to assemble RuBP from G3P. ADP and $NADP^+$ are formed during this process.

Two G3P molecules formed during the C_3 pathway are combined to form one glucose molecule. Glucose is then used to form sucrose, starch, or cellulose.

Section 7.4 Why Do Some Plants Use Alternate Pathways for Carbon Fixation?

To conserve water, leaf **stomata** close, causing leaf CO_2 levels to drop and leaf O_2 levels to rise. In C_3 **plants**, this causes the enzyme **rubisco** to combine O_2 with RuBP (**photorespiration**), which prevents carbon fixation and does not generate ATP.

C_4 **plants** minimize photorespiration by adding an additional step to the carbon-fixation process. In **mesophyll** cells, CO_2 is combined with phosphoenolpyruvic acid (PEP) to form the four-carbon molecule malate, which releases CO_2 after being transported to adjacent **bundle sheath cells**. This CO_2 is then fixed using the C_3 cycle (**Figure 7-11a**).

C_4 **pathways** use more energy than C_3 pathways. Thus, C_3 plants have an advantage in cool, wet, low-light environments, whereas C_4 plants have an advantage in hot, dry, well-lit environments.

CAM plants use a variation on the C_4 strategy, in which they open their stomata at night to capture CO_2 and temporarily store it as malate, and then fix it as carbohydrate during the day, using the light reactions of photosynthesis (**Figure 7-11b**). By keeping stomata closed during the heat of the day, they avoid most water loss. CAM plants include cacti and most other succulents.

FLASH CARDS

To use the flash cards, tear the page from the book and cut along the dashed lines. The key term appears on one side of the flash card, and its definition appears on the opposite side.

accessory pigment	C_4 plant
ATP synthase	Calvin cycle
bundle sheath cells	carbon fixation
C_3 pathway	carotenoid
C_3 plant	chemiosmosis
C_4 pathway	chlorophyll

A plant that relies on the C$_4$ pathway to fix carbon.

A colored molecule other than chlorophyll *a*, that absorbs light energy and passes it to chlorophyll *a*.

The cyclic series of reactions whereby carbon dioxide is fixed into carbohydrates during photosynthesis.

A channel protein in the thylakoid membranes of chloroplasts and the inner membrane of mitochondria that uses the energy of H$^+$ ions moving through the channel down their concentration gradient to produce ATP from ADP and inorganic phosphate.

The process by which carbon derived from carbon dioxide gas is captured in organic molecules during photosynthesis.

One of a group of cells that surround the veins of plants; in C$_4$ (but not in C$_3$) plants, bundle sheath cells contain chloroplasts.

A red, orange, or yellow pigment, found in chloroplasts, that serves as an accessory light-gathering pigment in thylakoid photosystems.

The reactions whereby carbon dioxide is fixed into phosphoglyceric acid during the Calvin cycle of photosynthesis.

A process of ATP generation in chloroplasts and mitochondria. The movement of electrons down an electron transport system is used to pump hydrogen ions across a membrane, thereby building up a concentration gradient of hydrogen ions; the hydrogen ions diffuse back across the membrane through the pores of ATP-synthesizing enzymes; the energy of their movement down their concentration gradient drives ATP synthesis.

A plant that relies on the C$_3$ pathway to fix carbon.

A pigment found in chloroplasts that captures light energy during photosynthesis; chlorophyll absorbs violet, blue, and red light but reflects green light.

The series of reactions in certain plants that fixes carbon dioxide into a four-carbon molecule, which is later broken down for use in the Calvin cycle of photosynthesis. This reduces wasteful photorespiration in hot, dry environments.

chlorophyll *a*

chloroplast

cuticle

electromagnetic spectrum

electron transport chain (ETC)

epidermis

light reactions

mesophyll

photon

photorespiration

photosynthesis

photosystem

The first stage of photosynthesis, in which the energy of light is captured as ATP and NADPH; occurs in thylakoids of chloroplasts.

The most abundant type of chlorophyll molecule in photosynthetic eukaryotic organisms and in cyanobacteria; chlorophyll *a* is found in the reaction centers of the photosystems.

Loosely packed, usually photosynthetic cells located beneath the epidermis of a leaf.

The organelle in plants and plantlike protists that is the site of photosynthesis; surrounded by a double membrane and containing an extensive internal membrane system that bears chlorophyll.

The smallest unit of light energy.

A waxy or fatty coating on the exposed surfaces of epidermal cells of many land plants, which aids in the retention of water.

A series of reactions in plants in which O_2 replaces CO_2 during the Calvin cycle, preventing carbon fixation; this wasteful process dominates when C_3 plants are forced to close their stomata to prevent water loss.

The range of all possible wavelengths of electromagnetic radiation, from wavelengths longer than radio waves, to microwaves, infrared, visible light, ultraviolet, x-rays, and gamma rays.

The complete series of chemical reactions in which the energy of light is used to synthesize high-energy organic molecules, usually carbohydrates, from low-energy inorganic molecules, usually carbon dioxide and water.

A series of electron-carrier molecules, found in the thylakoid membranes of chloroplasts and the inner membrane of mitochondria, that extract energy from electrons and generate ATP or other energetic molecules.

In thylakoid membranes, a cluster of chlorophyll, accessory pigment molecules, proteins, and other molecules that collectively capture light energy, transfer some of the energy to electrons, and transfer the energetic electrons to an adjacent electron transport chain.

In plants, the outermost layer of cells of a leaf, young root, or young stem.

pigment molecule

stoma (plural, stomata)

primary electron acceptor

stroma

reaction center

thylakoid

rubisco

An adjustable opening in the epidermis of a leaf or young stem, surrounded by a pair of guard cells, that regulates the diffusion of carbon dioxide and water into and out of the leaf.

The semifluid material inside chloroplasts in which the thylakoids are located; the site of the reactions of the Calvin cycle.

A disk-shaped, membranous sac found in chloroplasts, the membranes of which contain the photosystems, electron transport chains, and ATP-synthesizing enzymes used in the light reactions of photosynthesis.

A light-absorbing, colored molecule, such as chlorophyll, carotenoid, or melanin molecules.

A molecule in the reaction center of each photosystem that accepts an electron from one of the two reaction center chlorophyll *a* molecules and transfers the electron to an adjacent electron transport chain.

Two chlorophyll *a* molecules and a primary electron acceptor complexed with proteins and located near the center of each photosystem within the thylakoid membrane. Light energy is passed to one of the chlorophylls, which donates an energized electron to the primary electron acceptor, which then passes the electron to an adjacent electron transport chain.

In the carbon fixation step of the Calvin cycle, the enzyme that catalyzes the reaction between ribulose bisphosphate (RuBP) and carbon dioxide, thereby fixing the carbon of carbon dioxide in an organic molecule; short for ribulose bisphosphate carboxylase.

SELF TEST

1. In most land plants, photosynthesis occurs in cells of the _____ of the leaves, because these cells contain the largest numbers of chloroplasts.
   a. epidermis
   b. stomata
   c. cuticle
   d. mesophyll
   e. vascular bundles

2. Leaves include a number of structural modifications for the purpose of photosynthesis, including
   a. adjustable openings in the surface that permit the passage of CO_2, H_2O, and O_2.
   b. leaf surface coatings that accelerate the evaporation of water.
   c. photosynthetic mitochondria.
   d. the ability to primarily absorb the green wavelengths of light.

3. Which of the following is a true statement about photosynthesis?
   a. In photosynthesis, inorganic molecules such as carbon dioxide and water react to produce organic, energy-rich molecules such as glucose.
   b. In photosynthesis, oxygen is used to help break down glucose.
   c. Photosynthesis is an exergonic reaction.
   d. Photosynthesis is a process that is carried out primarily by autotrophic prokaryotic bacteria.

4. Light-independent, carbon-fixing reactions occur in
   a. guard cell cytoplasm.
   b. chloroplast stroma.
   c. the thylakoid membranes.
   d. the thylakoids.
   e. mitochondria.

5. Imagine that you are using instruments that are capable of measuring many different attributes of a living, sunlit leaf and its surrounding environment. What would you examine, as the most direct measure of the rate at which the Calvin cycle is operating?
   a. the rate of H_2O movement into the leaf, coming from the stem it is attached to
   b. the rate of O_2 production by the leaf
   c. the rate at which photons are being absorbed by the leaf
   d. the rate of ATP regeneration by the thylakoids
   e. the rate of CO_2 uptake by the leaf

6. Which substance or structure is found in both plants and animals?
   a. ATP
   b. thylakoid
   c. stroma
   d. stoma

7. What is the most correct statement about photosystems?
   a. All photosystem complexes within a chloroplast are identical and have the same function.
   b. The photosystems are located in the stroma of a chloroplast.
   c. Photosystems are arrays of light-absorbing pigments that eject highly energized electrons.
   d. Photosystem I provides a steady supply of electrons to photosystem II.
   e. Photosystems absorb all wavelengths of light equally.

8. During photosynthesis, electrons are continuously lost from the reaction center of photosystem II. What source is used to replace these electrons?
   a. sunlight
   b. oxygen
   c. water
   d. carbon dioxide

9. What is NOT true about accessory pigments?
   a. When accessory pigments absorb photons, they immediately pass high-energy electrons to the primary electron acceptors.
   b. At least one accessory pigment has a chemical structure similar to that of chlorophyll *a*.
   c. Accessory pigments tend to absorb light from regions closer to the middle of the visible light spectrum, whereas chlorophyll *a* absorbs light from the more extreme red and blue-violet ends of the visible spectrum.
   d. At least one accessory pigment is essential for maintaining human vision.

10. What is the most accurate (not necessarily complete) description of the different roles of photosystems I and II?
    a. Both PS I and PS II generate ATP, but PS II also generates NADPH.
    b. PS II captures light energy to generate ATP, whereas PS I captures chemical energy to generate NADPH.
    c. PS II generates both energized electrons and H^+ ions, whereas PS I generates NADPH.
    d. PS I splits water molecules to release energy, whereas PS II captures this energy and stores it in ATP.

11. Light absorption takes place in the
    a. thylakoid membrane.
    b. stroma.
    c. space within the thylakoid disk.
    d. mitochondria.

12. Suppose that you use a very small hypodermic syringe to inject a basic substance, such as an antacid, into the thylakoid spaces of a chloroplast in a well-lit leaf cell. What should happen next?
    a. The rate of ATP production by this chloroplast should increase.

b. There should be a reduction of the rate at which H^+ ions flow from the thylakoid space to the stroma.

c. Photosystem II should stop generating high-energy electrons.

d. The base should serve as a chemical substrate that accelerates production of NADPH.

e. The rate of glucose production by this chloroplast should be unchanged.

13. You have just discovered a new plant with red-orange leaves. Which wavelengths of visible light are NOT being absorbed by this pigment?
   a. green, blue, and violet
   b. green and yellow
   c. green
   d. red and orange
   e. blue and violet

14. Take a deep breath and slowly exhale. The oxygen in that breath is being used by your mitochondria in reactions that produce ATP from sugars and other food molecules you ate. Where did that oxygen come from originally?
   a. The oxygen in the atmosphere is produced by the breakdown of carbon dioxide during photosynthesis.
   b. Plants produce the oxygen via the process of photorespiration when they break down sugars in their mitochondria.
   c. The oxygen in the air is produced by the splitting of water during the light-dependent reactions of photosynthesis.
   d. Plants produce the oxygen via the process of respiration, in which they break down sugars at night when photosynthesis cannot occur.

15. In the light-dependent reactions of photosynthesis, the difference in hydrogen ion concentration across the thylakoid membrane is used to generate
   a. NADPH.   d. ATP.
   b. glucose.   e. oxygen.
   c. $FADH_2$.

16. Imagine that you are trying to set up a large fish tank and want to select a colored light that will show off the fish to best advantage but will also allow the green plants in the tank to grow and stay healthy. You decide to measure the efficiency of photosynthesis by looking at the production of oxygen bubbles on the leaves. At first, you use a white fluorescent lamp and see many oxygen bubbles on the leaves, indicating that photosynthesis is occurring normally. Next, you put a sheet of green cellophane between the white light and the water so that the light coming through to the tank appears green. Now, the bubbles will
   a. remain constant because photosynthesis can occur to some extent at all visible wavelengths of light.
   b. increase because chlorophyll is green, and photosynthesis will work best if leaves are exposed to green light.
   c. decrease because green light is poorly absorbed by chlorophyll.
   d. disappear entirely, because photosynthesis cannot occur at all in green light.

17. What is the role of electron transport in ATP synthesis?
   a. Electron transport creates a proton gradient that is used to supply energy for the synthesis of ATP.
   b. Electron transport chains donate the final electrons directly to ATP.
   c. Electron transport chains donate electrons to phosphate, which transfers them to ATP.
   d. Electron transport chains are necessary to split water to generate oxygen.

18. Which of the following is NOT required for the light-independent reactions of photosynthesis?
   a. ATP   d. CO_2
   b. stroma   e. H_2O
   c. NADPH   f. O_2

19. Examine the figure on the following page, which shows the conversions of various intermediate molecules during the Calvin cycle. If NADPH was not being provided via the light reactions, then when a chloroplast attempts to transform PGA into G3P, what would end up "missing" from the G3P?
   a. some oxygen atoms
   b. some carbon atoms
   c. some hydrogen atoms
   d. several ATP molecules

20. The dark reactions of photosynthesis
   a. convert carbon dioxide to sugar.
   b. generate ATP and NADPH.
   c. make oxygen.
   d. use the atoms in ATP to make sugar.

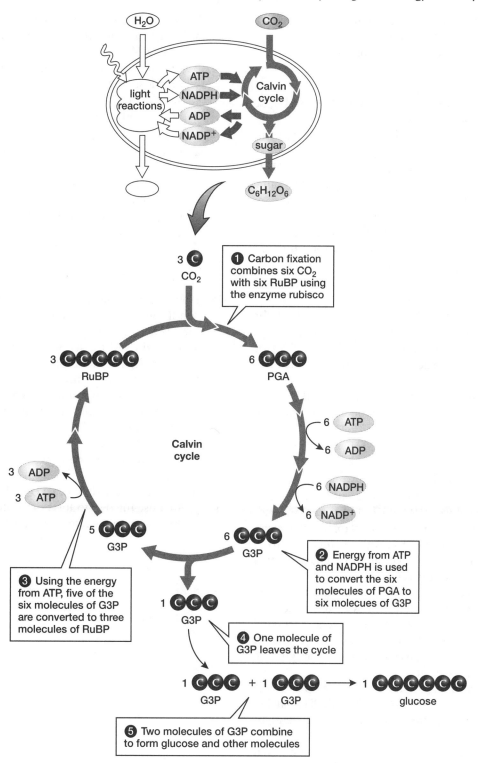

21. Dark reactions take place in the
    a. thylakoid membrane.
    b. space within thylakoid disks.
    c. mitochondria.
    d. stroma.

22. What is the linkage between the light and dark reactions?
    a. The light reactions are endergonic and the dark reactions are exergonic.
    b. The light reactions produce ATP and NADPH, and the dark reactions require ATP and NADPH.
    c. The light reactions pass electrons via an electron transport chain to the dark reactions.
    d. The light reactions produce carbon dioxide and the dark reactions use carbon dioxide.

23. Why is the use of carbon dioxide in the dark reactions important?
    a. The carbon dioxide produced by the light reactions is used by the dark reactions.
    b. The plant would not be able to get rid of carbon dioxide if the dark reactions did not use it.
    c. Fixation of carbon dioxide in the dark reactions converts the carbon into an organic form usable by all life forms.
    d. The dark reactions convert carbon dioxide directly into ATP to make energy.

24. Normally, in the Calvin cycle, one out of every six G3P molecules is shunted out of the cycle to produce glucose. What would happen, instead, if a chloroplast began to divert three out of every six G3P molecules to produce glucose, during each "turn" of the cycle, and to then export this glucose out of the cell?
    a. The enzyme, rubisco, would be converted into additional RuBP, allowing the cell to continue absorbing CO_2 from the atmosphere.
    b. The cell would lose its capacity to absorb CO_2, lowering its ability to produce glucose in the long run.
    c. The stroma would quickly develop a critical shortage of the ATP and NADPH that are being supplied by the light-dependent reactions.
    d. The chloroplast's supply of RuBP, the initial CO_2 acceptor molecule, would increase.

25. Photorespiration is the process by which
    a. plants produce energy at night.
    b. plant cells cool off in hot climates.
    c. sugar production is prevented in C_3 plants when CO_2 levels are low and O_2 levels are high.
    d. plants capture light energy and convert it into ATP.

26. Photorespiration occurs when
    a. oxygen is combined with ribulose bisphosphate.
    b. carbon dioxide is combined with ribulose bisphosphate.
    c. stomata are closed.
    d. both a and c.

27. How are C_4 plants different from C_3 plants?
    a. C_4 plants function better in low-light conditions.
    b. C_4 plants function better in relatively moist conditions.
    c. C_4 plants function better in bright-light conditions.
    d. C_4 plants function better in relatively dry conditions.
    e. Both c and d.

28. How have plants adapted to environmental conditions that would result in increased photorespiration?
    a. C_4 and CAM plants have eliminated use of the faulty enzyme rubisco.
    b. PEP combines with carbon dioxide only when oxygen concentrations are very low.
    c. A molecule transports temporarily stored carbon into cells where the normal C_3 photosynthetic pathway becomes favored due to the now higher concentrations of carbon dioxide.
    d. C_4 plants have developed chemical pathways that use the by-products of photorespiration to accelerate the rate of photosynthesis.

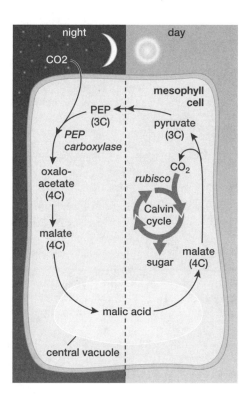

29. Examine the figure, showing how CAM plants temporarily store CO_2 as malic acid inside their central vacuoles. Succulent plants that are using the CAM pathway actually taste more acidic at certain times of the day or night than at others. Suppose that you bite into one. When should it have the most acidic taste?
   a. at 6 A.M., sunrise
   b. at 12 noon, midday
   c. at 5 P.M., late afternoon
   d. at 9 P.M., early evening

30. If we compare C_4 and CAM plants, what is one important aspect that is similar?
   a. Both types of plants initially store incoming CO_2 as part of a four-carbon molecule.
   b. Both types of plants take up CO_2 from the atmosphere during the daytime.
   c. Both types of plants use bundle sheath cells.
   d. Both types of plants typically have their stomata open at night, rather than during the day.

ANSWER KEY

1. d	16. c
2. a	17. a
3. a	18. f
4. b	19. c
5. e	20. a
6. a	21. d
7. c	22. b
8. c	23. c
9. a	24. b
10. c	25. c
11. a	26. d
12. b	27. e
13. d	28. c
14. c	29. a
15. d	30. a

CHAPTER 8 HARVESTING ENERGY: GLYCOLYSIS AND CELLULAR RESPIRATION

OUTLINE

Section 8.1 How Do Cells Obtain Energy?

Cells commonly obtain energy by breaking the chemical bonds of glucose molecules and using this energy to form **adenosine triphosphate (ATP)** (**Figure 8-2**).

The first step of glucose metabolism (**glycolysis**) occurs in the cytosol and breaks down glucose into pyruvate and two ATP molecules. If oxygen is absent (**anaerobic** conditions), **fermentation** converts pyruvate to either lactate or ethanol and CO_2.

If oxygen is present (**aerobic** conditions), eukaryotic cell **mitochondria** undergo **cellular respiration** to break pyruvate into CO_2 and H_2O, forming significantly more ATP than glycolysis (34 to 36 ATP).

Section 8.2 What Happens During Glycolysis?

Glycolysis occurs in two steps: (1) glucose activation and (2) energy harvest (**Figure 8-3**).

During glucose activation, two ATP molecules react with glucose to form fructose bisphosphate, which is an unstable "activated" molecule.

During energy harvest, fructose bisphosphate splits into two glyceraldehyde-3-phosphate (G3P) molecules that are then converted into two pyruvate molecules, which results in the production of four ATP and two **nicotinamide adenine dinucleotide (NADH)** carrier molecules.

Section 8.3 What Happens During Cellular Respiration?

Pyruvate is first broken down in the mitochondrial **matrix** in two stages: (1) acetyl CoA formation and (2) the **Krebs cycle** (**Figure 8-5**).

During acetyl CoA formation, each pyruvate reacts with coenzyme A, splitting it into CO_2 and forming acetyl CoA. NADH is also formed during this process.

During the Krebs cycle, each acetyl CoA combines with oxaloacetate to form citrate. Citrate is then processed by mitochondrial enzymes, regenerating oxaloacetate and producing two CO_2, one ATP, and four electron-carrier molecules (three NADH and one **flavin adenine dinucleotide [FADH$_2$]**).

The electron-carrier molecules produced in glycolysis and mitochondrial matrix reactions (NADH and FADH$_2$) deposit their electrons in the **electron transport chains (ETCs)** that are embedded in the inner mitochondrial matrix. The deposited electrons move along the chain, generating energy used to pump H^+ to the **intermembrane space** of the mitochondria, forming a H^+ gradient across the inner mitochondrial membrane. Electrons that reach the end of the ETC are accepted by hydrogen and oxygen to form water, clearing space on the ETC for the movement of additional electrons.

The resulting H^+ gradient across the inner mitochondrial membrane drives the **chemiosmotic** production of 32 or 34 ATP molecules by ATP-synthesizing enzymes (**Figure 8-6**). The total process of aerobic glucose metabolism, including the activities of glycolysis, the Krebs cycle, and the electron transport chain, yields 36 or 38 ATP molecules per molecule of glucose (**Figure 8-7**).

Section 8.4 What Happens During Fermentation?

Under anaerobic conditions, fermentation reactions convert pyruvate into either lactate or ethanol and CO_2. Both fermentation pathways produce NAD^+ electron-carrier molecules from NADH, which are required for glycolysis to continue to occur (**Figures 8-8 and 8-9**).

Lactate is produced in cells that are working so hard that they are carrying out glycolysis more rapidly than the resulting pyruvate is being oxidized by the Krebs cycle. In this case, **lactic acid fermentation** temporarily converts some pyruvate into lactic acid.

Some microbes such as yeasts carry out **alcoholic fermentation** when deprived of oxygen. In this case, the pyruvate created by glycolysis is converted to ethanol and CO_2.

FLASH CARDS

To use the flash cards, tear the page from the book and cut along the dashed lines. The key term appears on one side of the flash card, and its definition appears on the opposite side.

adenosine diphosphate (ADP)	chemiosmosis
adenosine triphosphate (ATP)	citric acid cycle
aerobic	electron transport chain (ETC)
alcoholic fermentation	fermentation
anaerobic	flavin adenine dinucleotide (FAD or FADH$_2$)
cellular respiration	glycolysis

A process of ATP generation in chloroplasts and mitochondria. The movement of electrons down an electron transport system is used to pump hydrogen ions across a membrane, thereby building up a concentration gradient of hydrogen ions; the hydrogen ions diffuse back across the membrane through the pores of ATP-synthesizing enzymes; the energy of their movement down their concentration gradient drives ATP synthesis.

A molecule composed of the sugar ribose, the base adenine, and two phosphate groups; a component of ATP.

A cyclic series of reactions, occurring in the matrix of mitochondria, in which the acetyl groups from the pyruvic acids produced by glycolysis are broken down to CO_2 accompanied by the formation of ATP and electron carriers; also called the Krebs cycle.

A molecule composed of the sugar ribose, the base adenine, and three phosphate groups; the major energy carrier in cells. The last two phosphate groups are attached by "high-energy" bonds.

A series of electron-carrier molecules, found in the thylakoid membranes of chloroplasts and the inner membrane of mitochondria, that extract energy from electrons and generate ATP or other energetic molecules.

Using oxygen.

Anaerobic reactions that convert the pyruvic acid produced by glycolysis into lactic acid or alcohol and CO_2, using hydrogen ions and electrons from NADH; the primary function of fermentation is to regenerate NAD^+ so that glycolysis can continue under anaerobic conditions.

A type of fermentation in which pyruvate is converted to ethanol (a type of alcohol) and carbon dioxide, using hydrogen ions and electrons from NADH; the primary function of alcoholic fermentation is to regenerate NAD^+ so that glycolysis can continue under anaerobic conditions.

An electron-carrier molecule produced in the mitochondrial matrix by the Krebs cycle; subsequently donates electrons to the electron transport chain.

Not using oxygen.

Reactions, carried out in the cytoplasm, that break down glucose into two molecules of pyruvic acid, producing two ATP molecules; does not require oxygen but can proceed when oxygen is present.

The oxygen-requiring reactions, occurring in mitochondria, that break down the end products of glycolysis into carbon dioxide and water while capturing large amounts of energy as ATP.

intermembrane space

matrix

Krebs cycle

mitochondrion

lactic acid fermentation

nicotinamide adenine dinucleotide
(NAD^+ or NADH)

The fluid contained within the inner membrane of a mitochondrion.

The fluid-filled space between the inner and outer membranes of a mitochondrion.

An organelle, bounded by two membranes, that is the site of the reactions of aerobic metabolism.

A cyclic series of reactions, occurring in the matrix of mitochondria, in which the acetyl groups from the pyruvic acids produced by glycolysis are broken down to CO_2 accompanied by the formation of ATP and electron carriers; also called the citric acid cycle.

An electron-carrier molecule produced in the cytoplasmic fluid by glycolysis and in the mitochondrial matrix by the Krebs cycle; subsequently donates electrons to the electron transport chain.

Anaerobic reactions that convert the pyruvic acid produced by glycolysis into lactic acid, using hydrogen ions and electrons from NADH; the primary function of lactic acid fermentation is to regenerate NAD^+ so that glycolysis can continue under anaerobic conditions.

SELF TEST

1. In eukaryotic cells, glycolysis occurs in the _____, and cellular respiration occurs in the _____.
   a. mitochondria; cytoplasm
   b. cytoplasm; mitochondria
   c. cytoplasm; chloroplasts
   d. chloroplasts; mitochondria

2. Photosynthesis and glucose metabolism are related because the
   a. products of photosynthesis are the raw materials for glucose metabolism.
   b. products of glucose metabolism are the raw materials for photosynthesis.
   c. products of photosynthesis are the same as the products of glucose metabolism.
   d. raw materials of photosynthesis are the same as the raw materials of glucose metabolism.
   e. both a and b.

3. The overall equation for glucose metabolism is $C_6H_{12}O_6 + 6 O_2 \rightarrow 6 CO_2 + 6 H_2O + ATP$ and heat. The carbon atoms in the CO_2 molecules in this equation come from _____ during reactions of _____.
   a. O_2; glycolysis
   b. O_2; the electron transport system
   c. O_2; the Krebs cycle
   d. $C_6H_{12}O_6$; glycolysis
   e. $C_6H_{12}O_6$; the electron transport system
   f. $C_6H_{12}O_6$; the Krebs cycle

4. Respiration is the process of gas exchange (breathing in oxygen and breathing out carbon dioxide); cellular respiration is the process of
   a. cellular gas exchange.
   b. cellular cooling.
   c. production of ATP via the electron transport system.
   d. cellular reproduction.

5. During the metabolic breakdown of glucose, most energy harvest occurs during the operations of
   a. glycolysis, in the cytoplasm.
   b. fermentation, in the cytoplasm.
   c. the Krebs cycle, in the mitochondria.

   d. the electron transport chain, in the mitochondria.
   e. chemiosmosis, involving ATP synthase.

6. If the metabolism of glucose had the same efficiency in our cells as the burning of gasoline has in car engines, our cells would generate about _____ molecules of ATP for each molecule of glucose used.
   a. 7   c. 23
   b. 15   d. 34

7. The energy-harvesting reactions of glycolysis produce four molecules of _____, two molecules of _____, and two molecules of _____.
   a. ATP; glyceraldehyde-3-phosphate; pyruvate
   b. ATP; NADH; pyruvate
   c. glucose; carbon dioxide; water
   d. pyruvate; glyceraldehyde-3-phosphate; water

8. What are the two parts of glycolysis?
   a. glucose activation and energy harvest
   b. fermentation and cellular respiration
   c. extracellular and intracellular events
   d. electron transport chain activity and CO_2 release

9. The production of which molecule marks the end of glycolysis and the beginning of cellular respiration?
   a. coenzyme A (CoA)   c. citrate
   b. acetyl CoA   d. pyruvate

10. Examine the figure below. Aerobic respiration regenerates the NAD^+ that is used during glycolysis. If the oxygen supply is cut off, and NAD^+ is no longer being generated by any means, how would this most directly affect the glycolysis process?
    a. The cell cannot convert glucose into fructose bisphosphate.
    b. Glycolysis will run in reverse to create more NAD^+, with pyruvate producing G3P and NAD^+ at the cost of two ATP molecules.
    c. The cell can no longer utilize the ATP supply that it already has on hand.
    d. The cell can no longer make pyruvate and NADH.

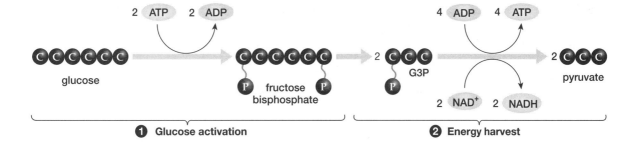

11. In glycolysis, if we compare two pyruvate molecules to one glucose molecule, it is NOT true that the two pyruvate molecules, combined,
    a. contain less stored chemical energy than the glucose.
    b. contain fewer hydrogen atoms than the glucose.
    c. contain the same number of carbon atoms as the glucose.
    d. contain more phosphorus atoms than the glucose.

12. In aerobic organisms growing in the presence of oxygen, the NADH produced by glycolysis ultimately donates its high-energy electrons to
    a. electron transport chains in the mitochondria.
    b. glucose.
    c. pyruvate.
    d. ATP.

13. Suppose that the reactions of the mitochondria of a green plant were completely inhibited. Which process would immediately stop?
    a. glycolysis
    b. fermentation
    c. photosynthesis
    d. ATP production
    e. lactate production

14. Oxygen that is breathed in during respiration is used in cellular respiration by being
    a. converted into CO_2 and exhaled.
    b. converted into ATP.
    c. the final electron acceptor of the electron transport system.
    d. used to produce glucose.

15. In eukaryotes, during the process of chemiosmosis, ATP is produced as hydrogen ions move from _____ to _____, passing through _____.
    a. the intermembrane compartment; the matrix; an ATP synthase
    b. the matrix; the intermembrane compartment; an ATP synthase
    c. the cytoplasm; the matrix; the electron transport system
    d. the matrix; the cytoplasm; the Krebs cycle

16. In eukaryotic cells, the enzymes for the Krebs cycle are located in the _____, and those for the electron transport system are located in the _____.
    a. cytoplasm; cell wall
    b. cytoplasm; mitochondrial matrix
    c. plasma membrane; cytoplasm
    d. mitochondrial matrix; inner mitochondrial membrane
    e. inner mitochondrial membrane; matrix

17. As high-energy electrons are passed from carrier to carrier along the electron transport system in cellular respiration, the electrons lose energy. Some of that energy is directly used to
    a. synthesize glucose.
    b. break down glucose.
    c. pump hydrogen ions across a membrane.
    d. synthesize ATP.

18. How is electron transport related in chloroplasts and mitochondria?
    a. Electron transport is used in each case to synthesize ATP, although the electron donors and final electron acceptors differ in the two systems.
    b. They have no similarities at all.
    c. The chloroplast system depends on intact membranes, whereas the mitochondrial does not.
    d. Only the mitochondrial electron transport system can synthesize ATP.

19. When glucose goes through glycolysis and all the way through the Krebs cycle, the six carbon molecules of glucose
    a. become pyruvate and G3P.
    b. become carbon dioxide.
    c. become joined to NADH.
    d. are used to make more glucose.

20. What are the most important direct products of the Krebs cycle?
    a. 34 to 36 ATP molecules generated per glucose molecule oxidized
    b. 2 acetyl CoA molecules generated per glucose molecule oxidized
    c. Many electron-carrier molecules, NADH, and $FADH_2$
    d. approximately 72 H^+ ions, which are moved into the intermembrane space, for every glucose molecule oxidized

21. Examine the figure at the top of the following page. What fact is NOT implied by it?
    a. ATP, once it is generated within a mitochondrion, must then be passed through two different membranes before reaching the cytoplasmic fluid of the cell.
    b. ATP is an acceptor of high-energy electrons.
    c. The intermembrane space has a lower pH than does the matrix of the mitochondrion.
    d. The inner mitochondrial membrane holds the proteins that form the electron transport chain.

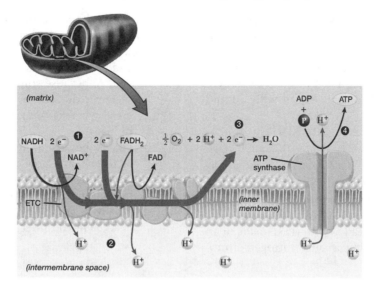

22. The oxygen atom in the H_2O that is produced as a waste product of cellular respiration comes from
    a. carbon dioxide (CO_2).
    b. glucose ($C_6H_{12}O_6$).
    c. oxaloacetate.
    d. oxygen (O_2).

23. A poison that interferes with the operation of the electron transport chain would have an effect most similar to
    a. rattlesnake venom, which can cause death in several hours by disrupting blood cells.
    b. the venom of the brown recluse spider, which can cause massive tissue damage.
    c. nerve gas, which can kill in seconds by blocking nerve function.
    d. carbon monoxide gas, which can kill in minutes or hours by blocking blood oxygen transport.
    e. hydrogen cyanide, which can kill in seconds or minutes.

24. Which of the following statements concerning fermentation is (are) true?
    a. Fermentation requires oxygen.
    b. Fermentation, like glycolysis, occurs in the cytoplasm of cells.
    c. Fermentation produces additional ATP.
    d. The end product of fermentation in human cells is ethanol.

25. At the beginning of most recipes for bread, you are instructed to dissolve the yeast in a mixture of sugar (sucrose) and hot water, in some cases with a small amount of flour. Within a short time, this yeast mixture begins to bubble and

foam, perhaps to the point of overflowing the container. What is happening?
    a. The bubbles are carbon dioxide that yeast produce as they break down the glucose and produce ATP via fermentation.
    b. The bubbles are oxygen produced by yeast as they grow.
    c. The bubbles are detergents that yeast produce to help them digest the proteins in the flour.
    d. The bubbles are water vapor produced as the hot water evaporates.

26. Which molecules are produced in glycolysis and used in fermentation?
    a. acetyl CoA and NADH
    b. pyruvate and NADH
    c. glucose, ATP, and NAD^+
    d. lactate, ATP, and CO_2
    e. pyruvate and ATP

27. How many molecules of ATP would be produced from 20 molecules of glucose at the end of fermentation?
    a. 10   d. 40
    b. 20   e. 100
    c. 30

28. What is the role of fermentation in glucose metabolism?
    a. Fermentation extracts the maximum ATP from glucose.
    b. Fermentation is necessary under aerobic conditions.
    c. Fermentation regenerates NAD^+ under anaerobic conditions.
    d. Fermentation allows the cell to use glucose under aerobic conditions.

29. Suppose that a person is suddenly deprived of oxygen. If he metabolically switches from aerobic respiration to fermentation, then to obtain the same amount of ATP as before, he must increase his rate of glycolysis by a factor of almost _____ times, and must also deal with a rapid accumulation of _____.
    a. 10; pyruvate   d. 20; $FADH_2$
    b. 10; ethanol   e. 40; lactic acid
    c. 20; lactic acid

30. Suppose that in a laboratory, one loaf of bread (A) is kneaded with yeast inside of an enclosure containing a normal atmosphere, while another loaf (B) is kneaded with yeast inside of an enclosure containing a pure nitrogen atmosphere at the same pressure and temperature. Both loaves are allowed to rise, inside of their respective enclosures, for 1 hour. Which loaf should rise the most and why?
    a. Loaf A, because some of its glucose is completely oxidized to CO_2 during the Krebs cycle, whereas none is completely oxidized in loaf B.
    b. Loaf A, because in loaf B, the yeast produce lactic acid and no CO_2 at all.
    c. Loaf B, because more acetyl CoA enters the mitochondria than with loaf A.
    d. Loaf B, because the rate of electron transport is accelerated under low-oxygen conditions.

ANSWER KEY

1. b	16. d
2. e	17. c
3. f	18. a
4. c	19. b
5. e	20. c
6. c	21. b
7. b	22. d
8. a	23. e
9. d	24. b
10. d	25. a
11. d	26. b
12. a	27. d
13. d	28. c
14. c	29. c
15. a	30. a

CHAPTER 9 THE CONTINUITY OF LIFE: CELLULAR REPRODUCTION

OUTLINE

Section 9.1 Why Do Cells Divide?

When cells divide, the parent cell passes hereditary information to its daughter cells in the form of **DNA** that contains **genes**, the instructions for making essential proteins (**Figure 9-1**).

Cell division is required for growth and development. In multicelled organisms such as animals, many cells **differentiate**, taking on specialized roles. Highly differentiated cells either lose the ability to divide further, or else can only divide to make more cells of the same type. **Stem cells** can divide to produce many different cell types.

Cell division is required for reproduction (**Figure 9-2**). **Asexual reproduction** (**cloning**) involves one parent producing offspring that are genetically identical to it. In **sexual reproduction**, two parents each make specialized reproductive cells (**gametes**) that fuse to produce unique offspring containing some genetic information from each parent.

Section 9.2 What Occurs During the Prokaryotic Cell Cycle?

The prokaryotic **cell cycle** consists of growth and DNA replication, followed by **binary fission**. Binary fission results in the formation of two cells with identical DNA (**Figure 9-3**).

Section 9.3 How Is the DNA in Eukaryotic Chromosomes Organized?

Eukaryotic **chromosomes** consist of linear DNA molecules bound to proteins. This binding allows otherwise long molecules to be wrapped as tightly coiled packages (**Figure 9-4**). Chromosomes are composed of **telomeres** (regions that stabilize the ends of a chromosome) and a **centromere** (which joins replicated DNA helices) (**Figure 9-5a**).

Duplicated chromosomes are composed of identical sister **chromatids** that can separate to form independent chromosomes (**Figures 9-5b** and **9-5c**).

Eukaryotic chromosomes typically occur in homologous pairs with similar genetic information. Cells with these pairs (most of them) are called **diploid**, whereas those with only one of each chromosome pair are called **haploid**. Cell chromosomes have both **autosomes** and **sex chromosomes** (**Figure 9-6**).

Section 9.4 What Occurs During the Eukaryotic Cell Cycle?

The eukaryotic cell cycle consists of **interphase** and cell division. During interphase, the cell grows and replicates its DNA over three phases (G_1, S, and G_2). In the G_1 phase, the cell grows and acquires materials needed for cell division. DNA synthesis occurs during the S phase. Growth is completed during the G_2 phase (**Figure 9-7**). Some cells differentiate once they are formed, and never divide again.

During cell division, a cell may duplicate itself asexually by **mitosis** (nuclear division) followed by **cytokinesis**. Mitosis can be used to maintain and repair tissues. Alternatively, cells of the ovaries and testes may form genetically unique gametes by **meiosis**, in anticipation of sexual reproduction.

Section 9.5 How Does Mitotic Cell Division Produce Genetically Identical Daughter Cells?

Mitotic cell division involves division of the nucleus (mitosis) and of the cytoplasm (cytokinesis). During interphase, chromosomes are duplicated, forming sister chromatids. These then enter the four stages of mitosis: **prophase**, **metaphase**, **anaphase**, and **telophase**. Cytokinesis usually follows telophase (**Figure 9-8**).

During prophase, chromosomes condense and their **kinetochores** attach to **spindle microtubules**. The nuclear membrane breaks down.

During metaphase, sister chromatids move to the equator of the cell.

During anaphase, sister chromatids separate and are pulled along their spindle microtubules to the cell poles.

During telophase, the chromosomes uncoil and nuclear envelopes re-form around each nucleus. During cytokinesis, the cytoplasm is divided into nearly equal halves by either contracting microfilaments (animal cells) or the fusion of vesicles along the center of the cell to form a **cell plate** (plant cells) (**Figures 9-8g** and **9-9**).

Section 9.6 How Is the Cell Cycle Controlled?

Complex protein interactions drive the cell cycle. Cells are usually in the G_1 phase and normally only begin preparations for entering mitosis after receiving a signal from a growth factor (**Figure 9-10**).

The major **checkpoints** by which the cell cycle is regulated occur (1) between G_1 and S, (2) between G_2 and mitosis, and (3) between metaphase and anaphase (**Figure 9-11**). Cancer can be a result of faulty checkpoint control.

Section 9.7 Why Do So Many Organisms Reproduce Sexually?

Mutations are the origin of genetic variability, forming **alleles** that can produce differences in structure and function (**Figure 9-12**).

Sexual reproduction can combine parental alleles in unique ways within offspring, creating variation that can improve the survival chances of an organism, as well as its ability to reproduce.

Section 9.8 How Does Meiotic Cell Division Produce Haploid Cells?

Meiosis produces haploid cells by separating **homologous chromosomes** so that each cell contains only one **homologue** (**Figure 9-13**).

Chromosomes are duplicated during interphase, and then the cell undergoes two specialized cell divisions (meiosis I and meiosis II) to produce four haploid cells (**Figure 9-15**).

Meiosis I: During prophase I, homologous duplicated chromosomes pair up, and **crossing over** occurs (**Figure 9-16**). During metaphase I, paired homologous chromosomes line up along the cell equator. During anaphase I, paired homologous chromosomes separate and are pulled to opposite poles of the cell. New nuclei form during telophase I. Because each daughter cell has only one of each pair of homologues, it is considered haploid.

Meiosis II: Both haploid daughter nuclei undergo divisions similar to mitosis. Nuclear membranes break down during prophase II. During metaphase II, the two chromatids of each chromosome move to the cell equator. During anaphase II, the two chromatids separate, each moving to opposite cell poles. During telophase II, nuclear membranes re-form, producing four haploid nuclei. Cytokinesis typically occurs after (or during) telophase II, producing four haploid cells.

Section 9.9 When Do Mitotic and Meiotic Cell Division Occur in the Life Cycles of Eukaryotes?

Eukaryotic life cycles have three parts: (1) fertilization causes the fusion of gametes, forming a diploid cell; (2) **meiotic cell division** occurs, creating haploid cells; and (3) the growth of multicellular bodies or asexual reproduction via mitosis occurs (**Figure 9-17**).

The proportion of time spent in these stages is dependent on the life cycle of the particular species. Haploid life cycles predominantly consist of haploid cells (**Figure 9-18**). Diploid life cycles predominantly consist of diploid cells (**Figure 9-19**). Alternation of generations life cycles have both diploid and haploid multicellular stages (**Figure 9-20**).

Section 9.10 How Do Meiosis and Sexual Reproduction Produce Genetic Variability?

Genetic variability occurs through the shuffling of homologous maternal and paternal chromosomes during metaphase I (**Figure 9-21**), crossing over during prophase I, and the fusion of gametes during fertilization.

FLASH CARDS

To use the flash cards, tear the page from the book and cut along the dashed lines. The key term appears on one side of the flash card, and its definition appears on the opposite side.

allele	cell division
anaphase	cell plate
asexual reproduction	centriole
autosome	centromere
binary fission	checkpoint
cell cycle	chiasma (plural, chiasmata)

Splitting of one cell into two; the process of cellular reproduction.

One of several alternative forms of a particular gene.

In plant cell division, a series of vesicles that fuse to form the new plasma membranes and cell wall separating the daughter cells.

In mitosis, the stage in which the sister chromatids of each chromosome separate from one another and are moved to opposite poles of the cell; in meiosis I, the stage in which homologous chromosomes, consisting of two sister chromatids, are separated; in meiosis II, the stage in which the sister chromatids of each chromosome separate from one another and are moved to opposite poles of the cell.

In animal cells, a short, barrel-shaped ring consisting of nine microtubule triplets; a pair of centrioles is found near the nucleus and may play a role in the organization of the spindle; centrioles also give rise to the basal bodies at the base of each cilium and flagellum that give rise to the microtubules of cilia and flagella.

Reproduction that does not involve the fusion of haploid gametes.

The region of a replicated chromosome at which the sister chromatids are held together until they separate during cell division.

A chromosome that occurs in homologous pairs in both males and females and that does not bear the genes determining sex.

A mechanism in the eukaryotic cell cycle by which protein complexes in the cell determine whether the cell has successfully completed a specific process that is essential to successful cell division, such as the accurate replication of chromosomes.

The process by which a single bacterium divides in half, producing two identical offspring.

A point at which a chromatid of one chromosome crosses with a chromatid of the homologous chromosome during prophase I of meiosis; the site of exchange of chromosomal material between chromosomes.

The sequence of events in the life of a cell, from one cell division to the next.

chromatid	daughter cell
chromosome	deoxyribonucleic acid (DNA)
clone	differentiate
cloning	diploid
crossing over	duplicated chromosome
cytokinesis	gamete

One of the two cells formed by cell division.

One of the two identical strands of DNA and protein that forms a duplicated chromosome. The two sister chromatids of a duplicated chromosome are joined at the centromere.

A molecule composed of deoxyribose nucleotides; contains the genetic information of all living cells.

A DNA double helix together with proteins that help to organize and regulate the use of the DNA.

The process whereby a cell becomes specialized in structure and function.

Offspring that are produced by mitosis and are, therefore, genetically identical to each other.

Referring to a cell with pairs of homologous chromosomes.

The process of producing many identical copies of a gene; also the production of many genetically identical copies of an organism.

A eukaryotic chromosome following DNA replication; consists of two sister chromatids joined at the centromeres.

The exchange of corresponding segments of the chromatids of two homologous chromosomes during meiosis I; occurs at chiasmata.

A haploid sex cell, usually a sperm or an egg, formed in sexually reproducing organisms.

The division of the cytoplasm and organelles into two daughter cells during cell division; normally occurs during telophase of mitosis.

gene	kinetochore
haploid	locus (plural, loci)
homologous chromosome	meiosis
homologue	meiotic cell division
interphase	metaphase
karyotype	mitosis

A protein structure that forms at the centromere regions of chromosomes; attaches the chromosomes to the spindle.

The unit of heredity; a segment of DNA located at a particular place on a chromosome that encodes the information for the amino acid sequence of a protein, and hence particular traits.

The physical location of a gene on a chromosome.

Referring to a cell that has only one member of each pair of homologous chromosomes.

In eukaryotic organisms, a type of nuclear division in which a diploid nucleus divides twice to form four haploid nuclei.

A chromosome that is similar in appearance and genetic information to another chromosome with which it pairs during meiosis; also called *homologue*.

Meiosis followed by cytokinesis.

A chromosome that is similar in appearance and genetic information to another chromosome with which it pairs during meiosis; also called *homologous chromosome*.

In mitosis, the stage in which the chromosomes, attached to spindle fibers at kinetochores, are lined up along the equator of the cell; also the approximately comparable stages in meiosis I and meiosis II.

The stage of the cell cycle between cell divisions in which chromosomes are duplicated and other cell functions occur, such as growth, movement, and acquisition of nutrients.

A type of nuclear division, used by eukaryotic cells, in which one copy of each chromosome (already duplicated during interphase before mitosis) moves into each of two daughter nuclei; the daughter nuclei are, therefore, genetically identical to each other.

A preparation showing the number, sizes, and shapes of all of the chromosomes within a cell.

mitotic cell division

sexual reproduction

mutation

spindle microtubule

nucleotide

stem cell

prophase

telomere

recombination

telophase

sex chromosome

A form of reproduction in which genetic material from two parent organisms is combined in the offspring; normally, two haploid gametes fuse to form a diploid zygote.

Mitosis followed by cytokinesis.

Microtubules organized in a spindle shape that separate chromosomes during mitosis or meiosis.

A change in the nucleotide sequence of the DNA of a gene.

An undifferentiated cell that is capable of dividing and giving rise to one or more distinct types of differentiated cell(s).

A subunit of which nucleic acids are composed; a phosphate group bonded to a sugar (deoxyribose in DNA), which is in turn bonded to a nitrogen-containing base (adenine, guanine, cytosine, or thymine in DNA).

The nucleotides at the end of a chromosome that protect the chromosome from damage during condensation, and prevent the end of one chromosome from attaching to the end of another chromosome.

The first stage of mitosis, in which the chromosomes first become visible in the light microscope as thickened, condensed threads and the spindle begins to form; as the spindle is completed, the nuclear envelope breaks apart, and the spindle microtubules invade the nuclear region and attach to the kinetochores of the chromosomes. Also, the first stage of meiosis: In meiosis I, the homologous chromosomes pair up, exchange parts at chiasmata, and attach to spindle microtubules; in meiosis II, the spindle re-forms and chromosomes attach to the microtubules.

In mitosis and both divisions of meiosis, the final stage, in which the spindle fibers usually disappear, nuclear envelopes re-form, and cytokinesis generally occurs. In mitosis and meiosis II, the chromosomes also relax from their condensed form.

The formation of new combinations of the different alleles of each gene on a chromosome; the result of crossing over.

The pair of chromosomes that usually determines the sex of an organism; for example, the X and Y chromosomes in mammals.

SELF TEST

1. Which of the following statements about DNA molecules is true?
   a. Genes consist of extremely long pieces of DNA, about 10 million nucleotides in length.
   b. Each DNA nucleotide contains both an adenine and a thymine base.
   c. When a parent cell divides to produce two daughter cells in mitosis, each daughter cell receives half of the genetic information that was in the original parent cell.
   d. The base adenine always pairs with thymine, whereas cytosine pairs with guanine.
   e. DNA is the carrier of hereditary information in eukaryotes, but not in prokaryotes.

2. Which of the following statements comparing sexual and asexual reproduction is true?
   a. Only sexual reproduction requires the copying of DNA.
   b. Sexual reproduction produces offspring receiving 50% of their DNA from each of two parents.
   c. Asexual reproduction can only occur with single-celled organisms such as bacteria and yeast cells, whereas more complex life-forms such as plants and animals always reproduce sexually.
   d. Organisms that reproduce asexually are always incapable of reproducing sexually.

3. Which of the following statements about the chromosomes of prokaryotic and eukaryotic cells is true?
   a. Both prokaryotic and eukaryotic cells have multiple chromosomes.
   b. The chromosome of a prokaryotic cell is a circular DNA double helix, but the chromosomes of eukaryotic cells are linear DNA double helices.
   c. The chromosomes of prokaryotic cells are located in their nuclei, but the chromosomes of eukaryotic cells are in the cytoplasm.
   d. Chromosomes of eukaryotic cells are attached to the plasma membrane, but the chromosome of prokaryotic cells floats free in the cytoplasm.

4. Which of the following statements about cell divisions in prokaryotes, such as bacteria, is NOT true?
   a. Even though the cell is very small, a single bacterium contains approximately 1 meter of DNA within its single, circular chromosome.
   b. Prokaryotes reproduce asexually, so offspring are usually genetically identical to their parent.
   c. If bacteria are provided with a growth medium that includes ideal conditions of food availability, temperature, pH, and so on, then they can reproduce extremely quickly.
   d. During cell division, prokaryotic DNA becomes temporarily attached to the plasma membrane.

5. A duplicated chromosome that has not yet entered mitosis contains all except which of the following?
   a. two long chains of alternating deoxyribose sugar-phosphate molecules
   b. two sister chromatids
   c. four telomeres
   d. two exact, identical copies of each gene locus

6. The term *haploid* refers to
   a. chromosomes that contain the same genes.
   b. cells that contain a pair of each type of chromosome.
   c. a complete set of chromosomes from a single cell that has been stained for microscopic examination.
   d. cells that contain only one of each type of chromosome.

7. If organism A has diploid cells, what is one advantage that it might have over organism B, which has haploid cells?
   a. Cellular division in organism A requires less time and/or investment of scarce cellular resources.
   b. Organism A can reproduce sexually, whereas a haploid organism such as organism B cannot.
   c. Organism A inherits a "backup" copy of each gene, providing some protection against inheritance of faulty versions of genes.
   d. Organism B cannot grow, because haploid cells are incapable of mitosis.

8. The copying of chromosomes occurs during
   a. G_1.   c. G_2.
   b. S phase.   d. mitosis.

9. The diploid number of chromosomes in human cells is 46. Thus, diploid human cell nuclei normally contain 46 distinct molecules of DNA, each organized into a chromosome. On the figure at the top of page 122, label each stage of interphase and mitosis, showing how many DNA molecules are present in a single cell nucleus at the beginning of each stage (three stages of interphase, and four of mitotic cell division).

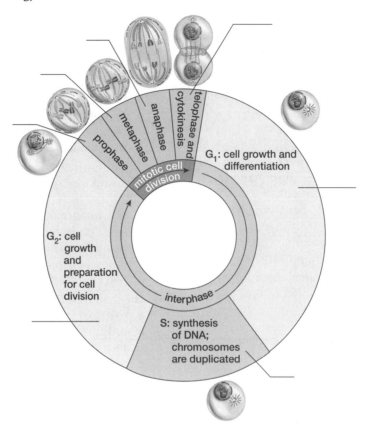

10. Which of the following statements about prophase of mitosis is NOT true?
    a. The nuclear membrane disintegrates.
    b. Chromosomes become condensed and, thus, visible with a microscope.
    c. Kinetochores separate from the centromeres.
    d. Centrioles migrate to opposite sides of the nucleus.
    e. Sister chromatids remain attached at their centromeres.

11. During anaphase, the cell becomes more ovoid because
    a. microtubules push the poles of the cell apart.
    b. the microfilament ring around the equator contracts.
    c. there are excess amounts of cytoplasm.
    d. the chromosomes line up at the equator.
    e. the centrioles migrate to opposite ends of the cell.

12. A clone is
    a. an unnatural creature fabricated by unscrupulous scientists.
    b. any cell or organism that is genetically identical to another.
    c. an exact duplicate of an organism.
    d. a sexually reproduced organism.

13. Nuclear envelopes re-form during
    a. anaphase.   c. metaphase.
    b. prophase.   d. telophase.

14. During prophase of mitosis
    a. the chromatids start to separate.
    b. chromosomes decondense.
    c. the nucleolus re-forms.
    d. centrioles are produced.
    e. the spindle microtubules attach to the chromosomes.

15. Cytokinesis following mitosis differs in animal and plant cells because
    a. plant cells have chloroplasts.
    b. plant cells have cell walls.
    c. animal cells have mitochondria.
    d. animal cells have more cytoplasm.

16. A chemical that is potentially useful for treating cancer would
    a. prevent recombination.
    b. prevent DNA synthesis.
    c. inhibit pairing of homologous chromosomes.
    d. induce mutations.

17. A cyclin-dependent kinase
    a. is a hormone that circulates in the blood, and binds to a receptor protein on a cell.
    b. strongly inhibits cell division, once cyclin has bonded to it.
    c. can have its effects on DNA copying easily blocked during chemotherapy for cancer, without unwanted side effects.
    d. stimulates DNA replication during interphase, once it has been activated.

18. An advantage of sexual reproduction is that it
    a. promotes genetic variability, thereby increasing the probability that an individual with new combinations of favorable traits may arise.
    b. ensures that individuals will inherit the most desirable genes from both parents.
    c. ensures that offspring are as similar as possible to their parents.
    d. allows a parent to pass along the greatest possible number of its own genes to its offspring.

19. Alternate forms of a particular gene are called _____; they arise as a result of _____.
    a. alleles; meiosis
    b. mutations; mitosis
    c. alleles; mutation
    d. clones; sexual reproduction

20. Meiosis comes from a Greek word that means "to decrease." What decreases during the process of meiosis?
    a. size of chromosomes
    b. number of cells
    c. length of the DNA double helices
    d. number of chromosomes

21. During the process of meiosis, DNA is replicated _____, followed by _____ nuclear division(s).
    a. twice; two   c. once; two
    b. twice; one   d. once; one

22. The random alignment of homologues at the equator during metaphase I is called
    a. crossing over.   c. random alignment.
    b. chiasmata.   d. independent assortment.

23. During meiosis I, _____ separate; during meiosis II, _____ separate.
    a. homologous chromosomes; sister chromatids
    b. sister chromatids; homologous chromosomes
    c. sister cells; gametes
    d. DNA double helices; DNA double helices

24. Cells that have only one of each type of chromosome, either the maternal or paternal homologue, are considered haploid. After which step in meiosis is this first seen?
    a. metaphase I   c. telophase I
    b. anaphase I   d. cytokinesis I

25. Genetic recombination (crossing over) produces
    a. new alleles.
    b. mutations.
    c. new combinations of alleles.
    d. longer chromosomes.

26. A zygote
    a. is the direct product of the fusion of two haploid cells.
    b. is haploid in some species, or diploid in others.
    c. always divides by mitosis to form a multi-celled organism, and never divides by meiosis instead.
    d. is, in humans, about half the size of the egg cell.

27. The key fact that sets the life cycle of plants apart from those of both the fungi and the animals is that
    a. plants have a multicelled, diploid stage.
    b. plants have a single-celled, haploid stage.
    c. plants have both multicelled diploid and multicelled haploid stages.
    d. plants have a multicelled haploid stage.
    e. plants have a single-celled, diploid stage.

28. Mitotic cell division allows for all of the following to occur except
    a. bodily repair.
    b. tissue and organ regeneration.
    c. production of sperm and eggs.
    d. growth of an individual.
    e. replacement of dead cells.

29. Refer to the example in the figure below. Imagine that an animal has a diploid (2n) number of four chromosomes (instead of the six shown in the figure). Assuming that crossing over does not occur during meiosis, how many genetically distinct versions of its gametes can this animal produce?
    a. 1   d. 8
    b. 2   e. 16
    c. 4

(a) The four possible chromosome arrangements at metaphase of meiosis I

(b) The eight possible sets of chromosomes after meiosis I

30. Which event(s) is (are) responsible for the genetic variability seen in meiosis?
    a. The particular daughter cell into which a parental chromosome moves, during metaphase I, is random.
    b. Homologous chromosomes exchange DNA with one another.
    c. Homologous chromosomes exchange RNA with one another.
    d. Both a and b.

ANSWER KEY

1. d
2. b
3. b
4. a
5. a
6. d
7. c
8. b
9.

10. c
11. a
12. b
13. d
14. e
15. b
16. b
17. d
18. a
19. c
20. d
21. c
22. d
23. a
24. d
25. c
26. a
27. c
28. c
29. c
30. d

CHAPTER 10 PATTERNS OF INHERITANCE

OUTLINE

Section 10.1 What Is the Physical Basis of Inheritance?

Inheritance occurs when characteristics of individuals are passed to their offspring. These characteristics are passed on in the form of **genes**, which are located on chromosomes.

Diploid cells contain pairs of homologous chromosomes that carry the same set of genes. Each gene is located at the same **locus** on its chromosome.

Differences in genes at the same locus represent different **alleles** of that gene (**Figure 10-1**). When a diploid organism has the identical allele, at the same locus, on both members of a pair of homologous chromosomes, the organism is called **homozygous** or **true-breeding**. If a pair of homologous chromosomes carries two different alleles at the same gene locus, the organism is called **heterozygous** or a **hybrid**.

Section 10.2 How Were the Principles of Inheritance Discovered?

Gregor Mendel discovered many of the principles of inheritance. He did this by being the first geneticist to choose the right organism, to design and perform the experiment properly, and to analyze the data correctly.

Section 10.3 How Are Single Traits Inherited?

Inheritance of traits depends on the allele composition of the parents. Each parent contributes one copy of every gene during gamete formation (**law of segregation, Figure 10-6**), resulting in the inheritance of a pair of alleles for each gene by the offspring.

Dominant alleles mask the expression of **recessive** alleles, resulting in offspring with the same **phenotype** but different **genotypes**. For example, homozygous dominant organisms have the same phenotype as those that are heterozygous (**Figure 10-7c**).

Each allele separates randomly during meiosis, allowing for the prediction of inherited traits according to the laws of probability. This can be done through the use of the **Punnett square method** (**Figure 10-8a**).

Section 10.4 How Are Multiple Traits Inherited?

As with the inheritance of single traits, the inheritance of multiple traits can be predicted according to the laws of probability using the Punnett square method (**Figure 10-11**).

If the genes for multiple traits are on separate chromosomes, their inheritance follows the **law of independent assortment**. This law states that alleles of one gene may be distributed to gametes independently of the alleles of other genes and is due to the random arrangement of homologous pairs during metaphase I (**Figure 10-12**).

Section 10.5 How Are Genes Located on the Same Chromosome Inherited?

Genes located on the same chromosome are linked and tend to be inherited together, that is, passed into the same gamete (**Figure 10-13**). However, some linked genes may be inherited separately (i.e., passed into different gametes) as the result of crossing over (**genetic recombination**), which occurs during prophase I (**Figure 10-14**).

Section 10.6 How Is Sex Determined?

The sex of many animals, including humans, is determined by **sex chromosomes**, often represented as X and Y. All other chromosomes in a cell are **autosomes**. Most females have two **X chromosomes**, as opposed to one X and one Y in males. Sex is determined by the **Y chromosome** carried in the sperm of the male (**Figure 10-16**).

Section 10.7 How Are Sex-Linked Genes Inherited?

The Y chromosome is small compared to the X, and most notably carries genes linked to male sex determination. Males fully express all alleles located on their single X chromosome. Females do not express a recessive allele on an X chromosome, as long as they also have a dominant allele for the same gene on their other X.

Many harmful recessive alleles of genes are located on the X chromosome. Males are much more likely to suffer from conditions related to these alleles. Red-green color blindness is an example of a **sex-linked** condition (**Figure 10-17**).

Section 10.8 Do the Mendelian Rules of Inheritance Apply to All Traits?

Not all traits follow Mendelian inheritance patterns. **Incomplete dominance** occurs when heterozygotes are an intermediate between the homozygous phenotypes (**Figure 10-18**). **Codominance** occurs when different alleles at one locus both contribute to the phenotype (e.g., blood types, **Table 10-1**). **Polygenic inheritance** occurs when some phenotypic traits are determined by several different genes at different loci (**Figure 10-19**). **Pleiotropy** occurs when genes have multiple effects on phenotype (e.g., the SRY gene). The environment can affect most, if not all, types of phenotypic expression (**Figure 10-20**).

Section 10.9 How Are Human Genetic Disorders Investigated?

Human genetic disorders can be investigated by analysis of the family history of genetic crosses (**pedigrees**) and use of molecular genetic technologies (**Figure 10-21**).

Section 10.10 How Are Human Disorders Caused by Single Genes Inherited?

Some genetic disorders, such as **albinism** and **sickle-cell anemia**, are caused by recessive alleles (**Figures 10-22** and **10-23**). The disorder is only phenotypically expressed in homozygotes. Heterozygotes do not express the disorder but are **carriers** for the recessive allele.

Some genetic disorders, such as **Huntington disease**, are caused by dominant alleles. Thus, inheriting only one dominant allele is sufficient for the offspring to express the disorder phenotypically.

Some genetic disorders, such as **hemophilia** and red-green color blindness, are sex-linked. The allele carrying the disorder is carried on the X chromosome; thus, males have a much greater chance of being affected because they lack a second X chromosome (**Figure 10-24**).

Section 10.11 How Do Errors in Chromosome Number Affect Humans?

Errors in chromosome number occur through **nondisjunction** of chromosomes during meiosis (**Figure 10-25**).

Some disorders, such as **Turner syndrome**, **trisomy X**, **Klinefelter syndrome**, and **Jacob syndrome**, are caused by an abnormal number of sex chromosomes (**Table 10-2**). These disorders result in distinctive physical characteristics derived from the sex chromosomes that are affected.

Some disorders, such as **trisomy 21 (Down syndrome)**, are caused by abnormal numbers of autosomes (**Figure 10-26**). These disorders usually result in spontaneous abortions, but for those fetuses that survive, severe mental and physical deficiencies occur. The likelihood of these disorders increases with the age of the mother and, to a lesser degree, the father.

FLASH CARDS

To use the flash cards, tear the page from the book and cut along the dashed lines. The key term appears on one side of the flash card, and its definition appears on the opposite side.

albinism	dominant
allele	Down syndrome
autosome	gene
carrier	genetic recombination
codominance	genotype
cross-fertilization	hemophilia

An allele that can determine the phenotype of heterozygotes completely, such that they are indistinguishable from individuals homozygous for the allele; in the heterozygotes, the expression of the other (recessive) allele is completely masked.

A recessive hereditary condition caused by defective alleles of the genes that encode the enzymes required for the synthesis of melanin, the principal pigment in mammalian skin and hair; albinism results in white hair and pink skin.

A genetic disorder caused by the presence of three copies of chromosome 21; common characteristics include mental retardation, distinctively shaped eyelids, a small mouth with protruding tongue, heart defects, and low resistance to infectious diseases; also called *trisomy 21*.

One of several alternative forms of a particular gene.

The unit of heredity; a segment of DNA located at a particular place on a chromosome that encodes the information for the amino acid sequence of a protein and, hence, particular traits.

A chromosome that occurs in homologous pairs in both males and females and that does not bear the genes determining sex.

The generation of new combinations of alleles on homologous chromosomes due to the exchange of DNA during crossing over.

An individual who is heterozygous for a recessive condition; a carrier displays the dominant phenotype but can pass on the recessive allele to offspring.

The genetic composition of an organism; the actual alleles of each gene carried by the organism.

The relation between two alleles of a gene, such that both alleles are phenotypically expressed in heterozygous individuals.

A recessive, sex-linked disease in which the blood fails to clot normally.

The union of sperm and egg from two individuals of the same species.

heterozygous	Jacob syndrome
homozygous	Klinefelter syndrome
Huntington disease	law of independent assortment
hybrid	law of segregation
incomplete dominance	linkage
inheritance	locus (plural, loci)

A set of characteristics typical of human males possessing one X and two Y chromosomes (XYY); most XYY males are phenotypically normal, but XYY males have a higher-than-average incidence of high testosterone levels, severe acne, and above-average height.

Carrying two different alleles of a given gene; also called *hybrid*.

A set of characteristics typically found in individuals who have two X chromosomes and one Y chromosome; these individuals are phenotypically males but are sterile and have several female-like traits, including broad hips and partial breast development..

Carrying two copies of the same allele of a given gene; also called *true-breeding*.

The independent inheritance of two or more traits, assuming that each trait is controlled by a single gene with no influence from gene(s) controlling the other trait; states that the alleles of each gene are distributed to the gametes independently of the alleles for other genes; this law is true only for genes located on different chromosomes or very far apart on a single chromosome.

An incurable genetic disorder, caused by a dominant allele, that produces progressive brain deterioration, resulting in the loss of motor coordination, flailing movements, personality disturbances, and eventual death.

The principle that each gamete receives only one of each parent's pair of alleles of each gene.

An organism that is the offspring of parents differing in at least one genetically determined characteristic; also used to refer to the offspring of parents of different species.

The inheritance of certain genes as a group because they are parts of the same chromosome. Linked genes do not show independent assortment.

A pattern of inheritance in which the heterozygous phenotype is intermediate between the two homozygous phenotypes.

The physical location of a gene on a chromosome.

The genetic transmission of characteristics from parent to offspring.

multiple alleles	Punnett square method
nondisjunction	recessive
pedigree	self-fertilization
phenotype	sex chromosome
pleiotropy	sex-linked
polygenic inheritance	sickle-cell anemia

A method of predicting the genotypes and phenotypes of offspring in genetic crosses.

Many alleles of a single gene, perhaps dozens or hundreds, as a result of mutations.

An allele that is expressed only in homozygotes and is completely masked in heterozygotes.

An error in meiosis in which chromosomes fail to segregate properly into the daughter cells.

The union of sperm and egg from the same individual.

A diagram showing genetic relationships among a set of individuals, normally with respect to a specific genetic trait.

One of the pair of chromosomes that usually determines the sex of an organism; for example, the X and Y chromosomes of mammals.

The physical characteristics of an organism; can be defined as outward appearance (such as flower color), as behavior, or in molecular terms (such as glycoproteins on red blood cells).

Referring to a pattern of inheritance characteristic of genes located on one type of sex chromosome (for example, X) and not found on the other type (for example, Y); in mammals, in almost all cases, the gene controlling the trait is on the X chromosome, so this pattern is also called X-linked. In X-linked inheritance, females show the dominant trait unless they are homozygous recessive, whereas males express whichever allele, dominant or recessive, that is found on their single X chromosome.

A situation in which a single gene influences more than one phenotypic characteristic.

A recessive disease caused by a single amino acid substitution in the hemoglobin molecule. Sickle-cell hemoglobin molecules tend to cluster together, distorting the shape of red blood cells and causing them to break and clog capillaries.

A pattern of inheritance in which the interactions of two or more genes determine phenotype.

test cross

Turner syndrome

trisomy 21

X chromosome

trisomy X

Y chromosome

true-breeding

A set of characteristics typical of a woman with only one X chromosome; women with Turner syndrome are sterile, with a tendency to be very short and to lack typical female secondary sexual characteristics.

A breeding experiment in which an individual showing the dominant phenotype is mated with an individual that is homozygous recessive for the same gene. The ratio of offspring with dominant versus recessive phenotypes can be used to determine the genotype of the phenotypically dominant individual.

The female sex chromosome in mammals and some insects.

See *Down syndrome.*

The male sex chromosome in mammals and some insects.

A condition of females who have three X chromosomes instead of the normal two; most such women are phenotypically normal and are fertile.

Pertaining to an individual all of whose offspring produced through self-fertilization are identical to the parental type. True-breeding individuals are homozygous for a given trait.

SELF TEST

1. Alleles are
   a. specific physical locations of genes on a chromosome.
   b. variations of the same gene (i.e., similar nucleotide sequences on homologous chromosomes).
   c. homozygotes.
   d. heterozygotes.

2. Humans have about 35,000 genes. How many alleles (either the same or different) of each of these genes are present in your muscle cells, disregarding genes on the X and Y chromosomes?
   a. 1   c. 23
   b. 2   d. 46

3. What is the MOST accurate description of how Gregor Mendel conducted his experiments?
   a. He cross-bred different species of pea plants to produce hybrids that had some qualities of each species.
   b. He carefully controlled breeding so that flowers were only allowed to self-pollinate.
   c. He made certain that pollinating honeybees had full access to all of his pea plants, so that fertilization could be carried out as quickly and thoroughly as possible.
   d. He removed the stamens of certain flowers, and then fertilized them by hand with pollen from certain other flowers.

4. Mendel's scientific approach was based in part on
   a. a thorough understanding of genetics and chromosome structure.
   b. precisely formulated experiments and careful numerical analysis of results.
   c. knowledge of evolutionary processes that was very advanced for its time.
   d. some of the first microscopic studies of meiosis, including crossing over and independent assortment of chromosomes.

5. The appearance of an organism that results from its particular complement of genes is called its
   a. genotype.
   b. dominance.
   c. phenotype.
   d. inheritance.

6. In pea plants, yellow pods are recessive to green pods. If you see yellow pods, then the genotype of that plant must be _____ for pod color.
   a. heterozygous
   b. homozygous recessive
   c. homozygous dominant
   d. both a and b
   e. both a and c

7. When Mendel conducted his experiments with purple- and white-flowered pea plants, he found that the F_1 generation did not contain any white-flowered plants. In the F_2 generation, the white-flowered plants were seen. Which of the following statements is true?
   a. The purple phenotype is always present in every generation.
   b. The white-flower allele is dominant to the purple-flower allele.
   c. The white-flower phenotype shows up only in alternate generations.
   d. Many factors are involved in this "masking" phenomenon.
   e. The white-flower allele is recessive to the purple-flower allele.

8. The mating of an individual that is true-breeding for the dominant phenotype with an individual that is true-breeding for the recessive phenotype will
   a. produce no offspring.
   b. give rise to offspring in which expression of the recessive allele is masked.
   c. produce only male organisms.
   d. produce offspring exhibiting both the dominant and recessive phenotypes.
   e. produce a generation consisting of individuals exhibiting only the recessive phenotype.

9. Mendel's law of segregation concludes that
   a. all gametes formed by an organism will have the same allele.
   b. each individual has two alleles for a particular gene.
   c. genes are found at the same loci on homologous chromosomes.
   d. dominant alleles may mask the expression of recessive alleles.
   e. each gamete will contain only one allele from the parent's pair.

10. In pea plants, the white-flower trait is recessive to the purple-flower trait, and the dwarf trait is recessive to the tall trait. Assume a pea plant that is tall with white flowers. Using T to designate the plant size gene and P to designate flower color, this would require a genotype of
    a. PPTt.   d. pptt or ppTt.
    b. ppTT or ppTt. e. PpTt.
    c. pptt.

11. In the mating between two individuals that are heterozygous for two traits,
    a. the offspring will have approximately the same number of individuals showing dominant phenotypes as the recessive phenotypes.
    b. the offspring will have only the dominant phenotype for each trait.

c. one-quarter of the offspring will exhibit one recessive phenotype and one-quarter will exhibit the other recessive phenotype.

d. there will be three times as many offspring with the recessive phenotype as with the dominant phenotype for one of the traits.

e. the offspring will have only the recessive phenotype for each trait.

12. In the results of a two-trait cross, an individual from the F₂ generation displays the recessive form for both traits. Which of the following statements describing the individual is true?

a. This overall phenotype is the most abundant among the offspring.

b. The individual must be heterozygous for both genes involved.

c. The individual must be homozygous recessive for at least one of the two genes involved.

d. The individual must have two recessive alleles for each of the genes in question.

e. This individual will be weaker than the individual showing the dominant form for both traits.

13. Which of the following can account for a situation in which Mendel's law of independent assortment fails to hold?

a. genes on the same chromosome

b. pleiotropy

c. self-fertilization

d. genes have undergone recombination

e. crossing over

14. Study the figure. Imagine that crossing over is performed incorrectly such that DNA is exchanged between sister chromatids rather than between homologous chromosomes. What is the effect on the gametes being produced?

a. 50% will contain purple (P) and round (l) alleles, and 50% will contain red (p) and long (L) alleles.

b. 50% will contain purple (P) and long (L) alleles, and 50% will contain red (p) and round (l) alleles.

c. 25% will contain purple (P) and long (L) alleles, 50% will contain purple (P) and round (l) alleles, and 25% will contain red (p) and round (l) alleles.

d. 100% will contain purple (P), red (p), long (L), and round (l) alleles.

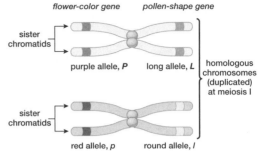

(a) Replicated chromosomes in prophase of meiosis I

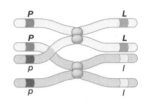

(b) Crossing over during prophase I

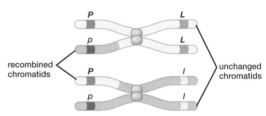

(c) Homologous chromosomes separate at anaphase I

(d) Unchanged and recombined chromosomes after meiosis II

15. Anne Boleyn, King Henry VIII's second wife, was beheaded because she did not provide him with a son as an heir. Explain why King Henry should have blamed himself and not his wife.

a. All the sperm that males produce contain an X chromosome, so their genetic contribution to the child determines its sex.

b. All the eggs that females produce contain an X chromosome, so their genetic contribution to the child does not determine its sex.

c. The eggs that females produce contain either an X or a Y chromosome, so their genetic contribution to the child is unrelated to its sex.

d. All sperm and eggs contain an X chromosome, but the father determines a child's sex because half his sperm also contain a Y chromosome.

16. How many pairs of autosomes do humans have?
    a. 2   d. 44
    b. 22   e. 46
    c. 23

17. If a woman who is heterozygous for the X-linked recessive condition Duchenne muscular dystrophy (DMD) mates with a normal man and produces children, the chance that any particular son will develop DMD is _____, whereas the chance that any particular daughter will develop DMD is _____.
    a. 50%; 50%   d. 50%; 0%
    b. 0%; 100%   e. 75%; 25%
    c. 100%; 0%

18. A man with red-green color blindness can inherit the allele from _____, and can pass it on to _____.
    a. either parent; sons only
    b. his father only; daughters only
    c. his mother only; daughters only
    d. either parent; sons and daughters
    e. his mother only; sons and daughters

19. Labrador retrievers may have yellow fur, chocolate-brown fur, or black fur. Coat color is determined by two genes that have two alleles each. One gene, called E, determines whether the dog's fur is dark or light, with the dark allele (E) dominant to the light allele (e). The other gene, called B, determines whether the dark fur will be black or chocolate, with the black allele (B) dominant to the chocolate allele (b). If you mate two chocolate labs, what color puppies can they have?
    a. Only chocolate puppies can be produced because both parents must be EEbb.
    b. Chocolate or yellow puppies can be produced because the parents can be either Eebb or EEbb.
    c. Chocolate, yellow, and black puppies will be produced because both parents must be EeBb.
    d. Chocolate and black puppies will be produced because both parents must be EeBb.

20. A single gene capable of influencing multiple phenotypes within a single organism is said to be
    a. codominant for that gene.
    b. incompletely dominant for that gene.
    c. polygenic for that gene.
    d. pleiotropic for that gene.

21. A couple brings home their new, nonidentical twin daughters, Joan and Jill. After several months, the father begins to suspect that there was a mix-up at the hospital, because Jill does not look much like either parent or like her sister. When the twins' blood tests come back, the father calls his lawyer to start a lawsuit against the hospital. The mother, father, and Joan have type A blood, but Jill has type O blood. Based on blood type, does the father have a case? Explain your answer. (The gene for blood type has three alleles: A, B, and O. The A and B alleles are codominant, and the O allele is recessive.)
    a. No, because parents with type A blood can have a child with type O blood.
    b. No, because parents with any blood type (A, B, AB, or O) can produce children with type O blood.
    c. Yes, because all of this couple's children will have type A blood.
    d. Yes, because people with type A blood can pass on only A alleles to their children.

22. Study the human pedigree shown in the figure on page 138. Notice that for generation III, the three siblings to the left are shown as definitely being carriers for the recessively inherited condition, whereas in generation II, four people are shown as having genotypes that are "unknown." The best explanation for this difference is that
    a. the two parents in generation I are both heterozygous carriers for the condition, who can therefore create any of three genetic outcomes by their mating, whereas the mother of the three generation III siblings actually has the condition, and therefore passes the recessive allele to 100% of her children.
    b. the four people in generation II are all too young to have yet developed the condition, whereas all of the people in generation III are older and have definitely developed the phenotypes that reveal their particular genotypes.
    c. the two parents in generation I are both heterozygous carriers for the condition and will therefore always have heterozygous children, whereas it is impossible for someone who actually has a recessive condition, such as the generation II mother, to produce children who are only carriers for the condition.
    d. conditions of this type are unable to "skip" a generation, but instead must appear in each and every generation for as long as they are being inherited.

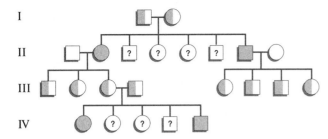

23. We are constructing a family pedigree to track an inherited, autosomal recessive condition, and to understand the risk that it poses to various family members. What fact could cause the greatest harm to the integrity of the study, possibly leading to faulty predictions of who is at risk of passing the allele for the condition to their children or of developing the condition themselves?

   a. The nature of this condition is that everyone who is homozygous for the allele will then develop the associated condition, at a very early age.

   b. The man who is named as the father, on the birth certificate of one of the family members, is not the true biological father.

   c. Two parents, both of Irish ancestry, have adopted two Chinese children.

   d. One particular family, within the larger pedigree, has had eight daughters but no sons.

24. An individual who is a "carrier" of a genetic disorder is

   a. heterozygous for the disorder, and the allele for the disorder is recessive.

   b. heterozygous for the disorder, and the allele for the disorder is dominant.

   c. homozygous for the disorder.

   d. homozygous for the disorder but is protected by the presence of an immunity gene.

25. Cystic fibrosis is a recessive trait. Imagine that your friend Roger has cystic fibrosis but that his parents do not. What do you know about Roger's alleles and those of his parents at the cystic fibrosis locus of their DNA?

   a. This information is insufficient to allow me to conclude anything about the cystic fibrosis alleles in the DNA of Roger's parents.

   b. This information is insufficient to allow me to conclude anything about the cystic fibrosis alleles in Roger's DNA.

   c. Roger is heterozygous and his parents are homozygous at the cystic fibrosis locus.

   d. Roger is homozygous and his parents are heterozygous at the cystic fibrosis locus.

26. A man with hemophilia inherited the hemophilia gene from

   a. his mother.

   b. his father.

   c. either his mother or his father.

   d. his paternal grandfather.

27. With a disorder such as Huntington disease, which has an autosomal dominant mode of inheritance, a heterozygous father who develops the disease at age 50 has a _____ chance of having already passed the allele for the disease to any particular son, and a _____ chance of having already passed it to any particular daughter.

   a. 100%; 0%

   b. 50%; 100%

   c. 100%; 100%

   d. 50%; 50%

   e. 0%; 50%

28. What happens if a baby has only one X chromosome and no Y?

   a. Such a deficiency is lethal, so the baby would be stillborn.

   b. This baby would be a female with Turner syndrome.

   c. The baby would be a male with Turner syndrome.

   d. The baby would have Klinefelter syndrome.

29. Which of the following is caused by an abnormal number of autosomes?

   a. Down syndrome

   b. Klinefelter syndrome

   c. Turner syndrome

   d. Marfan syndrome

30. Which condition, involving nondisjunction of sex chromosomes, can only be caused by a faulty sperm cell, rather than by a faulty egg cell?

   a. Turner syndrome (XO)

   b. trisomy X (XXX)

   c. Klinefelter syndrome (XXY)

   d. Jacob syndrome (XYY)

ANSWER KEY

1. b		16. b	
2. b		17. d	
3. d		18. c	
4. b		19. b	
5. c		20. d	
6. b		21. a	
7. e		22. a	
8. b		23. b	
9. e		24. a	
10. b		25. d	
11. c		26. a	
12. d		27. d	
13. a		28. b	
14. b		29. a	
15. b		30. d	

Chapter 11 DNA: The Molecule of Heredity

OUTLINE

Section 11.1 How Did Scientists Discover That Genes Are Made of DNA?

By the early 1900s, scientists had established that genetic information existed in units called genes, which are parts of chromosomes. Because chromosomes are composed of both protein and **DNA**, however, it was unclear which acted as the cell's hereditary blueprint.

Griffith's experiments on mice determined that genes could be transferred from one strain of bacteria to another, causing it to become deadly to mice (**Figure 11-1**).

Avery, MacLeod, and McCarty determined that this transformation was caused by DNA, thus they concluded that genes are made of DNA (**Figure 11-2**).

Section 11.2 What Is the Structure of DNA?

DNA is composed of **nucleotides**, each with a phosphate group, a sugar (deoxyribose), and one of four possible nitrogen-containing **bases: adenine (A)**, **guanine (G)**, **thymine (T)**, and **cytosine (C)** (**Figure 11-3**).

Nucleotides are arranged in **strands**, with the phosphate group of one nucleotide bonded to the adjacent sugar group of another, forming a **sugar-phosphate backbone**. Nitrogen-containing bases protrude from this sugar-phosphate backbone.

Two DNA nucleotide strands are held together by hydrogen bonds between nitrogen-containing bases of adjacent nucleotide strands, forming a **double helix** (**Figure 11-5**). Hydrogen bonds form between nitrogen-containing bases in a predictable way: adenine with thymine (A–T) and guanine with cytosine (G–C). These pairings are called **complementary base pairs**.

Section 11.3 How Does DNA Encode Information?

DNA encodes information based on the sequence of nucleotide bases in a strand.

The four bases, A, T, C and G, function like an alphabet that can be used to spell thousands of different words.

Section 11.4 How Does DNA Replication Ensure Genetic Constancy During Cell Division?

Prior to mitosis or meiosis, DNA synthesizes an exact copy of itself through a process called **DNA replication**.

During DNA replication, **DNA helicases** pull apart the parental DNA double helix, exposing its nitrogen-containing bases. **DNA polymerases** read the exposed nitrogen-containing bases and bond complementary **free nucleotides** to them, forming two new DNA strands (**Figure 11-6**).

In each new DNA double helix, one strand came from the parent DNA double helix, whereas the other is a daughter strand (**semiconservative replication, Figure 11-7**).

Section 11.5 How Do Mutations Occur?

Mutations are changes in the sequence of bases in DNA. Mutations are rare, occurring once every 100 million to 1 billion base pairs. This is due to DNA repair enzymes that proofread each daughter strand during and after its synthesis.

Mutations can occur from normal errors in base pairing, as well as by environmental conditions that can cause damage to DNA (such as UV rays in sunlight).

Nucleotide substitutions (point mutations) occur when the correct nucleotide in a pair is replaced instead of the incorrect one (**Figure 11-8a**). **Insertion mutations** occur when one or more nucleotide pairs are inserted into the DNA double helix (**Figure 11-8b**), whereas **deletion mutations** occur when pairs are removed from the DNA double helix (**Figure 11-8c**).

Inversions are mutations that occur when chromosome pieces are cut out, turned around, and reinserted (**Figure 11-8d**). **Translocations** are mutations that occur when a chunk of DNA is removed from a chromosome and moved to another one (**Figure 11-8e**).

Most mutations are harmful or neutral, but some can be beneficial and favored by natural selection, depending on the environment.

FLASH CARDS

To use the flash cards, tear the page from the book and cut along the dashed lines. The key term appears on one side of the flash card, and its definition appears on the opposite side.

adenine (A)	deoxyribonucleic acid (DNA)
bacteriophage	DNA helicase
base	DNA ligase
complementary base pairs	DNA polymerase
cytosine (C)	DNA replication
deletion mutation	double helix

A molecule composed of deoxyribose nucleotides; contains the genetic information of all living cells.

A nitrogenous base found in both DNA and RNA; abbreviated as A.

An enzyme that helps unwind the DNA double helix during DNA replication.

A virus that specifically infects bacteria.

An enzyme that bonds the terminal sugar in one DNA strand to the terminal phosphate in a second DNA strand, creating a single strand with a continuous sugar-phosphate backbone.

One of the nitrogen-containing, single- or double-ringed structures that distinguishes one nucleotide from another. In DNA, the bases are adenine, guanine, cytosine, and thymine.

An enzyme that bonds DNA nucleotides together into a continuous strand, using a preexisting DNA strand as a template.

In nucleic acids, bases that pair by hydrogen bonding. In DNA, adenine is complementary to thymine and guanine is complementary to cytosine; in RNA, adenine is complementary to uracil, and guanine to cytosine.

The copying of the double-stranded DNA molecule, producing two identical DNA double helices.

A nitrogenous base found in both DNA and RNA; abbreviated as C.

The shape of the two-stranded DNA molecule; similar to a ladder twisted lengthwise into a corkscrew shape.

A mutation in which one or more pairs of nucleotides are removed from a gene.

free nucleotides	point mutation
guanine (G)	semiconservative replication
insertion mutation	strand
inversion	sugar-phosphate backbone
mutation	thymine (T)
nucleotide	translocation
nucleotide substitution	

A mutation in which a single base pair in DNA has been changed.

Nucleotides that have not been joined together to form a DNA or RNA strand.

The process of replication of the DNA double helix; the two DNA strands separate, and each is used as a template for the synthesis of a complementary DNA strand. Consequently, each daughter double helix consists of one parental strand and one new strand.

A nitrogenous base found in both DNA and RNA; abbreviated as G.

A single polymer of nucleotides; DNA is composed of two strands wound about each other in a double helix; RNA is usually single stranded.

A mutation in which one or more pairs of nucleotides are inserted into a gene.

A chain of sugars and phosphates in DNA and RNA; the sugar of one nucleotide bonds to the phosphate of the next nucleotide in a DNA or RNA strand. The bases in DNA or RNA are attached to the sugars of the backbone.

A mutation that occurs when a piece of DNA is cut out of a chromosome, turned around, and reinserted into the gap.

A nitrogenous base found only in DNA; abbreviated as T.

A change in the nucleotide sequence of DNA in a gene.

A mutation that occurs when a piece of DNA is removed from one chromosome and attached to another chromosome.

A subunit of which nucleic acids are composed; a phosphate group bonded to a sugar (deoxyribose in DNA), which is in turn bonded to a nitrogen-containing base (adenine, guanine, cytosine, or thymine in DNA).

A mutation in which a single base pair in DNA has been changed.

SELF TEST

1. Identify the research that first provided the basis for the following statement: DNA is the genetic material.
   a. Watson and Crick proposed a new model for DNA structure.
   b. Avery, MacLeod, and McCarty isolated the material that transformed R-strain bacteria into S-strain bacteria.
   c. Chargaff found that DNA contains equal amounts of adenine and thymine, as well as equal amounts of cytosine and guanine.
   d. Wilkins and Franklin used X-ray diffraction to study DNA structure.

2. Before it was actually determined, many scientists had trouble believing that DNA was the genetic material. This is most likely because
   a. it was known that proteins could be passed from generation to generation.
   b. the number of different nucleotides in DNA is very small.
   c. destruction of proteins prevented genetic transformation.
   d. it was known that RNA could be passed from generation to generation.

3. An interpretation of Griffith's experiments is that
   a. fragments of DNA containing genes were taken up by the R-strain bacteria.
   b. genetic material must have been transferred from the R-strain bacteria into the S-strain bacteria.
   c. the genetic material must be protein.
   d. base-pairing accounts for the amounts of each base found.

4. The Hershey-Chase experiment, which provided additional evidence that DNA is the carrier of genetic information, took advantage of the fact(s) that
   a. DNA contains phosphorus but not sulfur, whereas protein contains sulfur but not phosphorus.
   b. bacteriophages are specialized types of bacteria.
   c. radioactive isotopes behave very differently, chemically, than their nonradioactive counterparts (for example, radioactive C^{14} behaves very differently than C^{12}).
   d. bacteriophage coats are denser than bacteria.

5. Why did Hershey and Chase choose bacteriophages as their study organism in their attempts to determine which overall class of biological molecule was the carrier of genetic information?
   a. Bacteriophages are chemically and reproductively self-sufficient, making it easy to grow them in vast numbers in the laboratory, as long as they are provided with a simple chemical nutrient mixture.
   b. Bacteriophages are the simplest organisms to still contain all of the basic types of biological molecules.
   c. Bacterial cell walls are impervious to bacteriophage DNA, forcing this DNA to remain outside of the bacteria where it is then easier to study.
   d. Bacteriophages only contain DNA and protein, which simplifies the job of determining whether DNA or protein is the carrier of genetic information.

6. Avery, MacLeod, and McCarty concluded that genes are made of DNA. However, scientific conclusions should always be regarded as tentative and subject to later change. Imagine that the four observations listed below had later been made. Which one would have been the most likely to cast doubt on the conclusion of Avery, MacLeod, and McCarty, regarding the role of DNA in heredity?
   a. A mouse species is found to have natural immunity against the S-strain of *Streptococcus pneumoniae* bacteria.
   b. Some proteins, common in the nuclei of human cells (and possibly in cells of other species), are very resistant to breakdown by protein-destroying enzymes.
   c. Other researchers, following Griffith's work, use the R- and S-strains of *Streptococcus pneumoniae* to finally develop an effective vaccine against it.
   d. Enzymes that destroy the DNA in chromosomes have no effect on the proteins that are also found in chromosomes.

7. The structure of DNA explained Chargaff's observations because
   a. the DNA molecule is a regular, repeating molecule.
   b. the twisting of the DNA required a certain number of bases.
   c. adenines were found to pair with thymines and cytosines with guanines.
   d. the sequence of bases is crucial to the storage of information.
   e. there are only four different bases.

8. The two strands of DNA that make up a double helix are
   a. identical to each other.
   b. held together by covalent bonds.
   c. oriented in the same direction.
   d. complementary to each other.

9. Imagine that you are studying a newly discovered bacterium from a hot spring in Yellowstone National Park. When you examine the

nucleotide composition of this organism, you find that 10% of the nucleotides in its DNA are adenine. What percentage of nucleotides are guanine? Explain.

a. 10%, because A pairs with G.

b. 90%, because A pairs with T, accounting for 10% of the bases, leaving the remaining 90% of them to be guanine.

c. 40%, because A pairs with T (accounting for 20% of the bases), leaving 80% of the nucleotides as G–C base pairs; half of 80% is 40%.

d. 80%, because A and T together account for 20% of the nucleotides, leaving 80% to be guanine.

10. Consider the backbone of the DNA double helix. Which of the following statements is NOT true?

a. It is composed of alternating sugar and phosphate groups.

b. The backbone is not straight, but twisted.

c. The DNA backbone forms the central core of the DNA molecule.

d. There is directionality to the backbone.

11. DNA structure can be described as a twisted ladder. Imagine you are climbing a model of DNA, just as if you were climbing a ladder. Which parts of a nucleotide are your feet touching as you climb?

a. sugars   c. phosphates
b. nitrogenous bases d. deoxyriboses

12. In a DNA helix, all of the following statements are true except:

a. the nitrogenous bases are covalently bonded to one another.

b. the nitrogenous bases are in the inner part of the helix.

c. the strands are in opposite orientation.

d. cytosine pairs with guanine.

13. Examine the figure at right, showing the molecular structures of the four DNA bases. Normally, A pairs with T, and C pairs with G. Imagine that, somehow, base-pairing behavior has changed, so that A now pairs with G, and C now pairs with T. What are the consequences for the overall shape of the DNA molecule?

a. The double helix is now constricted, along its entire length.

b. It constricts where A pairs with G.

c. It constricts at some base-pair locations, and bulges at others, all along its length.

d. It bulges where C pairs with T.

14. Information in DNA is carried in

a. the sugar-phosphate backbone of one DNA strand.

b. the base pairs between nucleotides in the two DNA strands.

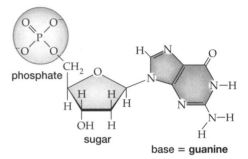

base = adenine

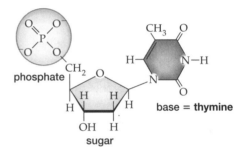

base = guanine

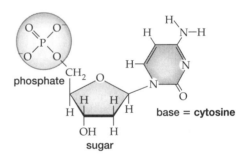

base = thymine

base = cytosine

c. the proteins that bind to the DNA double helix.

d. the order of the nucleotide bases in one DNA strand.

15. Suppose that you are spelling "words," using the four bases ("letters") A, T, C, and G. How many different three-letter "words" can you spell, using only these bases? Try it!

a. 4   c. 16
b. 8   d. 64

16. Multiple replication bubbles on a single eukaryotic chromosome

a. allow for rapid replication of eukaryotic DNA.

b. do not occur.

c. lead to many DNA strands being synthesized from the same chromosome simultaneously.

d. occur to ensure that the DNA is replicated faithfully.

17. Which of the following events occurs within a DNA replication bubble?

a. DNA polymerase reforms the hydrogen bonds joining the two parental DNA strands.

b. DNA helicase attaches the phosphate of a free nucleotide to the sugar of the previous nucleotide in the daughter strand.

c. DNA helicase unwinds the double helix at each replication fork within a replication bubble, breaking the hydrogen bonds between complementary bases.

d. DNA ligase creates new hydrogen bonds between the deoxyribose sugars and phosphates in the new daughter strands.

18. Human chromosomes range in size dramatically, with the smallest (sex chromosome Y) being many times smaller than the largest (autosomal chromosome 1). What is most directly responsible for determining the size of a chromosome?

a. the number of base pairs contained within its DNA molecule

b. the sizes of the protein molecules associated with it

c. the number of DNA molecules in it

d. whether it mainly contains recessive or dominant alleles of its genes

19. Which of the following options would result from the actions of DNA polymerase during DNA replication?

a. Two DNA polymerase molecules act to synthesize a long continuous daughter DNA strand from each parental strand; ligase is not needed.

b. Two DNA polymerase molecules act to synthesize a short segment of daughter DNA from each parental strand; ligase is used to connect these short segments of both daughter strands.

c. Two DNA polymerase molecules act to synthesize daughter DNA strands: one via a long continuous strand that moves in the same direction as the helicase, and a second polymerase synthesizes short segments of DNA that must be joined by DNA ligase.

d. Two DNA polymerase molecules act to synthesize daughter DNA strands: one via a long continuous strand that elongates in the opposite direction as the helicase moves, and a second polymerase synthesizes short segments of DNA that elongate in the same direction that the helicase moves and that must be joined by ligase.

20. Replication bubbles

a. consist of one moving replication fork and one fixed replication fork.

b. are always shrinking in size.

c. are only present once per chromosome.

d. consist of two replication forks moving in opposite directions.

21. Which of the following statements is false?

a. DNA replication involves uncoiling of the parental DNA molecule.

b. DNA replication produces a long, continuous strand and a series of short pieces.

c. DNA ligase is required in the synthesis of one strand.

d. Both parental strands end up in the same daughter strand after replication.

e. DNA polymerase molecules move toward the replication fork on both strands.

22. The purpose of DNA replication is to produce

a. two similar DNA double helices differing in a small number of specific sites.

b. two very different DNA double helices.

c. two identical DNA double helices.

d. one copy that is identical to the parental DNA molecule and one that is totally different.

e. a single-stranded DNA molecule from the double-stranded parent DNA.

23. Look at the two growing daughter strands of DNA in the figure below. Note the direction in which each strand is being assembled. Label the ends of each daughter strand to show which is its 3′ end and which is its 5′ end.

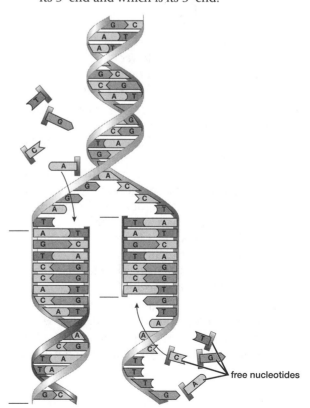
free nucleotides

24. Imagine that all of the phosphorous atoms in the DNA of a single circular chromosome of a bacterial cell are of the isotope P^{32}. Now, imagine that this cell is placed into an environment where the only available new phosphorus is of the isotope P^{31}. Assume that the cell does not contain or use any other phosphorous-containing molecules. Let the bacterium grow and divide in this environment, and let each daughter cell grow and divide again, producing four eventual daughter cells. What can we say about the amount of P^{32} that will exist in the daughter cells?
    a. Each daughter cell will contain 25% of the P^{32} that was in the original parent cell.
    b. One daughter cell will contain 100% of the original P^{32}, whereas the other three will contain none.
    c. Two daughter cells will each contain 50% of the original P^{32}, whereas the other two cells will each contain none.
    d. One daughter cell will contain 75% of the original P^{32}, whereas one will contain 25%, and the other two will each contain none.

25. Which is least likely to cause errors to accumulate in DNA?
    a. exposure to cigarette smoke
    b. ultraviolet light
    c. cold temperatures
    d. X-ray exposure

26. The approximate, overall error rate for DNA replication is one error for every _____ base pairs, which is _____ the error rate for DNA polymerase acting alone.
    a. 10,000 to 100,000; higher than
    b. 100 million to 1 billion; lower than
    c. 10,000 to 100,000; lower than
    d. 100 million to 1 billion; higher than
    e. 1 million to 10 million; equal to

27. Which is most likely to be caused by the ultraviolet radiation in sunlight?
    a. Adjacent thymine bases form cross-linked covalent bonds.
    b. The DNA molecule is degraded from each end, causing the progressive loss of bases from the ends of human chromosomes, thus steadily shortening them.
    c. DNA nucleotides can be deleted, added, changed, or rearranged.
    d. Base-pairing rules are changed, so that A pairs with C, and T with G.

28. A mutation has occurred. The length of the DNA molecule has not been changed. However, a short section of DNA that originally had the sequence of bases ATTCCG along strand 1, and their complementary bases along strand 2, has been changed to have the sequence of bases CGGAAT along strand 1, and their complementary bases along strand 2. What type of mutation has probably occurred here? (*Hint:* Draw the sequence of bases on both strands for both the before and after situations.)
    a. substitution
    b. insertion
    c. deletion
    d. translocation
    e. inversion

29. Suppose that due to a mutation in the synthesis of a human chromosome in a particular cell, DNA strand 1 has been erroneously constructed to have a "C" instead of a "T" at a particular position, whereas strand 2 has an "A" in the complementary position, forming an improper C–A "complementary" base pair. Next, the cell duplicates its DNA, in preparation for mitosis. It creates a daughter strand 2 that is complementary and joined to parental strand 1, and a daughter strand 1 that is complementary and joined to parental strand 2. What happens at the location of the mutation? (*Hint:* Make a sketch.)
    a. DNA replication remains perfect, so the two resulting daughter chromosomes are identical to each other, as well as to the parental chromosome.
    b. After DNA polymerase builds the two complementary daughter strands, daughter strand 2 has a "G" at this location, whereas daughter strand 1 has a "T" at the corresponding location.
    c. After the parental strands are separated by the action of DNA helicase, DNA polymerase corrects the error in parental strand 1, so it once again has a "T" at this position.
    d. After DNA polymerase builds the two complementary daughter strands, daughter strand 2 has a "T" at this location, whereas daughter strand 1 has a "C" at the corresponding location.

30. What is the LEAST accurate statement about mutations?
    a. In the long run, they make members of a species more genetically similar to each other.
    b. Mutations are what create different alleles of genes.
    c. Most mutations are harmful, and some can be quickly lethal.
    d. Some mutations are adaptive, improving the survival of organisms, at least under certain circumstances.

ANSWER KEY

1. b
2. b
3. a
4. a
5. d
6. b
7. c
8. d
9. c
10. c
11. b
12. a
13. c
14. d
15. d
16. a
17. c
18. a
19. c
20. d
21. d
22. c
23.

24. c
25. c
26. b
27. c
28. e
29. b
30. a

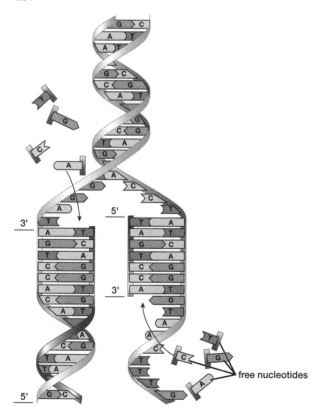

free nucleotides

Chapter 12 Gene Expression and Regulation

OUTLINE

Section 12.1 How Is the Information in DNA Used in a Cell?

The experiments of Beadle and Tatum demonstrated that one gene encodes the information for the synthesis of one enzyme (**Figure E12-1**).

Ribonucleic acid (RNA) carries the information from DNA to cytoplasmic **ribosomes** for protein synthesis. RNA is structurally different from DNA in that it is normally single stranded, has the sugar ribose, and replaces the use of the nitrogen-containing base thymine (T) with uracil (U) (**Table 12-1**).

There are three forms of RNA. **Messenger RNA (mRNA)** carries the gene code from DNA to ribosomes. **Ribosomal RNA (rRNA)** combines with proteins to make ribosomes. **Transfer RNA (tRNA)** carries amino acids to ribosomes (**Table 12-1** and **Figure 12-1**).

Protein synthesis is a two-step process. Within the nucleus, the DNA gene message is copied as mRNA by a process called **transcription** (RNA synthesis) (**Figure 12-2a**). The mRNA then binds to a ribosome where tRNA base pairs to mRNA, building the amino acid chain of a protein in the process of **translation** (**Figure 12-2b**).

The genetic code uses three mRNA bases to specify an amino acid. The bases of mRNA are read as triplets (**codons**), each coding for the insertion of a single amino acid in the growing protein. The mRNA code for a protein begins with a **start codon** (AUG) and ends with a **stop codon** (UAG, UAA, or UGA). The different combinations of bases in a codon code for the insertion of up to 20 different types of amino acids in a protein (**Table 12-3**), but each codon specifies only one amino acid.

Section 12.2 How Is the Information in a Gene Transcribed into RNA?

Transcription consists of three steps: (1) initiation, (2) elongation, and (3) termination (**Figure 12-3**).

Initiation occurs when transcription factors initiate the binding of the **promoter** region of the gene to **RNA polymerase**.

Elongation occurs when RNA polymerase travels down the **template strand** of DNA, assembling RNA nucleotides into a single mRNA strand.

Elongation continues until a termination signal DNA sequence is reached, causing RNA polymerase to release the assembled mRNA molecule and detach from DNA.

Section 12.3 How Is the Base Sequence of Messenger RNA Translated into Protein?

In prokaryotic cells, every nucleotide of a protein-coding gene codes for amino acids (**Figure 12-5**), whereas in eukaryotic cells these genes consist of **introns** and **exons**. Exons are regions that code for amino acids, whereas introns do not, so the introns must be excised from the original RNA transcript before translation can commence (**Figure 12-6**).

In eukaryotes, mRNA carries the genetic information to ribosomes in the cytoplasm where amino acid assembly occurs.

Translation consists of three steps: (1) initiation, (2) elongation, and (3) termination (**Figure 12-7**).

Initiation occurs when large and small ribosomal subunits come together at the first AUG codon of mRNA to form an initiation complex.

Elongation occurs when tRNA molecules deliver amino acids to the mRNA molecule. The specific amino acid delivered depends on the base pairing between the tRNA **anticodon** and the mRNA codon. Two tRNA molecules can bind to the ribosome at the same time, with the large subunit causing the formation of a peptide bond between the amino acids. One tRNA then detaches from the ribosome, and the ribosome moves over one codon. A new tRNA then binds to the complementary mRNA codon, the amino acid is bonded to the growing amino acid chain, and the process repeats.

Termination occurs when the ribosome encounters a stop codon, causing it to release the amino acid chain and then disassemble.

Section 12.4 How Do Mutations Affect Protein Function?

Mutations are changes in the sequence of bases in DNA. Mutations can occur from normal errors in base pairing, as well as from environmental conditions that can cause damage to DNA (such as UV rays in sunlight).

Nucleotide substitutions (point mutations) occur when the correct nucleotide in a pair is replaced instead of the incorrect one. **Insertion mutations** occur when one or more nucleotide pairs are inserted into the DNA double helix, whereas **deletion mutations** occur when pairs are removed from the DNA double helix (**Table 12-4**).

Inversions are mutations that occur when pieces of a chromosome are cut out, turned around, and reinserted. **Translocations** are mutations that occur when a chunk of DNA is removed from a chromosome and moved to another one.

Most mutations are harmful or neutral, but some can be beneficial and favored by natural selection, depending on the environment.

Section 12.5 How Are Genes Regulated?

Gene expression occurs when the information in a gene is transcribed and translated, forming a protein that causes an effect within the cell. The specific gene that is expressed is regulated by cell function, developmental stage, and the environment.

Prokaryotic DNA is organized as packages (**operons**) in which genes for related functions lie close to each other (**Figure 12-9a**). Operons are regulated as units so functionally related proteins can be synthesized and expressed simultaneously. An example of this prokaryotic gene regulation mechanism is illustrated by the **lactose operon** of *E. coli* (**Figures 12-9b 12-9c**).

Eukaryotic gene expression is regulated at many steps: (1) Cells can control the frequency of individual gene transcription. (2) One gene may be used to produce different mRNAs and proteins. (3) Cells may control mRNA stability, as well as its translation. (4) Proteins may require modification before they are functional. (5) The life span of a protein can be regulated (**Figure 12-10**).

Eukaryotic cells may regulate the transcription of genes, regions of chromosomes, or entire chromosomes by several mechanisms. Individual gene transcription can be altered by changing which transcription factors are present, thus affecting the ability of the promoter region to initiate transcription. Chromosomal regions that are condensed are not typically transcribed because they are inaccessible to RNA polymerase. Sometimes entire chromosomes may remain condensed (**Figures 12-11** and **12-12**).

FLASH CARDS

To use the flash cards, tear the page from the book and cut along the dashed lines. The key term appears on one side of the flash card, and its definition appears on the opposite side.

anticodon	insertion mutation
Barr body	intron
codon	inversion
deletion mutation	lactose operon
exon	messenger RNA (mRNA)
genetic code	mutation

A mutation in which one or more pairs of nucleotides are inserted into a gene.

A sequence of three bases in transfer RNA that is complementary to the three bases of a codon of messenger RNA.

A segment of DNA in a eukaryotic gene that does not code for amino acids in a protein (see also exon).

A condensed, inactivated X chromosome in the cells of female mammals, which have two X chromosomes.

A mutation that occurs when a piece of DNA is cut out of a chromosome, turned around, and reinserted into the gap.

A sequence of three bases of messenger RNA that specifies a particular amino acid to be incorporated into a protein; certain codons also signal the beginning or end of protein synthesis.

In prokaryotes, the set of genes that encodes the proteins needed for lactose metabolism, including both the structural genes and a common promoter and operator that control transcription of the structural genes.

A mutation in which one or more pairs of nucleotides are removed from a gene.

A strand of RNA, complementary to the DNA of a gene, that conveys the genetic information in DNA to the ribosomes to be used during protein synthesis; sequences of three bases (codons) in mRNA that specify particular amino acids to be incorporated into a protein.

A segment of DNA in a eukaryotic gene that codes for amino acids in a protein (see also intron).

A change in the nucleotide sequence of DNA in a gene.

The collection of codons of mRNA, each of which directs the incorporation of a particular amino acid into a protein during protein synthesis.

neutral mutation	regulatory gene
nucleotide substitution	repressor protein
operator	ribonucleic acid (RNA)
operon	ribosomal RNA (rRNA)
point mutation	ribosome
promoter	RNA polymerase

In prokaryotes, a gene encoding a protein that binds to the operator of one or more operons, controlling the ability of RNA polymerase to transcribe the structural genes of the operon.

A mutation that does not detectably change the function of the encoded protein.

In prokaryotes, a protein encoded by a regulatory gene, which binds to the operator of an operon and prevents RNA polymerase from transcribing the structural genes.

A mutation in which a single base pair in DNA has been changed.

A molecule composed of ribose nucleotides, each of which consists of a phosphate group, the sugar ribose, and one of the bases adenine, cytosine, guanine, or uracil; involved in converting the information in DNA into protein; also the genetic material of some viruses.

A sequence of DNA nucleotides in a prokaryotic operon that binds regulatory proteins that control the ability of RNA polymerase to transcribe the structural genes of the operon.

A type of RNA that combines with proteins to form ribosomes.

In prokaryotes, a set of genes, often encoding the proteins needed for a complete metabolic pathway, including both the structural genes and a common promoter and operator that control transcription of the structural genes.

A complex consisting of two subunits, each composed of ribosomal RNA and protein, found in the cytoplasm of cells or attached to the endoplasmic reticulum, that is the site of protein synthesis, during which the sequence of bases of messenger RNA is translated into the sequence of amino acids in a protein.

A mutation in which a single base pair in DNA has been changed.

In RNA synthesis, an enzyme that catalyzes the bonding of free RNA nucleotides into a continuous strand, using RNA nucleotides that are complementary to those of the template strand of DNA.

A specific sequence of DNA at the beginning of a gene, to which RNA polymerase binds and starts gene transcription.

start codon	transcription
stop codon	transfer RNA (tRNA)
structural gene	translation
template strand	translocation

The synthesis of an RNA molecule from a DNA template.

The first AUG codon in a messenger RNA molecule.

A type of RNA that binds to a specific amino acid, carries it to a ribosome, and positions it for incorporation into the growing protein chain during protein synthesis. A set of three bases in tRNA (the anticodon) is complementary to the set of three bases in mRNA (the codon) that codes for that specific amino acid in the genetic code.

A codon in messenger RNA that stops protein synthesis and causes the completed protein chain to be released from the ribosome.

The process whereby the sequence of bases of messenger RNA is converted into the sequence of amino acids of a protein.

In the prokaryotic operon, the genes that encode enzymes or other cellular proteins.

A mutation that occurs when a piece of DNA is removed from one chromosome and attached to another chromosome.

The strand of the DNA double helix from which RNA is transcribed.

SELF TEST

1. Which of the following statements about the functions of RNA is correct?
   a. The information for protein synthesis is carried by tRNA.
   b. rRNA is an intermediate in the synthesis of mRNA.
   c. rRNA is an important component of ribosomes.
   d. Translation requires tRNA and mRNA, but not rRNA.

2. Imagine that a probe sent to Mars brings back a sample that contains a very primitive life-form that appears similar to bacteria. Scientists are able to revive it and begin to grow it in culture. Much to their amazement, they discover that the organism has DNA and that the DNA encodes proteins. However, the DNA of these Martian microbes contains only two nucleotides, and these nucleotides contain bases that are not present in the DNA of organisms on Earth. If the Martian microbe uses triplet codons, what is the maximum number of different amino acids that it can have in its proteins? Explain.
   a. 9, because $3^2 = 3 \times 3 = 9$
   b. 16, because $4^2 = 4 \times 4 = 16$
   c. 8, because $2^3 = 2 \times 2 \times 2 = 8$
   d. 7, because there are 8 possible codons (2^3)

3. Which of the following is an accurate statement concerning the differences between DNA and RNA?
   a. RNA is usually double stranded, but DNA is usually single stranded.
   b. RNA has the sugar deoxyribose, but DNA has the sugar ribose.
   c. RNA contains three different nucleotides, but DNA contains four different nucleotides.
   d. RNA lacks the base thymine (which is found in DNA) and has uracil instead.

4. The process that uses the genetic information carried by mRNA to specify the sequence of amino acids in a protein is called
   a. replication.   d. synthesis.
   b. translation.   e. elongation.
   c. transcription.

5. Examine the genetic code table below. Which DNA triplets all code for threonine?
   a. ACU, ACC, ACA, and ACG
   b. TGU, TGG, TGT, and TGC
   c. TGA, TGG, TGT, and TGC
   d. UGA, UGG, UGU, and UGC

Table 12-3   The Genetic Code (Codons of mRNA)

First Base	U		C		A		G		Third Base
	\<Second Base\>								
U	UUU	Phenylalanine (Phe)	UCU	Serine (Ser)	UAU	Tyrosine (Tyr)	UGU	Cysteine (Cys)	U
	UUC	Phenylalanine	UCC	Serine	UAC	Tyrosine	UGC	Cysteine	C
	UUA	Leucine (Leu)	UCA	Serine	UAA	Stop	UGA	Stop	A
	UUG	Leucine	UCG	Serine	UAG	Stop	UGG	Tryptophan (Trp)	G
C	CUU	Leucine	CCU	Proline (Pro)	CAU	Histidine (His)	CGU	Arginine (Arg)	U
	CUC	Leucine	CCC	Proline	CAC	Histidine	CGC	Arginine	C
	CUA	Leucine	CCA	Proline	CAA	Glutamine (Gln)	CGA	Arginine	A
	CUG	Leucine	CCG	Proline	CAG	Glutamine	CGG	Arginine	G
A	AUU	Isoleucine (Ile)	ACU	Threonine (Thr)	AAU	Asparagine (Asp)	AGU	Serine (Ser)	U
	AUC	Isoleucine	ACC	Threonine	AAC	Asparagine	AGC	Serine	C
	AUA	Isoleucine	ACA	Threonine	AAA	Lysine (Lys)	AGA	Arginine (Arg)	A
	AUG	Methionine (Met) Start	ACG	Threonine	AAG	Lysine	AGG	Arginine	G
G	GUU	Valine (Val)	GCU	Alanine (Ala)	GAU	Aspartic acid (Asp)	GGU	Glycine (Gly)	U
	GUC	Valine	GCC	Alanine	GAC	Aspartic acid	GGC	Glycine	C
	GUA	Valine	GCA	Alanine	GAA	Glutamic acid (Glu)	GGA	Glycine	A
	GUG	Valine	GCG	Alanine	GAG	Glutamic acid	GGG	Glycine	G

6. What is the most correct statement about transcription and translation in eukaryotes?
   a. Transcription occurs inside the nucleus, whereas translation occurs in the cytoplasm.
   b. Transcription takes a message that is written in "DNA language" and converts it into a message that is written in "amino acid language," whereas translation then takes this message and converts it further into "RNA language."
   c. Transcription begins inside the nucleus but ends in the cytoplasm, whereas translation begins in the cytoplasm but ends inside the nucleus, so the two processes overlap.
   d. Functioning ribosomes, tRNA, free amino acids, and RNA polymerase are all abundant inside the nucleus, allowing both transcription and translation to rapidly occur there, thus increasing the efficiency and speed of protein production by the cell.
   e. Transcription occurs in the cytoplasm, whereas translation occurs inside the nucleus.

7. What does mRNA carry from the nucleus?
   a. ribosomes   c. amino acids
   b. information   d. tRNA

8. Unripe black walnuts contain a compound, juglone, that inhibits RNA polymerase. With which process would juglone most likely interfere?
   a. mutation rate   c. transcription
   b. DNA replication   d. translation

9. What determines which of the two DNA strands will serve as the template?
   a. Only one of the strands can be transcribed.
   b. The sequence of the bases in the RNA after transcription determines which will serve as the template.
   c. The RNA polymerase can transcribe only one strand of DNA.
   d. Only one strand will unwind.
   e. The orientation of the promoter and other regulatory sequences determines which will serve as the template.

10. The next base to be added to the RNA strand is determined by
    a. the previous base.
    b. the order of the backbone in the RNA.
    c. base pairing between the two DNA strands.
    d. base pairing between the template strand and the RNA nucleotides.

11. After an RNA polymerase has completed transcription, the enzyme
    a. is degraded.
    b. joins with another RNA polymerase to carry on transcription.

    c. begins transcribing the next gene on the chromosome.
    d. is free to bind to another promoter and begin transcription.
    e. will remain bound to the RNA.

12. Biology is generally very efficient in its use of resources. With this concept in mind, consider that when a gene is transcribed to make mRNA, only one DNA strand (the template strand) is used. What, then, is the best explanation of why cells use double-stranded DNA, rather than having their DNA exist in a single-stranded form?
    a. Because they have two complementary DNA strands, when they undergo mitosis, they can pass one complete parental strand to each daughter cell (semiconservative DNA replication).
    b. If the template strand is damaged or destroyed, the cell still has the other strand available as a backup copy of its genetic information.
    c. Double-stranded DNA contains twice as much information as single-stranded DNA, which means that genes can be more compact overall.
    d. Any single-stranded polypeptide of nucleic acids would be immediately broken down by enzymes, before it could serve any useful function.

13. Consider how eukaryotic genes contain both introns and exons, how the coding exons may be assembled into a variety of arrangements (alternative splicing), and how a single gene may thus code for several different proteins. What situation would cause this system to fail, producing many faulty proteins as a result?
    a. mRNA transcripts are exported from the nucleus to the cytoplasm, and translation happens in the cytoplasm.
    b. Introns are removed from the initial mRNA transcript, before it leaves the nucleus.
    c. Proteins that are used inside the nucleus are imported into the nucleus from the cytoplasm, via the nuclear pores.
    d. Functional ribosomes, tRNA molecules, and free amino acids exist inside the nucleus.

14. Which molecule is responsible for bringing the correct amino acid to the ribosome at the correct time?
    a. rRNA
    b. mRNA
    c. small subunit
    d. large subunit
    e. tRNA

15. Which of the following is the first step in translation?
    a. Bases of the tRNA anticodon bind with bases of the mRNA codon.
    b. A peptide bond forms between amino acids attached to the adjacent tRNAs on the ribosome.
    c. The ribosomal subunits are disassembled.
    d. Stop codons on the mRNA bind to special proteins rather than to tRNA molecules.

16. After translation is completed, the ribosome
    a. stays intact and finds another AUG codon to begin translation.
    b. moves back to the first AUG codon of the mRNA to begin again.
    c. joins with other ribosomes to continue translation.
    d. breaks into small and large ribosomal subunits.
    e. will remain bound to the mRNA.

17. The flow of genetic information in cells depends on specific base pairing between nucleotides. Which of the following correctly matches the type of base pairing with the process of translation?
    a. RNA base-pairs with DNA.
    b. rRNA base-pairs with mRNA.
    c. tRNA base-pairs with rRNA.
    d. tRNA base-pairs with mRNA.

18. Imagine that a triplet in the template strand of a gene has the sequence TAC. What sequence of the anticodon would decode this DNA triplet? Explain your answer.
    a. ATG, because the anticodon is complementary to the template strand
    b. AUG, because the anticodon is complementary to the template strand
    c. UAC, because the anticodon has the same sequence as the template strand (but it has U instead of T)
    d. TAC, because the anticodon has the same sequence as the template strand

19. Study the figure, showing the actions of transcription and translation in prokaryotes. Note that because prokaryotic DNA is not stored in a compartment separate from the ribosomes, ribosomes may immediately attach to the mRNA transcript. The figure shows three different transcripts, and several ribosomes that are creating proteins. Consider which transcripts and proteins are the oldest. With the transcripts, the oldest one is toward the _____ of the figure, whereas for the proteins, the oldest is to the _____ of the figure.

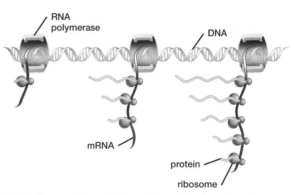

direction of transcription

RNA polymerase

DNA

mRNA

protein

ribosome

(b) Simultaneous transcription and translation in prokaryotes

    a. left; upper-left
    b. right; upper-left
    c. left; lower-right
    d. right; lower-right
    e. left; upper-right
    f. right; upper-right

20. Imagine that a DNA triplet is AAG. What are the corresponding mRNA codon, tRNA anticodon, and amino acid, respectively?
    a. TTC; AAG; leucine
    b. UUC; AAG; phenylalanine
    c. UUC; TTG; phenylalanine
    d. AAG; UUC; lysine
    e. UUC; AAG; lysine

21. Some people have two different colors of eyes. What is a possible explanation for this trait?
    a. A mutation occurred in the sperm that produced this individual's embryo.
    b. A mutation occurred in the egg that produced this individual's embryo.
    c. A chemical toxin inhibited production of pigment in one eye but not the other.
    d. During early stages of development, a mutation occurred in the cell that developed into one of the eyes, but not other cells in the embryo.

22. Which kind of point mutation would have the most dramatic effect on the protein coded for by that gene?
    a. a base substitution
    b. a base insertion near the beginning of the coding sequence
    c. a base insertion near the end of the coding sequence
    d. deletion of three bases near the start of the coding sequence

23. Suppose that a mutation changes a particular sequence of three DNA triplets, on the template strand, from ATATGTCCA to ATATGACCA. The overall length of the gene is unchanged by the mutation. This mutation may be best described as a(n) _____ mutation.
    a. neutral
    b. deletion
    c. point
    d. insertion and neutral
    e. neutral and point

24. Cystic fibrosis involves the production of faulty membrane transport proteins. People who inherit one normal and one defective allele for the *CFTR* gene have nearly normal health, but people who inherit two defective alleles have the disease. What can we say about the mode of inheritance of the disease, and the nature of the faulty proteins?
    a. The disease is a sex-linked recessive condition that mainly affects males, and the faulty protein, while not toxic, is also ineffective at doing its membrane transport job. The males who receive a single defective allele of the *CFTR* gene, on their single X chromosome, are unable to produce any of the correctly formed proteins.
    b. The disease is an autosomal recessive condition, and the faulty protein is actively toxic, with the people who are homozygous for the defective allele of the *CFTR* gene receiving an extremely high dose of the toxin.
    c. The disease is an autosomal recessive condition, and the faulty protein, although not toxic, is also ineffective at doing its membrane transport job. People who are homozygous for the defective allele of the *CFTR* gene are unable to produce any of the correctly formed protein.
    d. The disease is an autosomal dominant condition, and the faulty protein is actively toxic, with the people who are homozygous for the defective allele of the *CFTR* gene receiving an extremely high dose of the toxin.

25. The cells in your skin have a different shape and different function from the cells in your liver because the two types of cells have different
    a. DNA.
    b. proteins.
    c. lipids.
    d. carbohydrates.

26. Which of the following is not a step at which gene expression can be regulated in eukaryotic cells?
    a. rate of transcription
    b. rate of translation
    c. rate of DNA replication
    d. rate of enzyme activity

27. In mammals, males have one X chromosome and one Y chromosome, and females have two X chromosomes. How is the expression of genes on the X chromosome regulated so that there is equal expression of genes on the X chromosome in males and females?
    a. One X chromosome in females is inactivated so that females have only a single X chromosome capable of transcription.
    b. The genes on the X chromosome in males are transcribed twice as fast as in females.
    c. All of the X chromosomes are inactivated so that no genes are expressed from the X chromosome in either males or females.
    d. The Y chromosome contains balancing genes that help raise the levels of mRNA produced by the X chromosome in males.

28. Certain genes, sometimes called housekeeping genes, are expressed in all cells in your body. Other genes are expressed only in certain specialized cells. Which of the following genes is likely to be a housekeeping gene?
    a. hemoglobin   c. ribosomal proteins
    b. milk proteins   d. insulin

29. If androgen insensitivity, as described in the chapter, was caused by a faulty allele on the Y chromosome instead of on the X chromosome, could it still be an inherited condition? Assume that no advanced medical technologies, such as in vitro fertilization, are used to create children.
    a. Yes, because genes on the Y chromosome are always passed from father to son.
    b. Yes, because the genes that are involved in triggering testicular development and testosterone production are also carried on the Y chromosome.
    c. No, because a person would need to inherit the allele from the male parent, and yet the person who possesses the faulty allele would have a female phenotype and could not be a father.
    d. No, because a child who does not develop proper androgen receptor proteins will die either during infancy or fetal development.

30. Regarding the operation of the lactose (milk sugar) operon in *E. coli* bacteria, what is the most accurate statement?
    a. The repressor protein prevents RNA polymerase from binding to the promoter, thus preventing the transcription of a single gene.
    b. The repressor protein can block the transcription of several genes.
    c. The repressor protein is prevented from attaching to the operon, whenever lactose is not present.
    d. The repressor proteins and lactose molecules do not directly interact with each other.

ANSWER KEY

1. c
2. d
3. d
4. b
5. b
6. a
7. b
8. c
9. e
10. d
11. d
12. a
13. d
14. e
15. a

16. d
17. d
18. c
19. f
20. b
21. d
22. b
23. e
24. c
25. b
26. c
27. a
28. c
29. c
30. b

CHAPTER 13 BIOTECHNOLOGY

OUTLINE

Section 13.1 What Is Biotechnology?

Biotechnology is any use or alteration of organisms, cells, or biological molecules to achieve specific practical goals. Modern biotechnology uses genetic engineering as a direct method to modify genetic materials.

Some genetic engineers transfer **recombinant DNA** (which can be grown in bacteria, viruses, or yeasts) to other organisms so that the trait carried by the DNA can be expressed. Organisms that express DNA derived from other species are called **genetically modified organisms (GMOs).**

Section 13.2 How Does DNA Recombine in Nature?

Recombinant DNA occurs naturally by three processes: (1) sexual reproduction by crossing over; (2) bacterial transformation by the acquisition of DNA from **plasmids** or other bacteria (**Figure 13-1**); or (3) viral infection, when DNA fragments are transferred between or among species (**Figure 13-2**).

Section 13.3 How Is Biotechnology Used in Forensic Science?

Small amounts of DNA can be amplified into large amounts by the **polymerase chain reaction (PCR)** technique (**Figure 13-3**). This DNA can be cut into specific fragments, such as **short tandem repeats (STRs)**, using **restriction enzymes** (**Figure 13-4**).

DNA fragments (such as STRs) can be separated by **gel electrophoresis** (**Figure 13-5**) and then made visible by the use of **DNA probes** (**Figure 13-6**). STRs patterns are unique for each individual, creating a **DNA profile** that can be used to match crime scene DNA to that of a suspect (**Figure 13-7**).

Section 13.4 How Is Biotechnology Used in Agriculture?

Crop plants have been modified for herbicide or pest resistance. This can be accomplished by (1) using restriction enzymes to insert the desired gene into a plasmid, (2) transforming bacteria that are then used to infect plant cells, and (3) culturing them to form whole **transgenic** plants (**Figures 13-9 and 13-10**). Transgenic plants can also be used to produce a variety of different medicines.

Transgenic animals can also be produced (often using a disabled virus to transfer DNA) with enhanced growth, milk production, or the ability to produce human proteins.

Section 13.5 How Is Biotechnology Used to Learn About the Human Genome?

Biotechnology techniques were used to sequence the human genome. This knowledge is used to discover medically important genes, to examine evolutionary relatedness among organisms, and to study genetic variability among individuals.

Section 13.6 How Is Biotechnology Used for Medical Diagnosis and Treatment?

Restriction enzymes are used to produce DNA segments at a variety of lengths (called **restriction fragment length polymorphisms—RFLPs**) to identify alleles for a specific genetic disorder (e.g., sickle-cell anemia, **Figure 13-11**).

DNA probes can also be used to identify alleles for specific genetic disorders (e.g., cystic fibrosis, **Figure 13-12**).

Genetic engineering can be used to insert functional alleles into normal cells, stem cells, or egg cells to correct a genetic disorder.

Section 13.7 What Are the Major Ethical Issues of Modern Biotechnology?

One controversial aspect surrounding the use of genetically modified organisms is consumer safety. GMOs usually contain proteins that are harmless to mammals, but the potential exists for new allergens to be transferred to food that is normally harmless. This can be avoided by thorough testing prior to commercial distribution.

A second controversial aspect of GMOs is environmental protection. Modified plant genes can be transferred to wild plants, possibly disrupting ecosystems and agriculture. Also, mobile transgenic animals could displace wild populations.

Use of biotechnology on human embryos is controversial (**Figure 13-13**). The primary benefit is that genetic disorders can be cured, but it can also mean that embryos with genetic defects can be detected and the pregnancies terminated. Potentially, the genomes of human embryos can be modified to enhance their traits rather than to fix any serious defects.

FLASH CARDS

To use the flash cards, tear the page from the book and cut along the dashed lines. The key term appears on one side of the flash card, and its definition appears on the opposite side.

amniocentesis	genetic engineering
biotechnology	genetically modified organism (GMO)
chorionic villus sampling (CVS)	plasmid
DNA probe	polymerase chain reaction (PCR)
DNA profile	recombinant DNA
gel electrophoresis	restriction enzyme

The modification of the genetic material of an organism, usually using recombinant DNA techniques.

A procedure for sampling the amniotic fluid surrounding a fetus: A sterile needle is inserted through the abdominal wall, uterus, and amniotic sac of a pregnant woman, and 10 to 20 milliliters of amniotic fluid are withdrawn. Various tests may be performed on the fluid and the fetal cells suspended in it to provide information on the developmental and genetic state of the fetus.

An organism that has been produced through the techniques of genetic engineering, usually by the addition of genes from different species or modified genes from the same species.

Any industrial or commercial use or alteration of organisms, cells, or biological molecules to achieve specific practical goals.

A small, circular piece of DNA located in the cytoplasm of many bacteria; usually does not carry genes required for the normal functioning of the bacterium but may carry genes that assist bacterial survival in certain environments, such as a gene for antibiotic resistance.

A procedure for sampling cells from the chorionic villi produced by a fetus: A tube is inserted into the uterus of a pregnant woman, and a small sample of villi is suctioned off for genetic and biochemical analyses.

A method of producing virtually unlimited numbers of copies of a specific piece of DNA, starting with as little as one copy of the desired DNA.

A sequence of nucleotides that is complementary to the nucleotide sequence in a gene under study; used to locate a given gene during gel electrophoresis or other methods of DNA analysis.

DNA that has been altered by the addition of DNA from a different organism, typically from a different species.

The pattern of short tandem repeats of specific DNA segments; using a standardized set of 13 short tandem repeats, DNA profiles identify individual people with great accuracy.

An enzyme, normally isolated from bacteria, that cuts double-stranded DNA at a specific nucleotide sequence; the nucleotide sequence that is cut differs for different restriction enzymes.

A technique in which molecules (such as DNA fragments) are placed in wells in a thin sheet of gelatinous material and exposed to an electric field; the molecules migrate through the gel at a rate determined by certain characteristics, most commonly size.

restriction fragment length polymorphism (RFLP)

transformation

short tandem repeat (STR)

transgenic

A method of acquiring new genes, whereby DNA from one bacterium (normally released after the death of the bacterium) becomes incorporated into the DNA of another, living, bacterium.

A difference in the length of DNA fragments that were produced by cutting samples of DNA from different individuals of the same species with the same set of restriction enzymes; fragment length differences occur because of differences in nucleotide sequences, and hence in the ability of restriction enzymes to cut the DNA, among individuals of the same species.

Referring to an animal or a plant that contains DNA derived from another species, usually inserted into the organism through genetic engineering.

A DNA sequence consisting of a short sequence of nucleotides (usually 2 to 5 nucleotides in length) repeated multiple times, with all of the repetitions side by side on a chromosome; variations in the number of repeats of a standardized set of 13 STRs produce DNA profiles used to identify people by their DNA.

SELF TEST

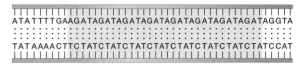

Eight side-by-side (tandem) repeats
of the same four-nucleotide sequence

1. Genetic engineering results in
   a. stem cells.
   b. viruses.
   c. DNA fragments.
   d. transgenic organisms.

2. The process of piecing together genes from different organisms results in
   a. transmorphic DNA.
   b. transgenic DNA.
   c. recombinant DNA.
   d. tandem repeat DNA.

3. A natural example of the use of recombinant DNA would be
   a. a cow that produces human growth hormone in milk.
   b. corn that is resistant to herbicides.
   c. crossing over during sexual reproduction.
   d. cotton that is resistant to fungi.

4. Transformation is the process by which foreign _____ is taken up from a cell's environment, permanently changing the characteristics of a cell and its offspring.
   a. DNA
   b. RNA
   c. protein
   d. plasmid

5. _____ would be a mechanism for development of antibiotic resistance in organisms.
   a. Transformation
   b. Natural selection
   c. Recombination
   d. Gene therapy

6. Which of the following describes an example of DNA recombination?
   a. any organism that utilizes sexual reproduction
   b. the exchange of fluids between a cell's interior (cytoplasm) and its exterior (extracellular fluid)
   c. a viral infection resulting in a common cold
   d. both a and c

7. Which of the following scenarios would be a good candidate for the use of the PCR technique?
   a. A small sample of blood is found at a crime scene and there are many suspects to eliminate.
   b. You need to insert a plasmid in bacteria to create an organism to produce growth hormone.
   c. You need to determine the restriction enzyme for the gene that codes for cystic fibrosis.
   d. You are trying to determine the DNA profile for a new species of mouse that was found in the rain forests of Chile.

8. The figure shows an example of which type of DNA technique?
   a. PCR
   b. DNA fingerprinting
   c. STRs
   d. RFLPs

9. Researchers must design short pieces of DNA that are complementary to DNA on either side of the segment to be amplified. These small pieces are called
   a. primers.
   c. plasmids.
   b. DNA probes.
   d. DNA fingerprints.

10. Which of the following sources could be used in DNA forensic analysis?
   a. semen
   b. saliva
   c. hair follicle
   d. blood
   e. all of the above

11. What are restriction enzymes?
   a. enzymes that are limited in how big they can become
   b. vitamins, such as vitamin A
   c. plasmids
   d. enzymes that cleave through a DNA helix wherever they encounter a specific sequence of nucleotides

12. The figure on page 174 shows an example of the utilization of DNA for
   a. inserting genes developed with plasmids to correct a genetic defect such as cystic fibrosis.
   b. conducting PCR analysis.
   c. determining parentage.
   d. determining genetic disorders of the fetus.

13. The primary goal of human gene therapy is to
   a. correct genetic disorders by inserting normal genes in place of defective ones.
   b. provide counseling to affected people to help them live with their disorders.
   c. treat the symptoms of genetic disorders.
   d. correct genetic disorders in developing embryos.

14. Why is stem cell technology a promising area of medical research?
   a. Stem cell DNA is easy to fingerprint.
   b. Under the right conditions, a stem cell could become any type of cell in the body.
   c. Stem cells can fight pathogens.
   d. Stem cells can be used to produce new medicines.

Amniocentesis

Chorionic villus sampling (by suction)

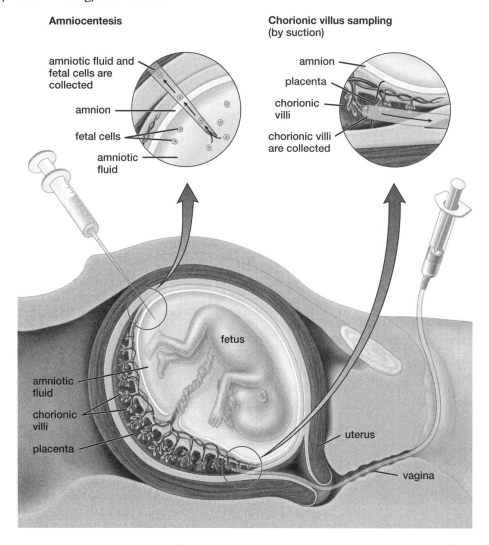

amniotic fluid and fetal cells are collected

amnion

fetal cells

amniotic fluid

amnion

placenta

chorionic villi

chorionic villi are collected

fetus

amniotic fluid

chorionic villi

placenta

uterus

vagina

15. How might human cloning permanently correct genetic defects?
 a. A viral vector would replace a defective gene and the resulting corrected nucleus would be injected into an egg lacking a nucleus and then allowed to develop in utero.
 b. Cloning many identical humans would increase the chance that mutation would correct the genetic defect.
 c. Replacement humans could be grown to be used when the original succumbs to a genetic disease.
 d. All of the above apply.

16. Transgenic organisms can
 a. contain only DNA from their same species.
 b. express DNA from another species.
 c. only be primates.
 d. only be bacteria.
 e. express only their own DNA.

17. The Human Genome Project
 a. is a method of recombinant DNA analysis used to determine the probability of genetic disorders in humans.

 b. is a project that uses DNA technology to determine the genetic code of humans.
 c. is a method used to determine any genetic abnormalities in a fetus.
 d. studies the nucleotide sequences of DNA using PCR and RFLP techniques.

18. An example of the purpose of the Human Genome Project would be
 a. to determine the evolutionary similarities of human and chimp DNA.
 b. to determine if a particular man did indeed father the child of a particular woman.
 c. to develop a cure for cystic fibrosis.
 d. to provide a long-term project in which to utilize the DNA technology that scientists have developed.

19. You can determine when transformation in bacteria has taken place by
 a. analyzing the appearance of the bacteria and monitoring any changes.
 b. using gel electrophoresis to fragment the DNA of the bacteria.

   c. determining if the desired product (insulin, growth hormone, etc.) is being produced by the bacteria.

   d. checking whether or not the organisms have reproduced.

20. In bacterial transformation, after a plasmid enters the new bacterium, it
   a. is destroyed.
   b. will replicate in the cytoplasm.
   c. will be incorporated into the bacterial chromosome.
   d. causes the bacterium to die.
   e. will be broken into small DNA pieces.

21. The polymerase chain reaction
   a. is a method of synthesizing human protein from human DNA.
   b. takes place naturally in bacteria.
   c. can produce billions of copies of a DNA fragment in several hours.
   d. uses restriction enzymes.
   e. is relatively slow and expensive compared with other types of DNA purification.

22. Imagine you are looking at a DNA fingerprint that shows an STR pattern of a mother's DNA and her child's DNA. Will all the bands on the Southern blot showing the child's DNA match those of the mother? Explain.
   a. Yes, because the child developed from an egg produced by the mother.
   b. Yes, because the DNA of mothers and children is identical.
   c. No, because a person's DNA pattern changes with age.
   d. No, because the father contributed half the child's DNA.

23. STR polymorphisms can be used to identify individuals because
   a. individuals are genetically unique.
   b. restriction enzymes cut different recognition sequences in different people.
   c. a given set of PCR primers works for only a small subset of individuals.
   d. the technique is more robust than PCR or DNA fingerprinting.

24. DNA fingerprinting can
   a. help determine the nucleotide sequence of a gene.
   b. utilize a system of RFLP and PCR analysis.
   c. help determine the parentage of a child.
   d. help determine the presence of a defective gene such as cystic fibrosis.

25. Why are genes inserted into bacterial plasmids during the gene-cloning process?
   a. Large amounts of the gene can be produced when the bacteria multiply.

   b. Genes can be cloned only for bacteria.
   c. The genes are more stable when inserted into bacterial plasmids.
   d. Bacterial plasmids require inserted genes to be functional.

26. Why might it be beneficial to produce genetically modified animals?
   a. We can populate our zoos with bizarre and interesting animals.
   b. We may be able to develop livestock with beneficial characteristics.
   c. We may be able to develop animals that can produce useful medicines.
   d. Both b and c.

27. Analysis of restriction fragment length polymorphisms (RFLPs) is a rapid way to examine differences in
   a. bacterial enzymes.
   b. the length of a DNA molecule.
   c. the DNA sequences of individuals.
   d. the number of genes on a chromosome.

28. What benefit does administration of genetically engineered human insulin provide that pig or bovine insulin does not?
   a. There is no benefit; the nucleotide sequence that determines the structure of insulin is the same in all three species.
   b. Allergic reactions were avoided by using the recombinant human insulin.
   c. Human insulin could be produced in large quantities using bacteria/yeast genetically engineered to carry the human insulin gene.
   d. Both b and c.

29. What are some of the problems associated with the use of genetic screens for cystic fibrosis (CF)?
   a. Positive testing of a man and a woman each carrying a copy of the CF gene guarantees their offspring will have the disease.
   b. Ethical dilemmas: Should individual carriers of the CF allele attempt to conceive? Should society and insurance carriers be forced to bear the financial costs?
   c. There is the possibility of genetic discrimination by insurance companies.
   d. Both b and c.

30. What ethical argument(s) has(have) been used against biotechnology?
   a. Genetically modified foods are dangerous to eat.
   b. Genetically modified organisms are harmful to the environment.
   c. The human genome should not be changed because the repercussions of doing so are unclear.
   d. All of the above.

ANSWER KEY

1. d	16. b
2. c	17. b
3. c	18. a
4. a	19. c
5. a	20. b
6. d	21. c
7. a	22. d
8. c	23. a
9. a	24. c
10. e	25. a
11. d	26. d
12. d	27. c
13. a	28. d
14. b	29. d
15. a	30. d

CHAPTER 14 PRINCIPLES OF EVOLUTION

OUTLINE

Section 14.1 How Did Evolutionary Thought Evolve?

Pre-Darwinian science stated that God simultaneously created all organisms, each being distinct and unchanging from the moment of its creation. This view was reflected in Aristotle's "ladder of Nature" (**Figure 14-2**).

By the 1700s, naturalists had explored many lands and discovered more species than they expected. They also noticed that physical and geographic comparisons of these species seemed inconsistent with the supposed "fixed" nature of life.

LeClerc suggested that God created a relatively small number of species and that modern species were "produced by Time"—that is, they evolved through natural processes.

The discovery and examination of **fossils** suggested that previously unknown organisms had gone extinct, and that fossil remains showed a remarkable progression over time (**Figure 14-4**).

The theory of catastrophism (proposed by Cuvier) suggested that these unknown species were eliminated in successive catastrophes that deposited their fossil remains in rock layers. Uniformitarianism (proposed by Lyell) contradicted this view, suggesting that natural, observable processes, repeated over long periods of time, could also produce the layers of rock in which fossils were found. This also suggested that Earth was many millions of years old.

Lamarck suggested that the **evolution** of organisms had occurred by the inheritance of acquired characteristics, in which organisms modify parts through use or disuse and pass these modifications on to offspring. In contrast, Darwin and Wallace proposed that organisms evolved over time through **natural selection**.

Section 14.2 How Does Natural Selection Work?

Darwin and Wallace suggest that (1) **populations** are variable (**Figure 14-6**), (2) variable traits can be inherited, (3) some individuals in a population survive and others do not, and then (4) the characteristics of successful individuals will be "naturally selected" and become more common over time.

Natural selection operates on individual organisms within populations, with certain individuals being favored to survive and reproduce more than other individuals. However, individual organisms do not evolve, but populations of species do.

Section 14.3 How Do We Know That Evolution Has Occurred?

The fossil record shows evidence of evolutionary change over time (**Figure 14-7**).

Homologous structures, **vestigial structures**, and **analogous structures** all provide evidence of descent with modification (**Figures 14-8, 14-9,** and **14-10**).

Similarities among vertebrate embryos suggest common ancestry (**Figure 14-11**).

Genetic and biochemical analyses suggest that all organisms are related to different degrees (**Figure 14-12**).

Section 14.4 What Is the Evidence That Populations Evolve by Natural Selection?

Artificial selection supports evolution by natural selection (**Figure 14-13**). Since humans have bred different forms of plants and animals over thousands of years, logic suggests that natural selection can produce the entire spectrum of organisms over hundreds of millions of years.

Natural selection occurs today, as observed in some guppy populations (**Figure 14-14**), insect pesticide resistance, and scientific experimentation with anole lizards (**Figure 14-15**).

FLASH CARDS

To use the flash cards, tear the page from the book and cut along the dashed lines. The key term appears on one side of the flash card, and its definition appears on the opposite side.

analogous structure	homologous structure
artificial selection	natural selection
convergent selection	population
evolution	vestigial structure
fossil	

Structures that may differ in function but that have similar anatomy, presumably because the organisms that possess them have descended from common ancestors.

Structures that have similar functions and superficially similar appearance but very different anatomies, such as the wings of insects and birds. The similarities are the result of similar environmental pressures rather than a common ancestry.

Unequal survival and reproduction of individuals with different phenotypes; causes better adapted phenotypes to become increasingly common in a population.

A selective breeding procedure in which only those individuals with particular traits are chosen as breeders; used mainly to enhance desirable traits in domestic plants and animals; may also be used in evolutionary biology experiments.

All the members of a species that occupy a particular area at the same time.

The independent evolution of similar structures among unrelated organisms as a result of similar environmental pressures; see *analogous structure*.

A structure that serves no apparent purpose but is homologous to functional structures in related organisms and provides evidence of evolution.

(1) The theory that all organisms are related by common ancestry and have changed over time; (2) any change in the proportions of different genotypes in a population from one generation to the next.

The remains of a dead organism, normally preserved in rock; may be petrified bones or wood; shells; impressions of body forms, such as feathers, skin, or leaves; or markings made by organisms, such as footprints.

SELF TEST

1. The idea put forward by Lamarck, that organisms could pass on to their offspring physical changes that the parents developed during their own lifetimes, is known as
   a. genetic drift.
   b. natural selection.
   c. artificial selection.
   d. adaptive radiation.
   e. inheritance of acquired characteristics.

2. Which of the following people did NOT influence Charles Darwin in his formulation of his theory of natural selection?
   a. Thomas Malthus   c. William Smith
   b. Charles Lyell   d. Gregor Mendel

3. Which of the following is NOT a major aspect of Darwin's theory of evolution?
   a. Life on Earth is quite old.
   b. Organisms are careful to produce no more offspring than the resources in the environment are capable of supporting.
   c. Contemporary species share a common descent.
   d. Species are formed and adapt by the process of natural selection.
   e. Evolution is gradual and continuous.

4. What does the idea of uniformitarianism suggest about the geological record and the age of Earth?
   a. Earth is 6,000 years old.
   b. Species are evidence of acts of divine creation.
   c. Earth's species were initially created all at the same time, created initially but many were destroyed by successive catastrophes.
   d. Natural processes (e.g., sedimentation due to river flow) occurring over long stretches of time rather than cataclysmic events account for the thick layers of rock where fossils are found.

5. What is a modern definition of evolution?
   a. the formation of a new species
   b. a change in the characteristics of a population over time
   c. natural selection for a novel trait
   d. a trait that increases the fitness of an individual relative to individuals without that trait

6. What was the role of Alfred Wallace in the development of evolutionary ideas?
   a. He developed the idea of evolution by natural selection before Darwin did, but he never published the idea, making it possible for Darwin to later claim all of the credit.
   b. He believed that the Earth was no more than 10,000 years old, and that all modern life-forms had evolved rapidly from earlier forms during that short time period.
   c. Although he accepted the idea that life-forms had evolved, he strongly opposed the idea that natural selection played an important role in evolution.
   d. He independently developed the idea of evolution by natural selection, and both he and Darwin published papers on the subject at the same time.

7. Why were Lyell's geological theories so significant?
   a. He demonstrated that the Earth's geological features, such as mountains and canyons, could have been formed by a single, global cataclysm.
   b. He showed that the Earth's features had been formed by gradual geological processes acting over very long time spans, thus proving the planet to be old enough that modern life-forms could have gradually evolved from earlier ancestors.
   c. He proved that fossils embedded in lower (deeper) layers of rock are younger than are the fossils in upper (shallower) layers of rock.
   d. He showed that modern types of geological forces, such as wind and waves, have only an insignificant role in shaping the Earth's surface.

8. Which of the following is the most essential requirement for natural selection to be an effective evolutionary force?
   a. Environmental conditions must be changing.
   b. Individuals are reproducing at a rapid rate.
   c. Each population is limited to a small size.
   d. A population exhibits some genetic variability.

9. Which of the following is an INCORRECT statement about mutation?
   a. Mutation introduces variation into a population.
   b. Mutations can be inherited from parents to offspring.
   c. Mutations may have no effect on the organism.
   d. Mutations that are favored by selection are more likely to occur.

10. Which of the following lists contains the four observations of natural selection?
    a. variation in population, heritable variation, selection, unlimited survival and reproduction
    b. uniform population, heritable traits, selection, differential survival or reproduction
    c. variation in population, environmental variation, selection, differential survival or reproduction
    d. rapid reproduction rate, constant population size, variation among individuals in a population, heritable variation

11. Examine the figure below, and notice that the four Galapagos finch species are each specialized to eat different foods. What is the best scientific explanation for how these islands (600 miles offshore from South America) became home to many distinctly different finch species?
   a. A single ancestral species of finch flew from the South American mainland to colonize the islands, and natural selection subsequently favored the evolution of many different species, each of which was specialized to eat a particular food source.
   b. All of these finch species also live in South America, and many different mainland species have independently colonized the islands, producing the variety of species that we see there.
   c. Ancient people must have artificially bred the different finch species and released them in the islands long before Darwin arrived there.
   d. There are far more diverse plant species available as food sources in the Galapagos Islands, compared to the very low-diversity tropical regions of nearby South America.

12. You are a biologist studying a natural population of mice, and you observe that in one area the proportion of darker-colored mice is greater than the proportion of lighter-colored mice. In another area, the opposite is true. You find that only the area with more dark mice has predators. Therefore, you hypothesize that darker mice are favored in areas with predators, perhaps because they are more difficult to see. If your hypothesis is true, what would you expect to happen in the few generations after predators are introduced to an area with a population of mice that previously did not have predators?
   a. The proportion of darker-colored mice will decrease.
   b. The proportion of darker-colored mice will not change.
   c. The proportion of darker-colored mice will increase.
   d. The predators will evolve to be able to see the darker mice better.
   e. both c and d.

(a) Large ground finch, beak suited to large seeds

(b) Small ground finch, beak suited to small seeds

(c) Warbler finch, beak suited to insects

(d) Vegetarian tree finch, beak suited to leaves

13. In which situation should evolution by natural selection be able to proceed more rapidly?
    a. Resources (food, water, etc.) that are needed by members of a population are plentiful, so that every individual can get what it needs with little effort.
    b. Members of a species reproduce sexually and rapidly, with a very short amount of time between generations.
    c. A species of fungus reproduces asexually (clonally) and seldom or never sexually.
    d. Mating occurs entirely at random, with potential mates not discriminating at all about who they mate with.
    e. The background level of cosmic radiation reaching the Earth's surface decreases, due to the Earth's shifting position in the galaxy.

14. Imagine that an environment contains constant levels of the resources that members of a species need to survive. What, then, could explain why the adults of some species produce extremely large numbers of offspring?
    a. Siblings in a large family evolve more rapidly than siblings in a small family, which allows these organisms to more rapidly improve their genetic match with their environment.
    b. With greater numbers of offspring per adult, the overall population size of the species can greatly increase, thus benefiting the entire species.
    c. Organisms are in competition with others of their own species, and those that produce more offspring may experience greater success in passing their genes on to future generations.
    d. Parents are trying to maximize the proportion of their own DNA that is being passed on to each of their particular offspring.

15. A human arm is homologous with a (an)
    a. crab forelimb.
    b. octopus tentacle.
    c. bird wing.
    d. sea star arm.

16. In Africa, a species of bird called the yellow-throated longclaw looks almost exactly like the meadowlark found in North America, but they are not closely related. This is an example of
    a. uniformitarianism.
    b. gradualism.
    c. vestigial structures.
    d. convergent evolution.

17. Structures that serve no apparent purpose but are homologous to functional structures in related organisms are called
    a. analogous structures.
    b. homologous structures.
    c. vestigial structures.
    d. convergent structures.

18. All vertebrate embryos resemble each other during the early stages of development. For example, fish, turtles, chickens, mice, and humans develop tails and gill slits during early stages of development. This suggests all of the following, EXCEPT that
    a. early embryonic development is conservative.
    b. all of these animals use their tails to swim, and gills to breathe, early in their embryological development.
    c. genes that modify the developmental pathways in vertebrates arose later in evolution.
    d. ancestral vertebrates possessed genes that directed the development of tails and gill slits, and all of their descendants still retain those genes.

19. All organisms share the same genetic code. This commonality is evidence that
    a. evolution is occurring now.
    b. convergent evolution has occurred.
    c. evolution occurs gradually.
    d. all organisms descended from a common ancestor.
    e. life began a long time ago.

20. Which of the following is an example of analogous structures?
    a. molar teeth in a vampire bat
    b. superficially similar structures in unrelated species
    c. internally similar structures (e.g., of birds and mammals) that are used for many different functions
    d. both a and c

21. Ostriches (native to Africa), rheas (native to South America), and emus (native to Australia) share only very ancient ancestral connections, and yet being large, flightless birds that are all fast runners adapted to grassland environments, they seem remarkably similar. Their apparent similarity is a result of
    a. convergent evolution.
    b. interbreeding between the species, producing hybrids.
    c. divergence of homologous structures.
    d. continuing inheritance of vestigial structures.

22. Consider the genetic similarity between humans and mice, as shown in the cytochrome *c* DNA sequence above. Cytochrome *c* is an extremely important metabolic enzyme and is found, in some form, in most species. How similar do you think the cytochrome *c* sequences in monkeys, sharks, and oak trees would be to that of humans? Choose the list that has the organisms ranked in the correct order, from the organism with the sequence most similar to the human sequence (on the left) to the organism with the least similar sequence (on the right).
    a. mouse, oak tree, shark, monkey
    b. monkey, shark, mouse, oak tree
    c. oak tree, shark, mouse, monkey
    d. mouse, monkey, oak tree, shark
    e. monkey, mouse, shark, oak tree

23. Early in their histories, Mars and Earth may have had similar environmental conditions. Also, several meteorites have been discovered on Earth that originated on Mars, but were knocked free of the planet by asteroid impacts there. Early in the history of the solar system, such asteroid impacts were much more frequent. These facts have led to the idea that life on Earth could have originated on Mars, or that alternatively, life on Mars (if any) could have originated on Earth. In either case, dormant bacteria might have ridden on a chunk of rock that was blasted free of one planet, and that eventually impacted on the other. If we find life still existing on Mars today, what would be the most convincing evidence that it originated independently of life on Earth, and shares no early ancestry with Earth life?
    a. Martian life uses a different genetic code (the correspondence between mRNA codons and amino acids; Table 12-3) than life on Earth does.
    b. Martian life tolerates the much higher levels of radiation that are found on Mars, levels that are harmful to Earthly life.
    c. The Martian versions of bacteria can gain energy by metabolizing minerals that are only found on Mars, and not on Earth.
    d. Studies of both modern Earth bacteria and the newly discovered Martian bacteria show that many species can survive, for many years, in the harsh vacuum of space.

24. Which of the following describes a case of artificial selection?
    a. the development of high-yielding varieties of wheat from ancestral wild grasses
    b. coloration changes in guppy populations in the absence of predators
    c. increased frequency of roaches that avoid sugar-baited poison traps
    d. the development and spread of antibiotic-resistant tuberculosis bacteria

25. Antibiotic prescriptions normally specify that one take the entire course of treatment, such as a fixed number of pills, rather than simply taking the medicine until one feels better. Why?
    a. It is wasteful to leave pills behind.
    b. You may suffer a relapse of the illness.

c. Bacteria carrying slight resistance will be killed by the full course but will persist with a lesser dose.

d. The full course of treatment will kill many other bacteria, in addition to the ones that are causing the current illness.

26. In the United States today, about half of the corn crop is genetically engineered with a protein that is toxic to corn borers, an insect pest of corn. Which of the following conditions is necessary for the corn borer to evolve resistance to the toxic protein?

a. The corn borer must have heritable variation in the resistance to the toxic protein. The resistant corn borers must survive and reproduce the same as nonresistant corn borers.

b. The corn borers must lack variation in the resistance to the toxic protein. The resistant corn borers must survive in the same way as nonresistant corn borers.

c. All corn borers must have resistance to the toxic protein. The resistant corn borers must survive better or reproduce more than other insects.

d. The corn borer must have heritable variation in the resistance to the toxic protein. The resistant corn borers must survive better or reproduce more than nonresistant corn borers.

27. Behavioral geneticists have found that the ability of house mice to squeak is determined by a single gene. If squeaky mice are more likely to be dropped by startled predators

a. the mouse population will consist of more and more squeakers over time.

b. the mouse population will consist of more and more nonsqueakers over time.

c. the occurrence of squeaking in the mouse population will not change.

d. the proportion of squeakers will not change, but the squeakers will squeak more loudly.

28. Biologists have shown that some species of fish have genetic variation in their tendency to bite an angler's bait and hook. With repeated fishing and harvesting of caught fish,

a. catching fish should become more difficult.

b. catching fish should become easier.

c. no change in catching success should occur.

d. an individual's ability to catch fish relative to other anglers will determine one's success.

29. Which of the following is NOT a situation in which humans are affecting and/or have affected the occurrence and/or rate of evolution?

a. Most dog breeds look very different from wolves.

b. Some ocean fish species are now becoming reproductive at smaller body sizes.

c. Many wild tropical grasses perform C_4 photosynthesis (Chapter 7), which helps them to grow in hot, dry climates.

d. Many roaches in Florida dislike the taste of the pesticide Combat®.

30. Which of the following is NOT a correct view of natural selection?

a. Natural selection improves the fit between a species and its environment.

b. Features of a species that are adaptive (beneficial) under current circumstances may later prove harmful under some altered, future circumstances.

c. For any given species, living in a range of habitats and through a time span of perhaps millions of years, there is always a single, most optimal, genotype that natural selection creates and then maintains until the species eventually becomes extinct.

d. The presence of variation (such as many different alleles of genes) in a population provides the raw material on which natural selection operates, increasing the frequency of some alleles and decreasing the frequency of others.

ANSWER KEY

1. e	16. d
2. d	17. c
3. b	18. b
4. d	19. d
5. b	20. b
6. d	21. a
7. b	22. e
8. d	23. a
9. d	24. a
10. d	25. c
11. a	26. d
12. e	27. a
13. b	28. a
14. c	29. c
15. c	30. c

CHAPTER 15 HOW POPULATIONS EVOLVE

OUTLINE

Section 15.1 How Are Populations, Genes, and Evolution Related?

Evolution is described as a change in the **allele frequencies** of a **population** over time, resulting in a change in that population's **gene pool** (**Figure 15-2**).

Allele frequencies in a population will remain stable over successive generations (forming an **equilibrium population**) only if the following conditions of the **Hardy–Weinberg principle** are met: (1) There are no **mutations**; (2) There is no **gene flow** between populations; (3) The population must be very large; (4) All mating must be random; (5) There must be no **natural selection**. These conditions are rarely met.

Section 15.2 What Causes Evolution?

Mutations are the original source of genetic variability and are usually the result of DNA copying errors during cell division. These mutations are random events and may be passed on to successive generations (**Figure 15-3**).

Gene flow describes the movement of alleles between populations. This can change population allele frequencies by increasing the genetic similarity of these populations.

Certain events have the potential to cause random changes in a population's allele frequency by a process called **genetic drift** (**Figure 15-5**). Small populations are particularly affected because these events can eliminate a disproportionate number of individuals bearing a particular allele (**Figure 15-6**). Both **population bottlenecks** and the **founder effect** are examples of situations in which genetic drift can occur (**Figures 15-7** and **15-8**).

Nonrandom mating can affect the genotypes of a population, typically by inbreeding (which can result in harmful homozygous genotypes) or assortative mating.

Not all genotypes are equally beneficial. Different genotypes result in different phenotypes, which affect the ability of an organism to survive in different ways. Thus, natural selection will favor the genotypes that produce beneficial phenotypes having the greatest **fitness**.

Section 15.3 How Does Natural Selection Work?

Natural selection acts on a population through differences in reproductive success resulting from phenotype variations. This, in turn, affects the genotypes that are found in the population.

Natural selection occurs due to interactions between organisms and the abiotic or biotic components of the environment. An example of abiotic influences on natural selection is a plant's ability to access soil minerals and soil water content. Biotic influences on natural selection include **competition** for resources, **coevolution** due to the interactions between predators and prey, and **sexual selection** (**Figures 15-11** and **15-12**).

Both natural and sexual selection can result in three patterns of evolutionary change (**Figure 15-13**). (1) **Directional selection** shifts character traits in a specific direction. (2) **Stabilizing selection** shifts character traits toward the average for a population. (3) **Disruptive selection** shifts character traits toward the phenotypic extremes for a population.

FLASH CARDS

To use the flash cards, tear the page from the book and cut along the dashed lines. The key term appears on one side of the flash card, and its definition appears on the opposite side.

adaptation	equilibrium population
allele frequency	fitness
coevolution	founder effect
competition	gene flow
directional selection	gene pool
disruptive selection	genetic drift

A population in which allele frequencies and the distribution of genotypes do not change from generation to generation.

A trait that increases the ability of an individual to survive and reproduce compared to individuals without the trait.

The reproductive success of an organism, relative to the average reproductive success in the population.

For any given gene, the relative proportion of each allele of that gene in a population.

The result of an event in which an isolated population is founded by a small number of individuals; may result in genetic drift if allele frequencies in the founder population are by chance different from those of the parent population.

The evolution of adaptations in two species due to their extensive interactions with one another, in which each species is a source of natural selection on the other.

The movement of alleles from one population to another owing to the migration of individual organisms.

Interaction among individuals who attempt to utilize a resource (for example, food or space) that is limited relative to the demand for it.

The total of all alleles of all genes in a population; for a single gene, the total of all the alleles of that gene that occur in a population.

A type of natural selection that favors one extreme of a range of phenotypes.

A change in the allele frequencies of a small population purely by chance.

A type of natural selection that favors both extremes of a range of phenotypes.

Hardy–Weinberg principle	population bottleneck
mutation	predation
natural selection	sexual selection
population	stabilizing selection

The result of an event that causes a population to become extremely small; may cause genetic drift that results in changed allele frequencies and loss of genetic variability.

A mathematical model proposing that, under certain conditions, the allele frequencies and genotype frequencies in a sexually reproducing population will remain constant over generations.

The act of eating another living organism.

A change in the base sequence of DNA in a gene; normally refers to a genetic change significant enough to alter the appearance or function of the organism.

A type of natural selection that acts on traits involved in finding and acquiring mates.

Unequal survival and reproduction of individuals with different phenotypes; causes better adapted phenotypes to become increasingly common in a population.

A type of natural selection that favors the average phenotype in a population.

All the members of a species that occupy a particular area at the same time.

SELF TEST

1. Within a population of 100 diploid individuals, for a particular autosomal gene locus, how many alleles does the gene pool contain?
   - **a.** 2
   - **b.** 50
   - **c.** 100
   - **d.** 200

2. A _____ is all the individuals of a species living in a particular place at a particular time.
   - **a.** bottleneck
   - **b.** species
   - **c.** gene pool
   - **d.** population

3. The Hardy–Weinberg principle is a model of population genetics that
   - **a.** shows that evolution will occur if all the conditions of the principle are met.
   - **b.** shows that evolution will occur if one or more of the conditions of the principle are violated.
   - **c.** describes how evolution occurs in nature.
   - **d.** both a and c.

4. The relative proportion of an allele in a population is the
   - **a.** allele frequency.
   - **b.** gene pool.
   - **c.** Hardy–Weinberg equilibrium.
   - **d.** population size.

5. An equilibrium population
   - **a.** evolves slowly.
   - **b.** evolves quickly.
   - **c.** evolves only by natural selection.
   - **d.** does not evolve.

6. Evolution is best defined as a change in
   - **a.** number of species.
   - **b.** physical traits.
   - **c.** DNA sequence.
   - **d.** allele frequencies.

7. Examine the figure below. Consider that out of 50 alleles in the mouse population, 20 are *B* and 30 are *b*. According to the box titled "A Closer Look at the Hardy–Weinberg Principle," this means that $p = 0.4$ and $q = 0.6$. In this case, what are the expected proportions of mice in this population with genotypes that are, respectively, (1) homozygous dominant, (2) heterozygous, and (3) homozygous recessive? *Hint:* Examine the proportions of these genotypes that are shown in the figure.
   - **a.** 0.16; 0.24; 0.36
   - **b.** 0.16; 0.48; 0.36
   - **c.** 0.36; 0.48; 0.16
   - **d.** 0.24; 0.48; 0.16
   - **e.** 0.36; 0.16; 0.48

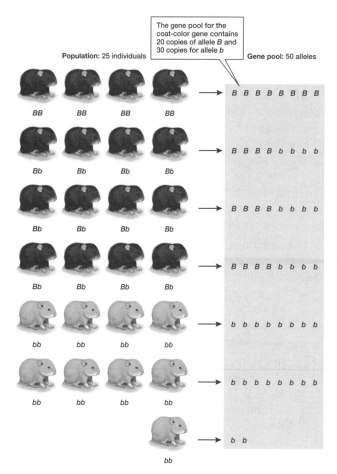

8. Suppose that in a randomly interbreeding population with 1,000 individuals, for a particular gene locus, there are 1,600 *A* alleles and 400 *a* alleles. How many individuals do we expect to be heterozygous, *Aa*?
   a. 0   d. 320
   b. 40   e. 640
   c. 160

9. Which of the following statements is NOT true?
   a. Natural selection is a mindless, mechanical process.
   b. Darwin had no knowledge of the mechanics of heredity.
   c. Heredity plays an important role in the theory of evolution by natural selection.
   d. Individual organisms evolve.
   e. The changes that we see in an individual as it grows and develops are not evolutionary changes.

10. A large population of birds on an island has a color distribution of 80% green birds and 20% yellow birds. Most of the population is killed by an unusually strong storm; color distribution of the survivors is 60% green birds and 40% yellow birds. The change in the color distribution is probably due to
    a. natural selection.   c. genetic drift.
    b. gene flow.   d. mutation.

11. Mutations
    a. are responsible for rapid evolution in populations.
    b. are defined as changes in single DNA nucleotides.
    c. provide the basic "raw material" for evolution.
    d. are usually beneficial to the organism.

12. In a mainland bird population, most individuals are black in color, and gray is a rare variation. A small group of these birds is carried by strong winds to a distant island, where they establish a new population. After a few generations, the gray phenotype is very common. What is most likely to be responsible for this?
    a. natural selection   d. nonrandom mating
    b. mutation   e. founder effect
    c. gene flow

13. Gene flow
    a. cannot influence the evolution of a population.
    b. prevents the spread of alleles through a species.
    c. causes populations to diverge from each other.
    d. makes populations more genetically similar.

14. Natural selection acts (through predation) against banded water snakes on certain Lake Erie islands, favoring the uniformly light-colored snakes. The banded form is very common on the nearby mainland. Yet banded snakes are maintained in the island populations and not eliminated completely. This is probably due to
    a. mutation.   c. genetic drift.
    b. gene flow.   d. natural selection.

15. Which of the following is characterized by random changes in allele frequencies?
    a. natural selection
    b. genetic drift
    c. gene flow
    d. nonrandom mating

16. Imagine that several small populations are isolated from each other. Each is subjected to genetic drift, which will tend to
    a. increase genetic variability both within and between populations.
    b. decrease genetic variability both within and between populations.
    c. increase genetic variability within populations, but decrease genetic variability between populations.
    d. decrease genetic variability within populations, but increase genetic variability between populations.

17. Examine the figure below. Assume that in generation 0, all members have the genotype *Bb* (they are heterozygous for this gene locus). For each of the four different computer simulations, label the figure to show one set of the possible genotypes for all of the members of generation 6. For two of the simulations, there can be more than one correct answer.

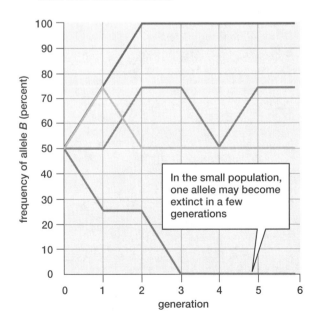

In the small population, one allele may become extinct in a few generations

18. When alleles increase in frequency because they allow the organism to better survive and reproduce in its environment, this is a result of
    a. natural selection.   c. genetic drift.
    b. gene flow.   d. mutation.

19. Which of the following does NOT change allele frequencies in a population?
    a. mutation   d. random mating
    b. gene flow   e. natural selection
    c. genetic drift

20. California has about 3,000 sea otters living along its coastline. All are descended from a small group of about 50 otters that were overlooked by fur hunters during the 1800s. The California sea otter population thus has little genetic variability. Suppose that we try to increase genetic variability by transplanting Alaskan otters into Californian waters, in the hope that they will interbreed with the California otters. In this case, we would be attempting to mitigate the effects of _____ by introducing the process of _____.
    a. a population bottleneck; genetic drift
    b. natural selection; gene flow
    c. mutations; sexual selection
    d. a population bottleneck; gene flow
    e. natural selection; sexual selection

21. Suppose that seven people embark on an intended 3-hour boat tour, but are unexpectedly shipwrecked on a small island that can support no more than about 20 people at a time. Their distant descendants (the 10th generation of castaways) are not rescued until 200 years later. What prediction can we make regarding the genetic variability of these descendants?
    a. They will be more genetically similar to each other than the original castaways were to each other, because many alleles will have been lost from the population due to genetic drift.
    b. Because of the opportunity for new mutations to create many new alleles, this population will show more genetic variability than the original group of castaways did.
    c. Because the island supports a long-term population somewhat larger than the original group of seven, the 10th generation descendants should have greater genetic variability.
    d. Genetic variability in the 10th generation will be about the same as in the first, because the population is large enough for the effects of genetic drift to be negligible.

22. Which of the following is NOT a threat posed by a very small population size (population bottleneck)?
    a. increasing rates of mutations, which are usually harmful
    b. mating between close relatives that creates lethal homozygous recessive genotypes in some offspring
    c. loss of alleles and thus genetic diversity in the population, due to genetic drift
    d. a very small population has limited genetic diversity, even before genetic drift occurs
    e. inability of such a population to deal with changing environmental conditions

23. Suppose that a species is widespread over a continent, and that it has also colonized an island. If environmental conditions on the island are sufficiently different from those on the mainland, the island population could, through many generations, become genetically adapted to these distinct conditions, and could eventually evolve into a new species. The process of _____ is most likely to prevent this evolution of a new species from occurring.
    a. natural selection
    b. mutation
    c. gene flow
    d. sexual selection
    e. genetic drift

24. In the Pacific Northwest, rough-skinned newts produce extremely high levels of tetrodotoxin, making them a fatal meal for almost any would-be predator. The sole exception is the common garter snake of the same region, which preys on the newts and which is extremely resistant to the effects of tetrodotoxin. The evolution of such high levels of toxicity in the newts, and of corresponding levels of resistance to the toxin in their predators, provides an example of
    a. coevolution.
    b. genetic drift.
    c. disruptive selection.
    d. stabilizing selection.

25. Which of the following phrases describes the most important concept of natural selection?
    a. survival of the fittest
    b. differences in reproductive success
    c. some individuals of a species enjoy a longer life span than others
    d. both a and c

26. We study a population of bighorn sheep in the Rocky Mountains, seeing how successful each adult is at producing offspring, during a 3-year period. We collect DNA samples so that we fully understand the parentage of each new offspring that is produced. We find that with the adult males, the numbers of offspring range from 0 for the least successful males up to 35 for the most successful. With the adult females, numbers of offspring produced range from 2 for the least successful female up to 5 for the most successful. Which of the following is NOT a plausible explanation for such a disparity in the range of reproductive success, comparing males and females?
    a. Males fight for dominance, and the more dominant males mate with the most females.
    b. Females differ greatly in how many offspring they may have in a year, with some having only one or two in a season, whereas others may have as many as seven or eight at a time.
    c. Every fertile female is likely to be impregnated by a dominant male, whereas most males are subordinate and thus are seldom able to mate.
    d. The sheep live long enough that an especially strong male can remain dominant for several consecutive years.

27. Selection against individuals at both ends of a phenotypic distribution for a character (i.e., favoring those in the middle or average of the distribution) is an example of
    a. the founder effect.
    b. disruptive selection.
    c. artificial selection.
    d. stabilizing selection.
    e. directional selection.

28. Body size varies among individuals in a species of lizard in the genus *Aristelliger*. Small lizards have a hard time defending territory, and thus mating, but large lizards are more likely to preyed on by owls. Therefore, natural selection favors individuals with an average body size. This is an example of
    a. directional selection.
    b. disruptive selection.
    c. stabilizing selection.
    d. female mate selection.

29. Black-bellied seedcrackers have either small beaks (better for eating soft seeds) or large beaks (better for hard seeds). There are no seeds of intermediate hardness; therefore, _____ acts on beak size in seedcrackers.
    a. directional selection
    b. stabilizing selection
    c. coevolution
    d. disruptive selection

30. Long necks make it easier for giraffes to reach leaves high on trees, while also making them better fighters in "neck-wrestling" contests. In both cases, _____ appears to have made giraffes the long-necked creatures they are today.
    a. random mating
    b. directional selection
    c. stabilizing selection
    d. genetic drift
    e. disruptive selection

ANSWER KEY

1. d
2. d
3. b
4. a
5. d
6. d
7. b
8. d
9. d
10. c
11. c
12. e
13. d
14. b
15. b
16. d
17.

18. a
19. d
20. d
21. a
22. a
23. c
24. a
25. b
26. b
27. d
28. c
29. d
30. b

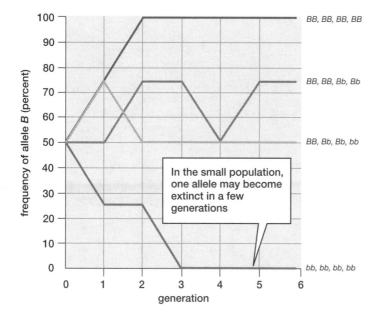

CHAPTER 16 THE ORIGIN OF SPECIES

OUTLINE

Section 16.1 What Is a Species?

The biological species concept defines a **species** as all groups of actually or potentially interbreeding natural populations that are also reproductively isolated from other populations.

The biological species concept has limitations: (1) It does not help to distinguish between species that use strictly asexual reproduction and (2) the mating habits of some species are difficult to observe, sometimes making the existence of **reproductive isolation** difficult to demonstrate.

Similarities or differences in appearance do not always prove whether different organisms belong to the same species (**Figures 16-1** and **16-2**).

Section 16.2 How Is Reproduction Isolation Between Species Maintained?

Isolating mechanisms prevent species from interbreeding by maintaining reproductive isolation (**Table 16-1**).

Premating isolating mechanisms prevent mating between species. These mechanisms include geographical isolation (**Figure 16-3**), ecological isolation (**Figure 16-4**), temporal isolation (**Figure 16-5**), behavioral isolation (**Figure 16-6**), and mechanical isolation (**Figure 16-7**).

Postmating isolating mechanisms prevent the formation of vigorous, fertile hybrids between species. These mechanisms include gametic incompatibility, hybrid inviability, and hybrid infertility (**Figure 16-8**).

Section 16.3 How Do New Species Form?

Speciation occurs when gene flow between two populations is reduced or eliminated and the populations diverge genetically.

Speciation can occur through geographical separation, a process called **allopatric speciation** (**Figure 16-9**).

Speciation can also occur through ecological isolation, a process called **sympatric speciation** (**Figure 16-10**).

Some new plant species can develop within a single generation, as a result of **polyploidy**, a mutation that causes the multiplication of a plant's entire chromosome set.

In some cases, many new species can form in a relatively short period of time, such as when a species encounters a wide variety of unoccupied habitats. This is called **adaptive radiation** (**Figures 16-11** and **16-12**).

Section 16.4 What Causes Extinction?

Extinction occurs when all the members of a species die off. This can occur from localized distribution (**Figure 16-13**), overspecialization (**Figure 16-14**), competition among species, and habitat destruction.

FLASH CARDS

To use the flash cards, tear the page from the book and cut along the dashed lines. The key term appears on one side of the flash card, and its definition appears on the opposite side.

adaptive radiation	premating isolating mechanism
allopatric speciation	reproductive isolation
extinction	speciation
isolating mechanism	species
polyploidy	sympatric speciation
postmating isolating mechanism	

Any structure, physiological function, or behavior that prevents organisms of two different populations from exchanging gametes.

The rise of many new species in a relatively short time; may occur when a single species invades different habitats and evolves in response to different environmental conditions in those habitats.

The failure of organisms of one population to breed successfully with members of another; may be due to premating or postmating isolating mechanisms.

The process by which new species arise following physical separation of parts of a population (geographical isolation).

The process of species formation, in which a single species splits into two or more species.

The death of all members of a species.

The basic unit of taxonomic classification, consisting of a group of populations that evolves independently. In sexually reproducing organisms, a species can be defined as a population or series of populations of organisms that interbreed freely with one another under natural conditions but that do not interbreed with members of other species.

A morphological, physiological, behavioral, or ecological difference that prevents members of two species from interbreeding.

The process by which new species arise in populations that are not physically divided; the genetic isolation required for sympatric speciation may be due to ecological isolation or chromosomal aberrations (such as polyploidy).

Having more than two sets of homologous chromosomes.

Any structure, physiological function, or developmental abnormality that prevents organisms of two different populations, once mating has occurred, from producing vigorous, fertile offspring.

SELF TEST

1. Which of the following is true of the biological species concept?
   a. Asexually reproducing organisms use the same criteria as sexually reproducing organisms.
   b. Naturally occurring populations must actually interbreed to be considered the same species.
   c. Different appearance is sufficient justification to categorize overlapping, naturally occurring populations as different species.
   d. Paleontologists, studying extinct life-forms, would find it especially difficult to use the biological species concept to determine whether several dinosaur fossils all belong to the same species.

2. Which of the following is MOST characteristic of populations of different species?
   a. Members of the two populations resemble each other.
   b. Members of the two populations can be distinguished by their appearance.
   c. The two populations are geographically separated from each other.
   d. The two populations are adapted to different habitats.
   e. A fertile female from one population mated with a fertile male from the other population produces no offspring.

3. Biologists sometimes combine two groups previously considered different species into the same species because the two species have evolved to
   a. be physically divergent, but they are found to interbreed in nature.
   b. lose premating reproductive-isolating mechanisms.
   c. lose postmating reproductive-isolating mechanisms.
   d. be more physically similar.
   e. expand their ranges and now encounter each other.

4. Under the biological species concept, the main criterion for identifying a species is
   a. anatomical distinctiveness.
   b. behavioral distinctiveness.
   c. geographic isolation.
   d. reproductive isolation.

5. An early study (#1) of myrtle warblers and Audubon's warblers notes that they have different appearances, live in (mostly) different ranges, and thus concludes that they are different species. A later study (#2) observes that they freely interbreed in those areas where their geographic ranges overlap, and concludes that the myrtle and Audubon's warblers are actually different populations of the same species. Which of the following (hypothetical) situations would constitute the strongest evidence that the conclusion of study #1 is actually the correct one?
   a. Audubon's warbler lives mainly in the western parts of the United States and Canada, whereas the myrtle warbler lives mainly in the eastern parts of the United States and Canada, so most Audubon's warblers have no opportunities to mate with myrtle warblers, and vice versa.
   b. The mating calls of male myrtle and Audubon's warblers are similar, but not identical.
   c. A further study of the offspring of interbreeding myrtle and Audubon's warblers finds that these offspring have little or no reproductive success of their own, compared to the offspring of either two myrtle warblers or of two Audubon's warblers.
   d. Most Audubon's warblers build nests in the branches of western coniferous trees, whereas most myrtle warblers build nests in the branches of eastern deciduous trees.

6. In many species of fireflies, males flash to attract females. Each species has a different flashing pattern. This is probably an example of
   a. temporal isolation.
   b. geographic isolation.
   c. behavioral isolation.
   d. ecological isolation.

7. Incompatibilities that prevent the formation of vigorous, fertile hybrids between species are called
   a. premating isolation mechanisms.
   b. postmating isolation mechanisms.
   c. geographical isolation.
   d. genetic divergence.

8. Which of the following is NOT a premating reproductive-isolating mechanism?
   a. ecological isolation
   b. gametic incompatibility
   c. behavioral isolation
   d. mechanical incompatibility
   e. temporal isolation

9. A female fig wasp will carry fertilized eggs from a mating that took place within a fig, then find another fig of the same species, enter it, lay eggs, and die. Her offspring will then hatch, develop, and mate within the fig. Because each species of fig wasp reproduces only in its

particular species of fig, each wasp will have the opportunity to mate only with other wasps of its own species. This would be considered
a. ecological isolation.
b. geographical isolation.
c. behavioral isolation.
d. temporal isolation.
e. mechanical incompatibility.

10. Bishop pines and Monterey pines coexist in nature. In the laboratory, they produce fertile hybrids; but in the wild, they do not interbreed. This is because each species releases pollen at different times of the year. This is an example of
a. temporal isolation.
b. behavioral isolation.
c. mechanical isolation.
d. geographical isolation.

11. Examine the table below, which summarizes the various mechanisms of reproductive isolation. Consider that in a zoo setting, lions and tigers will sometimes interbreed, producing, for example, ligers (healthy but sterile hybrid offspring of a male lion and a female tiger). In the artificial zoo situation, hybrid infertility serves as the reproductive barrier of last resort, keeping the two species separate. However, in nature, other barriers also function. Which mechanism of reproductive isolation is the LEAST likely to operate in nature to keep the lion and tiger species separate?
a. behavioral isolation
b. gametic incompatibility
c. geographical isolation
d. temporal isolation
e. ecological isolation

12. In tropical rain forests, environmental conditions of temperature, rainfall, and day length are similar year-round and, thus, flowering plants may bloom at any time of the year. Suppose that two very similar, closely related species of flowering plants live in the same region, but one of them blooms in May, whereas the other blooms in September. The species are prevented from hybridizing by
a. behavioral isolation.
b. gametic incompatibility.
c. geographical isolation.
d. temporal isolation.
e. ecological isolation.

13. Mammal species A lives in southern Florida. Mammal species B lives in western Cuba. Neither can swim well. The species appear nearly identical, and perhaps would interbreed (or attempt to do so) if given the chance, but we can say with the greatest confidence that they do NOT interbreed because of
a. temporal isolation.
b. ecological isolation.
c. geographical isolation.
d. behavioral isolation.

14. On occasion, brown and polar bears in Alaska will interbreed and produce offspring. Of the following statements, which does NOT describe a mechanism that would play at least some role in keeping the two species separate over the long run?
a. Polar bears mostly live in the extreme north of Alaska and on the floating Arctic ice, whereas brown bears mostly live in central and southern Alaska.
b. Polar bears mainly eat seals and other marine mammals, whereas brown bears eat many things including salmon, young caribou, and plants.

Mechanisms of Reproductive Isolation

Premating isolating mechanisms: factors that prevent organisms of two species from mating
- **Geographic isolation:** The species cannot interbreed because a physical barrier separates them.
- **Ecological isolation:** The species do not interbreed even if they are within the same area because they occupy different habitats.
- **Temporal isolation:** The species cannot interbreed because they breed at different times.
- **Behavioral isolation:** The species do not interbreed because they have different courtship and mating rituals.
- **Mechanical incompatibility:** The species cannot interbreed because their reproductive structures are incompatible.

Postmating isolating mechanisms: factors that prevent organisms of two species from producing vigorous, fertile offspring after mating
- **Gametic incompatibility:** Sperm from one species cannot fertilize eggs of another species.
- **Hybrid inviability:** Hybrid offspring fail to survive to maturity.
- **Hybrid infertility:** Hybrid offspring are sterile or have low fertility.

c. Under circumstances that occasionally occur in the wild, members of one species are receptive to the courtship advances of members of the other species.

d. The hybrid bears are not as well suited for "making a living" (that is, claiming territory, obtaining food, having success in their own mating) as either the true polar bears or the true brown bears are in their own respective habitats.

15. Having multiple sets of chromosomes beyond the diploid number is
   a. polyploidy.
   b. genetic drift.
   c. gene flow.
   d. aneuploidy.

16. Which of the following ecological barriers was a major factor in the evolution of *Rhagoletis pomonella*?
   a. a river separating the apple orchard and the forest with hawthorn trees
   b. a difference in canopy height between the two trees
   c. the greater appeal to females of males that smell like apples
   d. the preference of females for ovipositing (laying eggs) on different fruits

17. Which of the following is NOT an example of speciation?
   a. A small group from a large mainland population colonizes a remote island.
   b. A river that has long divided two populations of mice is diverted by an earthquake, and the two mouse populations come into contact and breed. The hybrid offspring, however, are sterile.
   c. In a bird population, there is disruptive selection for habitat: One group adapts to the treetops, whereas another adapts to the lower branches and ground. The two groups rarely interbreed; but when they do, the hybrid offspring do not live long because they have a mixture of both kinds of adaptations and are not adapted to either habitat.

d. Over a period of several million years, a deer-like species evolves from being very small and feeding on grasses and small shrubs to being much larger and feeding on the lower branches of trees.

e. Due to meiotic error, a diploid plant capable of self-fertilization produces a self-fertilizing tetraploid offspring.

18. Populations of two species living in the same areas (e.g., chorus frogs and wood frogs living in the same ponds of Ohio woodlots), which evolved from a single ancestral species inhabiting the same areas, are said to be
   a. allopatric.
   b. sympatric.
   c. convergent.
   d. divergent.

19. A mechanism by which many plants but few animal species have evolved sympatrically is through
   a. polyploidy.
   b. genetic drift.
   c. gene flow.
   d. aneuploidy.

20. Which of the following is the first step in the process of allopatric speciation?
   a. interspecies contact
   b. geographic isolation
   c. independent evolution of two species
   d. reproductive isolation

21. Which of the following is (are) likely to promote sympatric speciation?
   a. gene flow
   b. geographic isolation
   c. ecological isolation
   d. chromosomal aberrations
   e. both c and d

22. If a haploid egg from a diploid plant is fertilized by a haploid sperm from a diploid plant, and the fertilized egg duplicates its chromosomes but does not divide into two cells, the resulting cell may then divide normally, producing an individual plant that will be
   a. haploid.
   b. diploid.
   c. triploid.
   d. tetraploid.

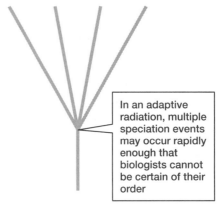

present

time

Each line represents a species

Forks represent speciation events

past

(a) Evolutionary tree

In an adaptive radiation, multiple speciation events may occur rapidly enough that biologists cannot be certain of their order

(b) Evolutionary tree representing adaptive radiation

23. Examine the figure above, showing (a) a standardized evolutionary tree and (b) an evolutionary tree with adaptive radiation. Read each of the following four numbered descriptions of evolutionary situations, and for each, determine whether (a) or (b) shows the most appropriate evolutionary tree for the situation. (1) A single squirrel species inhabits a river valley, and over 1 million years, the river changes course several times, each time isolating one population of squirrels from the others. (2) A species that cannot cross water lives in hills near the coast of a continent, and over time, the sea level gradually rises, thus turning into widely separated islands, each with its own unique environmental conditions, the various large hills that are inhabited by the species. (3) A species of finch (a bird) from South America successfully colonizes a recently formed archipelago of volcanic islands and finds many different available food sources that are best exploited by a wide variety of beak styles. (4) Dinosaurs, the dominant animals on land, are completely eradicated by a sudden catastrophe, and many different ecological niches now lie vacant and available to be taken over by the (initially) small mammals that have survived the catastrophe.
    a. 1:a; 2:a; 3:b; 4:b
    b. 1:b; 2:a; 3:b; 4:a
    c. 1:a; 2:b; 3:b; 4:a
    d. 1:b; 2:b; 3:a; 4:a

24. About how long should we expect it to take for the single fruit fly species *Rhagoletis pomonella* to diverge into two distinctly different species, one specialized to use hawthorn fruits and the other specialized to use apples?
    a. 10 years   c. 5,000 years
    b. 150 years   d. 20 million years

25. During the last Ice Age, conifers in the mountainous, western U.S. states, such as noble firs in Oregon, lived in widespread areas at lower elevations. Then, as the climate warmed up to the modern conditions, these trees migrated uphill, in some cases to small, isolated mountaintops. Some of these populations are now widely separated from others. What is the LEAST likely potential occurrence, if the planet eventually returns to an Ice Age climate in the next 100,000 years?
    a. Gene flow is restored between populations.
    b. Allopatric speciation has led by then to the development of one or more new tree species.
    c. The trees occupy a new and more connected range, at lower elevations.
    d. The overall range of the trees shifts about 1,000 kilometers to the north.

26. When a species has no surviving members, it is said to be
    a. temporally isolated.
    b. extinct.
    c. ecologically isolated.
    d. adaptively radiating.

27. Which of the following events is the leading cause of extinction?
    a. localized species distribution
    b. species overspecialization
    c. competition among species
    d. habitat change

28. Which species is at LEAST risk of extinction?
    a. the species that has many geographically isolated populations, all of them small
    b. the species that is composed of one large, continuous, genetically variable population
    c. the species that lives only in a tree that is itself endangered

d. the species whose major food source is an insect population that is declining because of pesticide use

e. a native plant species that lives where a newly introduced non-native plant has adapted and grows more quickly

29. Out of all the species that have ever lived on the planet, about what percentage have become extinct?

   a. 50%   d. 99%
   b. 80%   e. 99.9%
   c. 90%

30. About 2.5 million years ago, North and South America became joined by the Panamanian isthmus. Prior to this joining, the continents had been separated for a very long time. What was a major consequence of this event?

   a. Habitats on each continent were very suitable for their native mammals but unsuitable for species from the other continent, so very few mammals from North America were capable of colonizing South America, and vice versa.

   b. Mammals from each continent moved into the other, with the South American mammals mainly displacing and driving extinct many of their North American counterparts, whereas few if any North American mammal species succeeded in colonizing South America.

   c. Many mammal species from each continent succeeded in colonizing the other, whereas very few species became extinct, with the end result that both continents saw a large increase in their levels of mammalian biodiversity.

   d. Mammals from each continent moved into the other, with the main result that North American mammals displaced and drove extinct many of their South American counterparts.

ANSWER KEY

1. d
2. e
3. a
4. d
5. c
6. c
7. b
8. b
9. a
10. a
11. b
12. d
13. c
14. c
15. a
16. d
17. a
18. b
19. a
20. b
21. e
22. d
23. a
24. c
25. d
26. b
27. d
28. b
29. e
30. d

Chapter 17 The History of Life

OUTLINE

Section 17.1 How Did Life Begin?

In medieval times, people thought the **spontaneous generation** of new life-forms was common. Pasteur and others disproved this idea (**Figure 17-1**). However, life still has to have originated from nonliving molecules at least once in the history of the planet.

Prior to life on Earth, organic molecules (e.g., nucleic acids, amino acids, and lipids) may have formed when lightning, ultraviolet light, and heat reacted with components of the prebiotic atmosphere (**Figure 17-2**).

RNA many have been the earliest form of nucleic acid and may have functioned as an enzyme (**ribozyme, Figure 17-3**), replicating itself from nucleotides available in the waters of early Earth.

Aggregations of proteins and lipids may have formed vesicles that surrounded ribozymes, forming a semblance of a living cell (a **protocell**).

Section 17.2 What Were the Earliest Organisms Like?

Early Precambrian life consisted of **prokaryotes**, bacteria that were probably anaerobic (because of the lack of available atmospheric oxygen) and absorbed required organic molecules from the surrounding environment. Over time, depletion of available organics likely explained the development of photosynthesis, which allowed some bacteria to synthesize their own food using available inorganic molecules.

Photosynthetic organisms began releasing oxygen, but initially it reacted with iron and other elements. After the accessible iron had been oxidized, oxygen began to accumulate in the atmosphere, beginning about 2.3 billion years ago. The availability of oxygen gas likely resulted in the development of aerobic metabolism, which produces more cellular energy than anaerobic mechanisms.

The first **eukaryotes** evolved 1.7 billion years ago. Membranous organelles may have been derived from infolding of the plasma membrane. The **endosymbiont hypothesis** suggests that mitochondria and chloroplasts evolved from bacteria that were engulfed, but not digested, by predatory cells (**Figures 17-5 and 17-6**).

Section 17.3 What Were the Earliest Multicellular Organisms Like?

Multicellular life evolved in the form of eukaryotic algae 1.2 billion years ago. The algae could anchor themselves along the shore and take advantage of the abundant light and nutrients. Larger size also protected algae from predators.

Multicellular invertebrate animals first appeared about 610 million years ago. For these animals, multicellularity allowed faster locomotion, which imparted more effective prey capture and escape from predation. These behaviors established environmental pressures for faster locomotion, as well as sensory and nervous system development (**Figure 17-7**).

Section 17.4 How Did Life Invade the Land?

Land plants first evolved 475 million years ago. Terrestrial existence required the development of body support, as well as adaptations to prevent desiccation and allow reproduction out of water. However, terrestrial existence allowed access to abundant light and nutrients, in the absence of predators.

Some plant adaptations to life on land eventually included the development of more rigid structures, waterproof coatings, roots, and reproduction using pollen and seeds.

Arthropods were the first animals to move onto land, an event that occurred soon after the evolution of land plants, primarily because of the suitability of their **exoskeleton** for a terrestrial existence. Land plants provided them with abundant food sources.

The first terrestrial vertebrates evolved from **lobefin** fishes, which possessed fleshy fins and a primitive lung that were used to facilitate life on land (**Figure 17-9**).

The development of improved legs and lungs allowed the evolution of **amphibians** from lobefin fishes 350 million years ago. Although better suited for terrestrial life than their predecessors, amphibians needed to remain close to water to reproduce and sustain cutaneous respiration.

Reptiles evolved from amphibians and were less dependent on the proximity of water, as a result of a variety of adaptations, including waterproof scales and eggs and better developed lungs.

Mammals and birds evolved from separate reptile ancestors, developing body insulation in the form of hair and feathers, respectively.

Section 17.5 What Role Has Extinction Played in the History of Life?

New species evolved and existing species became extinct throughout the history of life. Generally, the overall number of species has been increasing.

Mass extinctions periodically occur, resulting in the extinction of many species in a short period of time (**Figure 17-11**). Mass extinctions can occur from climate change (i.e., continental movement from **plate tectonics**, **Figure 17-12**) and catastrophic events such as massive volcanic eruptions or meteorite impacts.

Section 17.6 How Did Humans Evolve?

Apes and humans evolved from tree-dwelling **primate** mammals (**Figure 17-13**) with binocular vision, grasping hands, and a relatively large brain.

The **hominin** lineage diverged from that of the apes between 5 and 8 million years ago; the oldest hominin fossils were found in Africa (**Figure 17-14**).

The first well-known hominins, the australopithecines, lived 4 million years ago. They could stand and walk upright and had larger brains than their ancestors.

The genus *Homo* diverged from the australopithecines 2.5 million years ago (**Figure 17-15**), and was accompanied by advances in tool technology (**Figure 17-16**). There are several lineages of the genus *Homo* (**Figure 17-19**), which include the Neanderthals (*H. neanderthalensis*) and modern humans (*H. sapiens*).

Human cultural evolution now occurs far more rapidly than our biological evolution. Some important cultural innovations include the development of a wide variety of tools, the development of widespread agriculture, and the Industrial Revolution.

FLASH CARDS

To use the flash cards, tear the page from the book and cut along the dashed lines. The key term appears on one side of the flash card, and its definition appears on the opposite side.

amphibian	hominin
arthropod	lobefin
conifer	mammal
endosymbiont hypothesis	mass extinction
eukaryote	plate tectonics
exoskeleton	primate

A human or a prehistoric relative of humans; the oldest known hominin is *Sahelanthropus* beginning with the Australopithecines, whose fossils are more than 6 million years old.

A member of the chordate clade Amphibia, which includes the frogs, toads, and salamanders, as well as the limbless caecelians.

A member of the fish order Sarcopterygii, which includes coelacanths and lungfishes. Ancestors of today's lobefins gave rise to the first amphibians and, thus, ultimately to all tetrapod vertebrates.

A member of the animal phylum Arthropoda, which includes the insects, spiders, ticks, mites, scorpions, crustaceans, millipedes, and centipedes.

A member of the chordate class Mammalia, which includes vertebrates with hair and mammary glands.

A member of a group of nonflowering vascular plants whose members reproduce by means of seeds formed inside cones and retain their leaves throughout the year.

The extinction of an extraordinarily large number of species in a comparatively short period of time. Mass extinctions have occurred periodically throughout the history of life.

The hypothesis that certain organelles, especially chloroplasts and mitochondria, arose as mutually beneficial associations between the ancestors of eukaryotic cells and captured bacteria that lived within the cytoplasm of the pre-eukaryotic cell.

The theory that Earth's crust is divided into irregular plates that are converging, diverging, or slipping by one another; these motions cause continental drift, the movement of continents over Earth's surface.

An organism whose cells are eukaryotic; plants, animals, fungi, and protists are eukaryotes.

A member of the mammal order Primates, characterized by the presence of an opposable thumb, forward-facing eyes, and a well-developed cerebral cortex; includes lemurs, monkeys, apes, and humans.

A rigid external skeleton that supports the body, protects the internal organs, and has flexible joints that allow for movement.

prokaryote

ribosome

protocell

spontaneous generation

reptile

An RNA molecule that can catalyze certain chemical reactions, especially those involved in the synthesis and processing of RNA itself.

An organism whose cells are prokaryotic; bacteria and archaea are prokaryotes.

The proposal that living organisms can arise from nonliving matter.

The hypothetical evolutionary precursor of living cells, consisting of a mixture of organic molecules within a membrane.

A member of the chordate group that includes the snakes, lizards, turtles, alligators, birds, and crocodiles.

SELF TEST

1. Which of the following was NOT a common component of Earth's early atmosphere?
   a. hydrogen   d. ammonia
   b. oxygen   e. carbon dioxide
   c. methane

2. Which of the following, if it had existed at the time, would have prevented the development of conditions favorable for the evolution of life after Earth was formed?
   a. complex organic molecules
   b. energy in the form of lightning and energy from the sun
   c. unbound, or free, oxygen molecules
   d. carbon dioxide and sulfur dioxide

3. For life to generate spontaneously today, by following the same series of events that produced life in the first place almost 4 billion years ago, which of the following conditions is NOT a requirement?
   a. Oxygen gas must be excluded from the region in which the spontaneous generation is to occur.
   b. Electricity and heat must be excluded.
   c. Bacteria and other microbes must be excluded.
   d. Simple inorganic molecules such as H_2, H_2O, CH_4, and NH_3 must be present.
   e. DNA must be allowed to form and replicate before other large organic molecules do.

4. What would NOT have been true about a protocell?
   a. It would have been able to absorb materials such as nucleotides from the surrounding environment.
   b. It would have periodically undergone mitosis, with duplication of DNA followed by the division into two identical protocells.
   c. It would have contained ribozymes.
   d. Its vesicle would have provided protection to its internal contents, much like what happens with cell membranes today.

5. Which of the following included the "age of dinosaurs"?
   a. Precambrian   c. Paleozoic
   b. Mesozoic   d. Cenozoic

6. The earliest living organisms were
   a. multicellular and photosynthetic.
   b. eukaryotic and single celled.
   c. prokaryotic and anaerobic.
   d. oxygen generating and eukaryotic.

7. The reason that both mitochondria and chloroplasts have a double membrane is that the inner membrane belonged to the original bacteria and the outer membrane
   a. was formed from a predatory cell membrane during phagocytosis.
   b. arose from the bacteria being surrounded by endoplasmic reticulum.
   c. was created by a mutation that later duplicated the bacterial cell membrane.
   d. was synthesized by the Golgi apparatus.

8. The MOST fundamental difference between prokaryotes and eukaryotes is that eukaryotes have
   a. a cell wall.
   b. a plasma membrane.
   c. a membrane-bound nucleus.
   d. genetic material.
   e. multiple cells.

9. Study the graph below, showing the radioactive decay of potassium-40 (^{40}K) to argon-40 (^{40}Ar). Suppose that you analyze an old volcanic rock, and find that it contains five times as many ^{40}Ar atoms as it does ^{40}K atoms. Which of the following statements, about the age of this rock (the amount of time that has passed, since it cooled from first being molten lava), is the MOST accurate?
   a. The rock is between 0 and 1.25 billion years of age.
   b. The rock is between 1.25 billion and 2.5 billion years of age.
   c. The rock is between 2.5 billion and 5 billion years of age.
   d. The rock is between 5 billion and 10 billion years of age.

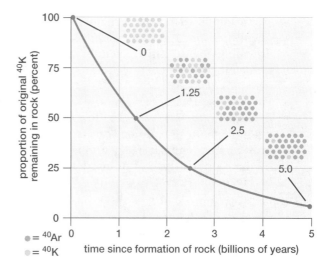

10. What was NOT a consequence of the advent of oxygen production by photosynthetic life-forms?
    a. Large amounts of iron metal were oxidized, producing the reddish iron ore that we see today.
    b. Many anaerobic microbes died, from oxygen poisoning.
    c. The earlier photosynthetic pathway that used the hydrogen atoms from hydrogen sulfide (H_2S) instead of from H_2O (as with oxygen-generating photosynthesis) became obsolete, and all bacteria that had used this earlier style of photosynthesis became extinct.
    d. Aerobic respiration evolved, as a way to produce many more ATP molecules from the complete oxidation of organic molecules.

11. The first multicellular organisms were thought to be
    a. arthropods.   c. cnidarians.
    b. conifers.   d. algae.

12. Which of the following characteristics is believed to be an evolutionary advantage of multicellular eukaryotic organisms?
    a. Multicellular organisms are easily ingested by single-celled predators.
    b. Multicellular organisms are able to let certain cells specialize to carry out functions that single-celled organisms cannot.
    c. Multicellularity allows cells to become much larger and able to exchange carbon dioxide and oxygen efficiently, despite decreasing the size of the organism.
    d. both a and c

13. What is the most accurate statement about the early development of animal life?
    a. Some of the earliest animals were land dwelling.
    b. The trilobites were large predators that frequently captured and ate nautiloids, thus providing the selective pressure that was needed to produce the more modern chambered *Nautilus*.
    c. The earliest fossils of vertebrate animals, including early fishes, are seen in rocks dating from about 700 million years ago.
    d. Most phyla of animals had developed either by the Cambrian period or else by shortly before it.

14. The earliest animals to have an amniotic egg, allowing eggs to develop out of water, were
    a. lobefin fish.   c. reptiles.
    b. amphibians.   d. birds.

15. The exoskeleton of early, marine-dwelling arthropods can be considered a preadaptation for life on land because the shell
    a. can support an animal's weight against gravity.
    b. allows a wide diversity of body types.
    c. resists drying.
    d. absorbs light.
    e. both a and c.

16. The first animals to "invade" the land were probably
    a. arthropods.   c. amphibians.
    b. fish.   d. reptiles.

17. The earliest land plants were restricted to wetland habitats because they
    a. were single celled.
    b. did not have a root system.
    c. could not stay upright.
    d. had swimming sperm rather than sperm encased in pollen grains.

18. Present-day coal deposits are actually remains of _____ that lived during the _____.
    a. marine invertebrates; Paleozoic
    b. plants; Carboniferous
    c. reptiles; Jurassic
    d. mammoths; Cenozoic

19. Which statement about the reptiles is NOT correct?
    a. Very large reptiles, such as large dinosaurs, could in part maintain stable, high temperatures because large bodies lose heat slowly to the environment.
    b. In cool weather, some small reptiles experience a drop in body temperature, and they become inactive.
    c. One solution to the problem of body heat loss from a small body is to evolve insulation in the form of feathers, as the reptilian ancestors of modern birds did.
    d. The dinosaurs probably had body temperatures and activity levels similar to those of the mammals, because the earliest mammals were also the ancestors of the dinosaurs.
    e. Reptiles are much more completely adapted to life on land than the amphibians are.

20. Each continent began to develop its own unique flora and fauna about 135 million years ago, when
    a. Pangaea had not yet formed.
    b. Pangaea had unified as one.
    c. Laurasia and Gondwanaland were formed.
    d. Laurasia and Gondwanaland broke up.
    e. the continents reached their current configuration.

21. Earth's continents were once unified into a single supercontinent called
    a. Pangaea.
    b. Gondwanaland.
    c. Laurasia.
    d. Gaia.
    e. Atlantis.

22. Which statement BEST describes the pattern of extinction throughout biological history?
    a. The rate of extinction has steadily increased over the entire history of life.
    b. The rate of extinction has steadily decreased over the entire history of life.
    c. There has been very little extinction except during mass extinction events.
    d. There has been a relatively constant turnover of species with occasional mass extinction events.

23. Study the figure below, showing the number of marine animal families living in the oceans, through the past 600 million years. Considering that a taxonomic family can contain many species, what does the figure imply?
    a. With the end-Permian mass extinction, far more than 50% of animal species in the oceans were lost.
    b. The end-Cretaceous mass extinction (which saw the end of the dinosaurs on land) was also the most severe marine mass extinction event on record.
    c. The end-Ordovician mass extinction destroyed about 20% of the marine animal species that existed at that time.
    d. Mass extinctions occur relatively quickly, and are followed by a subsequent, 10- to 20-million-year period of continuing net loss of marine families, occurring at a slower rate.

24. Which statement about primates is true?
    a. Most primate species have a primarily tree-dwelling lifestyle.
    b. At least one group of primates has evolved to lose opposable digits.
    c. At least one group of primates has evolved to lose binocular vision.
    d. All modern primates are strictly herbivorous.

25. Given the following list, which is the correct numbered sequence from earliest to most recent hominid development?
    1. bipedalism
    2. cave paintings
    3. first stone tools made
    4. first migration out of Africa
    5. buried dead and "religious" rites
    a. 1, 2, 3, 4, 5
    b. 3, 1, 2, 4, 5
    c. 1, 3, 4, 5, 2
    d. 3, 2, 1, 5, 4
    e. 1, 4, 3, 5, 2

26. Which of the following features distinguishes the hominid line from the lineage of apes?
    a. binocular vision
    b. large brain
    c. hands with opposable thumbs
    d. dependence on an upright posture for locomotion

27. The hominids belonging to the genus *Australopithecus* first evolved in
    a. Asia.
    b. Africa.
    c. Australia.
    d. Europe.

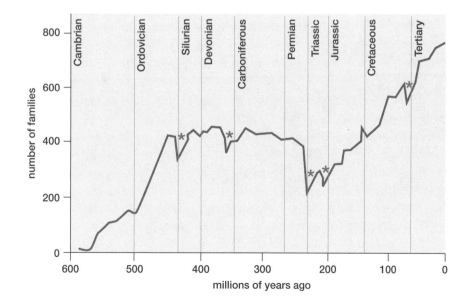

28. What is the relationship between hominins and primates?
    a. Hominins are the particular branch, within the larger category of primates, that led to the evolution of humans.
    b. The large class of mammals is divided into many orders, such as proboscidea (elephants), cetacea (whales), and chiroptera (bats), and it also includes two separate but closely related orders for the hominins and primates.
    c. Primates include monkeys and apes but not lemurs, whereas hominins are a subgroup of the primates that include all members of genus *Australopithecus* but not the members of genus *Homo*.
    d. Out of the overall large category of primates, the hominins are the subgroup including the species that spend most of their time living in trees and very little time on the ground.

29. Which species in the genus *Homo* was the first to arrive in Australia?
    a. *Homo habilis*
    b. *H. erectus*
    c. *H. neanderthalensis*
    d. *H. floresiensis*
    e. *H. sapiens*

30. What is the greatest number of hominin species that has ever existed simultaneously?
    a. 1
    b. 2
    c. 3
    d. 4 or 5

ANSWER KEY

1. b	16. a
2. c	17. d
3. e	18. b
4. b	19. d
5. b	20. d
6. c	21. a
7. a	22. d
8. c	23. a
9. c	24. a
10. c	25. c
11. d	26. d
12. b	27. b
13. d	28. a
14. c	29. e
15. e	30. d

CHAPTER 18 SYSTEMATICS: SEEKING ORDER AMIDST DIVERSITY

OUTLINE

Section 18.1 How Are Organisms Named and Classified?

Taxonomy is the branch of biology that names and classifies organisms. Each **species** is identified by a **scientific name**, composed of its **genus** name and species name. The first letter of the genus name is always capitalized, and the entire scientific name is either italicized or underlined (e.g., *Sialia sialis*). Scientific names allow for precise scientific communication about species (**Figure 18-1**).

Different species are placed into a series of nested hierarchical categories, based on their resemblance to other species. Following the modern form of the Linnaean classification system, in order of increasing inclusiveness, these categories are **domain, kingdom, phylum, class, order, family**, genus, and species. One domain can contain several kingdoms, one kingdom can contain several phyla, and so on.

Systematics is the practice of studying the **phylogeny**, or evolutionary relationships, between species. A **clade** consists of all of the species that are descended from a single, shared ancestral species. The species within a clade share a closer evolutionary history with each other than any of them does with other species that are not part of the clade (**Figure 18-2**). Tools for studies of phylogeny include comparisons of anatomy (**Figure 18-3**) and of genetics via **DNA sequencing** techniques (**Figure 18-4**).

Section 18.2 What Are the Domains of Life?

All life is classified into three domains that represent the main branches of the tree of life (**Figure 18-6**). These domains are **Bacteria, Archaea** (**Figure 18-5**), and **Eukarya**. Although the Bacteria and Archaea domains both consist of prokaryotic organisms, they are as evolutionarily and genetically distinct from each other as either domain is from the eukaryotes.

Among the Eukarya, commonly accepted kingdoms include the Fungi, Plantae, and Animalia (**Figure 18-7**), each of which is a clade. The Protist kingdom, as it is currently defined, includes a great diversity of other eukaryotic organisms that do not form a clade.

Section 18.3 Why Do Classifications Change?

Classification schemes are revised as new information is discovered.

The biological species concept, although useful, is not applicable to some organisms (primarily asexually reproducing organisms), which makes them difficult to classify. In these cases, alternative definitions of species can be useful.

Section 18.4 How Many Species Exist?

Biodiversity refers to the total range of species diversity on Earth.

The total number of named species is currently 1.5 million, with 700 to 10,000 new species named annually. Scientists believe there may be as many as 100 million species on Earth.

The greatest numbers of species that remain to be discovered and named probably live in tropical rain forests. However, at the rate at which these forests are being destroyed, most such species may become extinct before we ever even learn of their existence.

FLASH CARDS

To use the flash cards, tear the page from the book and cut along the dashed lines. The key term appears on one side of the flash card, and its definition appears on the opposite side.

Archaea	domain
Bacteria	Eukarya
biodiversity	family
clade	genus
class	kingdom
DNA sequencing	order

The broadest category for classifying organisms; organisms are classified into three domains: Bacteria, Archaea, and Eukarya.

One of life's three domains; consists of prokaryotes that are only distantly related to members of the domain Bacteria.

One of life's three domains; consists of all eukaryotes (plants, animals, fungi, and protists).

One of life's three domains; consists of prokaryotes that are only distantly related to members of the domain Archaea.

In Linnaean classification, the taxonomic rank composed of related genera. Closely related families make up an order.

The diversity of living organisms; measured as the variety of different species, the variety of different alleles in a gene pool, or the variety of different community interactions in an ecosystem.

In Linnaean classification, the taxonomic rank composed of related species. Closely related genera make up a family.

A group that includes all the organisms descended from a common ancestor, but no other organisms; a monophyletic group.

In Linnaean classification, the taxonomic rank composed of related phyla. Related kingdoms make up a domain.

In Linnaean classification, the taxonomic rank composed of related orders. Closely related classes form a phylum.

In Linnaean classification, the taxonomic rank composed of related families. Related orders make up a class.

The process of determining the order of nucleotides in a DNA molecule.

phylogeny

species

phylum

systematics

scientific name

taxonomy

The basic unit of taxonomic classification, consisting of a group of populations that evolves independently. In sexually reproducing organisms, a species can be defined as a population or series of populations of organisms that interbreed freely with one another under natural conditions but that do not interbreed with members of other species.

The evolutionary history of a group of species.

The branch of biology concerned with reconstructing phylogenies and with naming clades.

In Linnaean classification, the taxonomic rank composed of related classes. Related phyla make up a kingdom.

The branch of biology concerned with naming and classifying organisms.

The two-part Latin name of a species; consists of the genus name followed by the species name.

SELF TEST

1. The HIV-1 virus that causes most cases of AIDS originated in _____, located in _____.
   a. chickens; China
   b. domesticated pigs; Mexico
   c. cows; India
   d. chimpanzees; West Africa

2. Anatomical features of species may not always be useful for determining species relationships because of
   a. convergent evolution.
   b. homologous structures.
   c. adaptation.
   d. common ancestry.

3. What is the most inclusive of the major taxonomic categories?
   a. genus
   b. order
   c. phylum
   d. kingdom
   e. domain

4. What is the least inclusive of the major taxonomic categories?
   a. species
   b. order
   c. phylum
   d. division
   e. class

5. The science of reconstructing evolutionary history is
   a. ethology.
   b. archaeology.
   c. anthropology.
   d. systematics.
   e. systems analysis.

6. In the following sequence, what is missing from the hierarchy of major taxonomic categories (list in order): species—_____—family—_____—class.
   a. genus; domain
   b. genus; order
   c. kingdom; order
   d. order; genus
   e. order; division
   f. phylum; genus

7. Classification categories are arranged into a _____, in which each successive category is increasingly narrow and specifies a group whose common ancestor is increasingly recent.
   a. ladder
   b. hierarchy
   c. stairway
   d. checkerboard

8. The more taxonomic categories two organisms share, the more closely related those organisms are in an evolutionary sense. Which scientist's work led to this insight?
   a. Aristotle
   b. Darwin
   c. Linnaeus
   d. Whittaker
   e. Woese

9. The correct sequence of the taxonomic hierarchy is
   a. domain, kingdom, class, phylum, order, family, genus, species.
   b. domain, kingdom, phylum, order, class, family, genus, species.
   c. domain, kingdom, phylum, class, family, order, species, genus.
   d. kingdom, domain, order, phylum, class, family, genus, species.
   e. domain, kingdom, phylum, class, order, family, genus, species.

10. Humans belong to the order
    a. Primates.
    b. Hominidae.
    c. Homo.
    d. Mammalia.

11. It is possible to imagine the various levels of taxonomic classification as a "family tree" for an organism. If the kingdom is analogous to the trunk of the tree, which taxonomic category would be analogous to the large limbs coming off of that trunk?
    a. class
    b. family
    c. order
    d. phylum
    e. subfamily

12. Which of the following criteria could NOT be used to determine how closely related two types of organisms are?
    a. similarities in the presence and relative abundance of specific molecules
    b. DNA sequence
    c. the presence of homologous structures
    d. developmental stages
    e. occurrence of both organisms in the same habitat

13. What is the correct format for the technical, binomial name for the grizzly bear?
    a. Ursus Arctos
    b. *Ursus Arctos*
    c. Ursus arctos
    d. *Ursus arctos*
    e. ursus arctos
    f. *ursus arctos*

14. Examine the figure below. Which of the following do NOT constitute a monophyletic group?
    a. birds and crocodiles
    b. snakes, lizards, turtles, and birds
    c. crocodiles, birds, snakes, and lizards
    d. snakes and lizards

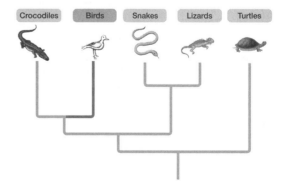

15. Examine the figure at right, comparing human and chimpanzee chromosomes. Notice that human chromosome 2 corresponds to two separate chromosomes (which are known as 2p and 2q) in chimpanzees (as well as in bonobos, gorillas, and orangutans). Thus, humans have a haploid chromosome number of 23, whereas the chimpanzee haploid number is 24. Imagine that the most recent shared ancestor of humans and chimpanzees had a haploid number of 24, but that later in the human evolutionary pathway, chromosomes 2p and 2q became fused together, end to end. What could a detailed examination of the structure of human chromosome 2 reveal? (*Note:* For each numbered chromosome pair, "H" stands for *human*, and "C" stands for *chimpanzee*.)

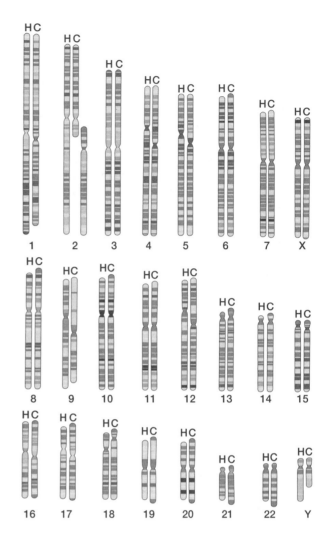

  a. one functional centromere, two functional telomeres, two nonfunctional telomeres, and one nonfunctional centromere
  b. two functional centromeres, one functional telomere, one nonfunctional telomere, and two nonfunctional centromeres
  c. one functional centromere, one functional telomere, two nonfunctional telomeres, and two nonfunctional centromeres
  d. two functional centromeres, two functional telomeres, one nonfunctional telomere, and one nonfunctional centromere
  e. one functional centromere, two functional telomeres, one nonfunctional telomere, and two nonfunctional centromeres

16. Which of the following pairs is the most distantly related?
  a. archaea and bacteria
  b. protists and fungi
  c. plants and animals
  d. fish and starfish
  e. fungi and plants

17. What is a domain?
  a. the broadest of all of the taxonomic categories
  b. the taxonomic category that contains insects, birds, and mammals
  c. the taxonomic category that contains all living species of humans
  d. a synonym for the taxonomic category "kingdom"

18. Which taxonomic group contains eukaryotic organisms that are typically unicellular?
  a. Animalia   d. Plantae
  b. Fungi   e. Protista
  c. Archaea

19. Humans belong to the domain
  a. Primates.   d. Eukarya.
  b. Hominidae.   e. Mammalia.
  c. Homo.

20. Which kingdom consists of eukaryotic organisms that are multicellular and autotrophic and possess cell walls?
  a. Animalia   d. Plantae
  b. Fungi   e. Protista
  c. Bacteria

21. Which group of protists constitutes a monophyletic group?
  a. the ciliates, dinoflagellates, apicomplexans, and diatoms
  b. the diplomonads, euglenids, kinetoplastids, ciliates, and apicomplexans
  c. the Phaeophyta, Oomycota, ciliates, dinoflagellates, apicomplexans, and diatoms
  d. the amoebozoans, Chlorophyta, and Rhodophyta

22. Which group is the most closely related to members of kingdom Plantae (the Bryophyta, Pteridophyta, gymnosperms, and angiosperms)?
  a. the Phaeophyta
  b. the Rhodophyta
  c. the Chlorophyta
  d. the amoebozoans

23. The greatest genetic variation in humans is found in _____, supporting the idea that modern humans have had their longest history in _____.
    a. Africa; Africa
    b. Africa; Asia
    c. Asia; Asia
    d. Europe; Asia
    e. Europe; Europe

24. The phylogenetic species concept stresses _____ as the criterion for assigning individuals to the same species.
    a. anatomical similarity
    b. recent common ancestry
    c. potential for interbreeding
    d. similarity in behavior

25. Which of the following is NOT used by taxonomists to distinguish one species from another?
    a. genetic characteristics
    b. similar features resulting from convergent evolution acting on distantly related species
    c. distinguishing features seen in an organism during its development
    d. external anatomical features

26. Why do biologists now consider there to be three species of elephants, rather than only two (the Asian elephant and African elephant)?
    a. The range of the Asian elephants has been so fragmented by human agricultural land conversion that some populations of Asian elephants have been completely cut off from each other, causing the loss of gene flow and consequent divergence into at least two species.
    b. Close examination of some elephant herds in Zimbabwe has revealed that these elephants are more genetically similar to the extinct mammoths than they are to other elephant populations in the rest of Africa.
    c. Human selective breeding of domesticated Asian elephants (which have been used for millennia as working animals) has led to these elephants becoming so genetically distinct from wild elephants that they are considered a new species (in the same way that dogs are distinct from wolves).
    d. The African elephants consist of two distinct groups, forest elephants and savanna elephants, and these groups seldom encounter each other, seldom if ever interbreed, and are genetically distinct from each other.

27. There are approximately 1.5 million named species today. Of these, approximately 5% are prokaryotes and protists. Scientists believe that
    a. this is a good representation of the true percentage of all species that are either prokaryotes or protists.
    b. this is an overestimation of the true percentage of all species that are either prokaryotes or protists.
    c. this is an underestimation of the true percentage of all species that are either prokaryotes or protists.
    d. all possible protist and prokaryote species have been discovered already.

28. Which one of the following habitats appears to have the greatest number of species?
    a. seafloor
    b. deserts
    c. tropical rain forests
    d. grasslands

29. With the search for new species, which types of organisms, and places, have been the most thoroughly studied?
    a. large organisms such as mammals and birds, in temperate places such as the United States
    b. small organisms that are inconspicuous, yet crucial for ecosystem functioning, such as mycorrhizal soil fungi and nutrient-cycling soil bacteria
    c. tree tops in tropical rain forests, which are home to vast numbers of as-yet unidentified arthropod species such as beetles
    d. fish and other marine organisms, especially in the deep oceans

30. What is the most accurate statement about the phylogeny of the HIV virus (and other, similar viruses)?
    a. Humans initially became infected with HIV-1 by contact with chimpanzees, and after the virus mutated within the human population, humans later passed the altered, HIV-2 version to monkeys and macaques.
    b. Humans have become infected separately with at least two different versions of the virus (HIV-1 and HIV-2), probably by eating chimpanzees and monkeys.
    c. Humans were infected once, through contact with chimpanzees, and the virus has since mutated and spread within the human population to produce four distinct strains.
    d. The HIV and SIV (simian immunodeficiency virus) viruses had independent origins, but have become very similar due to convergent evolution, because their respective human and simian hosts have extremely similar physiologies.

ANSWER KEY

1. d	16. a
2. a	17. a
3. e	18. e
4. a	19. d
5. d	20. d
6. b	21. c
7. b	22. c
8. b	23. a
9. e	24. b
10. a	25. b
11. d	26. d
12. e	27. c
13. d	28. c
14. b	29. a
15. a	30. b

CHAPTER 19 THE DIVERSITY OF PROKARYOTES AND VIRUSES

OUTLINE

Section 19.1 Which Organisms Are Members of the Domains Archaea and Bacteria?

Members of the domains **Bacteria** and **Archaea** are distinguished by fundamental differences in their structural and biochemical features. For example, bacteria have peptidoglycan in their cell walls (archaea do not), and both groups differ in the composition of their plasma membranes, ribosomes, and RNA polymerases.

Prokaryotic classification within domains is difficult and has historically been performed on the basis of shape, means of locomotion, pigmentation, nutrient requirements, appearance of colonies, staining properties, and genetic similarity. Comparison of DNA sequences is providing a new tool for determining the evolutionary relationships between species.

Prokaryotes can be found in a wide range of sizes and have common characteristic shapes, including spherical, rodlike, and corkscrew (**Figure 19-1**).

Section 19.2 How Do Prokaryotes Survive and Reproduce?

Prokaryotes have different ranges of mobility. Some adhere to surfaces or drift in their liquid surroundings, but others actively move by the use of **flagella** (**Figure 19-2**).

Some bacterial cell walls are surrounded by a protective slime (**biofilm**) that helps them adhere to surfaces (**Figure 19-3**).

Under harsh environmental conditions, many rod-shaped bacteria can form **endospores** that allow genetic material (and some enzymes) to survive for long periods of time in metabolic dormancy (**Figure 19-4**).

Prokaryotes are specialized for survival in specific habitats (**Figure 19-5**) and exhibit many different methods of energy acquisition, including anaerobic, aerobic, chemosynthetic, and photosynthetic (**Figure 19-6**) metabolisms.

Prokaryotes typically reproduce by asexual **binary fission** (**Figure 19-7**), but may exchange genetic material by a process called **conjugation**. During this process, circular DNA **plasmids** are transferred from donor to recipient prokaryotes (**Figure 19-8**).

Section 19.3 How Do Prokaryotes Affect Humans and Other Organisms?

Prokaryotes play a number of important roles in the lives of eukaryotes. Many herbivorous animal digestive systems could not extract nutrients from plant matter without the help of bacteria to break down plant cellulose. **Nitrogen-fixing bacteria** provide plants with a source of soil nitrogen by converting atmospheric nitrogen into ammonia (**Figure 19-9**). Other prokaryotes decompose dead organic matter, acting as nature's recyclers, and yet others can be used for **bioremediation** projects by breaking down pollution resulting from human activity.

Some bacteria are **pathogenic** and can cause a number of human diseases, including tetanus, the plague, strep throat, pneumonia, and various types of food poisoning (**Figure E19-1**). The majority of bacteria, however, are harmless.

Section 19.4 What Are Viruses, Viroids, and Prions?

Viruses are small, nonliving, host-specific parasites consisting of a protein coat that surrounds genetic material (**Figure 19-11**). They reproduce only by invading a **host** cell and, by using the cell's own enzymes, directing it to make more viruses, which exit the cell when it ruptures (**Figure E19-2**). Many viruses are pathogenic.

Viroids are circular strands of RNA that enter the cell nucleus and direct new viroid production. Viroids are known to affect only plant cells.

Prions are mutated proteins that may act enzymatically to form more prions, thus disrupting cell function (**Figure 19-14**). Prions have been implicated in a number of neurological diseases, including bovine spongiform encephalopathy (mad cow disease).

FLASH CARDS

To use the flash cards, tear the page from the book and cut along the dashed lines. The key term appears on one side of the flash card, and its definition appears on the opposite side.

anaerobe	bioremediation
Archaea	conjugation
Bacteria	endospore
bacteriophage	flagellum
binary fission	host
biofilm	nitrogen-fixing bacterium

The use of organisms to remove or detoxify toxic substances in the environment.

An organism whose respiration does not require oxygen.

In prokaryotes, the transfer of DNA from one cell to another via a temporary connection; in single-celled eukaryotes, the mutual exchange of genetic material between two temporarily joined cells.

One of life's three domains; consists of prokaryotes that are only distantly related to members of the domain Bacteria.

A protective resting structure of some rod-shaped bacteria that withstands unfavorable environmental conditions.

One of life's three domains; consists of prokaryotes that are only distantly related to members of the domain Archaea.

A long, hairlike extension of the plasma membrane; in eukaryotic cells, it contains microtubules arranged in a 9 + 2 pattern. The movement of flagella propel some cells through fluids.

A virus specialized to attack bacteria.

The prey organism on or in which a parasite lives; is harmed by the relationship.

The process by which a single bacterium divides in half, producing two identical offspring.

A bacterium that possesses the ability to remove nitrogen (N_2) from the atmosphere and combine it with hydrogen to produce ammonium (NH_4^+).

A community of prokaryotes of one or more species, in which the prokaryotes secrete and are embedded in slime that adheres to a surface.

pathogenic

viroid

plasmid

virus

prion

A particle of RNA that is capable of infecting a cell and of directing the production of more viroids; responsible for certain plant diseases.

Capable of producing disease; refers to an organism with such a capability (a pathogen).

A noncellular parasitic particle that consists of a protein coat surrounding genetic material; multiplies only within a cell of a living organism (the host).

A small, circular piece of DNA located in the cytoplasm of many bacteria; normally does not carry genes required for the normal functioning of the bacterium but may carry genes that assist bacterial survival in certain environments, such as a gene for antibiotic resistance.

A protein that, in mutated form, acts as an infectious agent that causes certain neurodegenerative diseases, including kuru and scrapie.

SELF TEST

1. Bacteria and archaea differ in all of the following EXCEPT in
   a. the structures and compositions of their plasma membranes.
   b. the organizations of their ribosomes.
   c. the particular chemicals used in their cell walls.
   d. their overall sizes and appearances.

2. Following classification into unique domains, bacteria and archaea have proven
   a. easy to classify into kingdoms because of their small size and structural simplicity.
   b. easy to classify into kingdoms because of their biochemical differences.
   c. difficult to classify into kingdoms because of their small size and structural simplicity.
   d. difficult to classify into kingdoms because of their biochemical differences.

3. Classification of prokaryotes may use many kinds of traits, including all EXCEPT which of the following?
   a. cell shape
   b. means of locomotion
   c. nutrient sources
   d. presence or absence of a nucleus
   e. staining properties

4. What are the three basic shapes of most bacteria and archaea?
   a. corkscrews, pentagons, and triangles
   b. pentagons, triangles, and spheres
   c. triangles, spheres, and rods
   d. spheres, rods, and corkscrews

5. During bacterial conjugation, the transferred item is a
   a. gene.   d. lipid.
   b. chromosome.   e. protein.
   c. plasmid.

6. Which of the following traits allows some bacteria to survive extreme conditions for millions of years?
   a. aerobic respiration
   b. endospore formation
   c. conjugation
   d. binary fission

7. Sexual reproduction performed by eukaryotes is similar to bacterial conjugation because
   a. two haploid cells fuse.
   b. genetic variability increases.
   c. crossing over occurs.
   d. genetically different diploid offspring form.

8. The simple form of cell division by which prokaryotic cells reproduce is called
   a. binary fission.
   b. conjugation.
   c. endospore formation.
   d. mitosis.

9. Hospitals must sterilize surgical instruments at a very high temperature and pressure because some bacteria can survive harsh conditions by making
   a. zygotes.   c. seeds.
   b. endospores.   d. fruiting bodies.

10. The bacterium on the left has attached to the one on the right, using a sex pilus. What happens next?
    a. The two bacterial cells will fuse, mix all of their DNA, and then separate from each other, with each cell taking away a 50–50 mix of the DNA from the two original cells.
    b. All of the DNA in the cell on the right will be made identical to the DNA in the cell on the left, thus eliminating any harmful mutations.
    c. The cell on the left will donate a small plasmid to the other, causing a small change to the genome of the recipient cell.
    d. The cell on the left will now "reel in" and engulf the cell on the right, thus acquiring its genes.

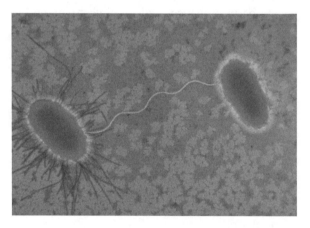

11. A bacterium can move by using one or more flagella that
    a. whip back and forth.
    b. whip at one end, and spin like a propeller at the other.
    c. spin like a propeller.
    d. grab hold of nearby stationary objects, and then pull the bacterium toward them.
    e. push against nearby stationary objects, thus moving the bacterium away from them.

12. Cyanobacteria, carrying out photosynthesis in essentially the same way that plants do, would require all of the following EXCEPT
    a. water.
    b. hydrogen sulfide.
    c. carbon dioxide.
    d. light.
    e. chlorophyll.

13. Imagine that a single bacterium, of a type capable of rapid reproduction, encounters an ideal environment (temperature and pH, as well as the concentrations of food, water, oxygen, etc.). After 2 full hours have passed, how many bacteria would there probably be?
    a. 4   d. 64
    b. 7   e. 256
    c. 16

14. Which of the following pairs of organism and disease is INCORRECT?
    a. virus: AIDS
    b. prion: kuru
    c. bacterium: syphilis
    d. archaean: gonorrhea

15. Which of the following is a bacterial disease?
    a. influenza   c. AIDS
    b. common cold   d. tetanus

16. Which of the following enables plants to obtain a usable form of nitrogen?
    a. bacteria, which "fix" a biologically usable form of nitrogen from the otherwise unusable N_2 form that is common in the atmosphere
    b. viruses, which carry the genetic instructions for the proteins that carry our nitrogen fixation
    c. prions, which transform plant proteins into new forms capable of carrying out nitrogen fixation
    d. photosynthesis, which directly provides new nutrients to plants

17. The majority of bacteria are
    a. pathogenic.   d. eukaryotic.
    b. photosynthetic. e. harmful.
    c. harmless.

18. Which of the following roles do bacteria NOT play in the environment?
    a. producers
    b. decomposers
    c. partners in mutualistic symbiosis
    d. consumers
    e. pathogens

19. Given that *Escherichia coli* bacteria are normally present in the guts of humans and many other animals, and thrive in that kind of environment, what is also generally true about *E. coli*?
    a. They are lovers of extreme heat, similar to those that thrive in boiling hot pools in Yellowstone National Park.
    b. They are photosynthetic, like the cyanobacteria found in lakes and oceans.
    c. They sometimes occur in streams and rivers.
    d. They nearly always cause incurable, deadly diseases in humans.

20. The bacteria that cause botulism, *Clostridium botulinum*, and those that cause tetanus, *Clostridium tetani*, are both anaerobes. Given what you know about tetanus, what is the best explanation for why improperly canned food can sometimes cause botulism?
    a. Even if only a few *C. botulinum* bacteria are still living in the food, once the can is opened, they will become active, will multiply to great numbers, and will rapidly spoil the food.
    b. If endospores of *C. botulinum* are present in the food that is being canned, and if they survive the heat of the canning process, they will resume metabolic activity, feed, and multiply in the anaerobic environment inside the sealed can.
    c. If mice or rats eat such food after the can has been opened, their fleas can then spread the botulism-causing bacteria to nearby people.
    d. Unless a can is completely filled with food, there will be enough air left in the can to support the rapid multiplication of the *C. botulinum* bacteria that produce the extremely deadly toxin.

21. What is NOT an accurate description of any virus, viroid, or prion?
    a. a nonliving, infectious agent or pathogen that can cause disease in organisms
    b. a small sequence of nucleotides or amino acids that is devoid of its own cell membrane
    c. a substance that can cause disease only after entering a host cell and taking over the nuclear machinery
    d. a small organism that carries out small amounts of metabolic activity, and actively propels itself through its environment, but that still relies on other living cells to be replicated

22. Which of the following is NOT true?
    a. Viruses cannot reproduce outside a host cell.
    b. Viroids are important pathogens of animals, seaweeds, and plants.
    c. Viroids lack a protein coat.
    d. Some viruses cause cancer.
    e. Prions are infectious proteins.

23. Viruses that attack bacteria are called
    a. prions.
    b. bacteriophages.
    c. herpes viruses.
    d. mosaic viruses.
    e. viroids.

24. Which components make up a virus?
    a. protein particles only
    b. RNA only
    c. DNA or RNA and a protein coat
    d. DNA and RNA and a protein coat
    e. DNA in a nucleus, RNA, ribosomes, plasma membrane, and cell wall

25. In HIV infection, after reverse transcription creates viral DNA from the viral RNA, the viral DNA
    a. is translated into new viral proteins.
    b. is packaged together with new viral proteins to form new viruses.
    c. enters the host cell's nucleus and is integrated into the host's chromosome.
    d. is used as a template to make new host cell DNA.
    e. remains in the cytoplasm where new viruses are assembled.

26. Viruses make new viral copies (reproduce) in
    a. water.
    b. the soil.
    c. a host cell.
    d. the air.

27. After viral DNA and viral envelope proteins are assembled into new viruses, they exit the host cell via
    a. transcription.
    b. exocytosis.
    c. mitosis.
    d. endocytosis.

28. What type of practice has caused the spread of prion-caused diseases, in certain cases involving humans and livestock?
    a. carnivorous diets, including cannibalism
    b. overuse of antibiotics that are administered too frequently to humans and that are often incorporated into animal feed
    c. forcing livestock and humans to live on strictly vegetarian diets
    d. using artificially produced fertilizers to grow genetically engineered plants, which are then fed to humans and to livestock

29. Study the different stages involving the herpes virus "life" cycle in the figure below. The common antiviral drug acyclovir is a modified form of the guanine base found in DNA. During the replication of viral DNA, DNA polymerase can add a nucleotide containing acyclovir to a growing DNA strand, in a position that would normally be occupied by guanine (that is, opposite from cytosine; see Figure 11-6). However, because acyclovir is shaped differently from guanine, DNA polymerase is then unable to add more nucleotides to the strand, so the viral DNA strand is not fully replicated. Examining the diagram with these facts in mind, label the figure to show the exact stage of viral replication with which acyclovir interferes.

30. Suppose that someone wanted to develop a biological weapon that would kill many people quickly. Which sort of microorganism, or other substance organism or "organism" would be the most suitable?
    a. the human immunodeficiency virus
    b. the prions that cause bovine spongiform encephalopathy
    c. *Bacillus anthracis* bacteria
    d. *Borrelia burgdorferi* bacteria

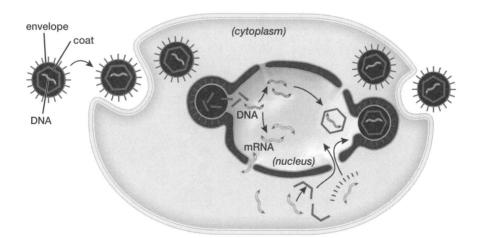

ANSWER KEY

1. d
2. c
3. d
4. d
5. c
6. b
7. b
8. a
9. b
10. c
11. c
12. b
13. d
14. d
15. d

16. a
17. c
18. d
19. c
20. b
21. d
22. b
23. b
24. c
25. c
26. c
27. b
28. a
29. See figure below
30. c

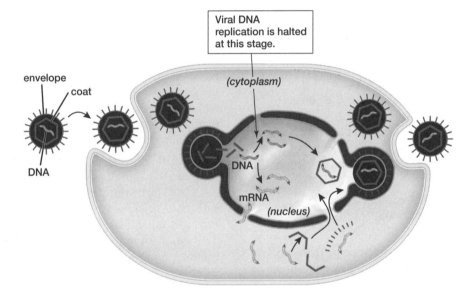

Chapter 20 The Diversity of Protists

OUTLINE

Section 20.1 What Are Protists?

Protists are a generalized group of eukaryotes that are not plants, animals, or fungi. Although most protists are single celled, some form colonies and some are multicellular. The various protist groups do not form a clade, and are mainly grouped together for convenience.

Protists exhibit several modes of nutrition, including predatory heterotrophs that use **pseudopods** or other structures to ingest food (**Figure 20-1**), absorptive feeders that are often parasites inside other organisms, and photosynthesis, seen in several different lineages that are collectively called the **algae**.

Protists reproduce through both asexual and sexual methods. Some species use strictly asexual reproduction, whereas others reproduce asexually at some times and sexually at others (**Figure 20-2**).

Protists are major producers in marine ecosystems, providing food for other marine organisms and contributing oxygen to the atmosphere. However, some parasitize humans, cause plant diseases, and release toxins into the water in coastal areas.

Section 20.2 What Are the Major Groups of Protists?

Excavates are protists that lack mitochondria. They include the **diplomonads** such as the parasitic *Giardia*, which have two nuclei and several flagella (**Figure 20-3**), and the flagellated, anaerobic **parabasalids**, all of which live inside animals (**Figure 20-4**).

Euglenozoans have distinctively shaped mitochondria. One group includes the **euglenids**, which lack a rigid covering and swim by means of flagella. Most of the euglenids are photosynthetic (**Figure 20-5**). Another group includes the flagellated **kinetoplastids**, many of which are parasites, such as *Trypanosoma* (**Figure 20-6**).

Stramenopiles are a genetically classified group of diverse protists. They include the filamentous **water molds**, some of which are extremely harmful to plants (**Figure 20-7**); the **diatoms**, which are single-celled members of the **phytoplankton** with silica shells (**Figure 20-8**); and the brown algae, which are multicellular algae that sometimes form large kelps and other seaweeds (**Figure 20-9**).

Alveolates have distinct cavities beneath their shells. They include the mostly photosynthetic **dinoflagellates**, which swim using two flagella (**Figures 20-10** and **20-11**) and are often important mutualists with other organisms such as corals, but which also sometimes form toxins; the parasitic **apicomplexans**, which cause malaria (**Figure 20-12**); and the **ciliates**, which are complex unicellular protists that move by beating large numbers of **cilia** (**Figures 20-13** and **20-14**).

Rhizarians have elaborate shells, and use thin pseudopodia to move and to capture food. They include the **foraminiferans**, which have shells made of chalk (**Figure 20-15a**), and the **radiolarians**, which have glassy shells (**Figure 20-15b**).

Amoebozoans move by the use of pseudopodia in both aquatic and terrestrial environments. They include the **amoebas**, which have thick pseudopods and no shells (**Figure 20-16**), and the **acellular** and **cellular slime molds**, which are forest-floor decomposers (**Figures 20-17** and **20-18**).

Red algae are multicellular, photosynthetic seaweeds, with red pigmentation that masks their chlorophyll (**Figure 20-19**).

Green algae are unicellular, colonial, or multicellular photosynthetic protists, most of which live in ponds and lakes (**Figure 20-20**). They are the closest living relatives of plants.

FLASH CARDS

To use the flash cards, tear the page from the book and cut along the dashed lines. The key term appears on one side of the flash card, and its definition appears on the opposite side.

acellular slime mold	ciliate
alga (plural, algae)	cilium (plural, cilia)
alveolate	diatom
amoeba	dinoflagellate
amoebozoan	diplomonad
apicomplexan	euglenid
cellular slime mold	euglenozoan

A member of a protist group characterized by cilia and a complex unicellular structure. Ciliates are part of a larger group known as the alveolates.

A type of organism that forms a multinucleate structure that crawls in amoeboid fashion and ingests decaying organic matter; also called *plasmodial slime mold*. Acellular slime molds are members of the protist clade Amoebozoa.

A short, hairlike projection from the surface of certain eukaryotic cells that contains microtubules in a 9 + 2 arrangement. The movement of cilia may propel cells through a fluid medium or move fluids over a stationary surface layer of cells.

Any photosynthetic protist.

A member of a protist group that includes photosynthetic forms with two-part glassy outer coverings; important photosynthetic organisms in fresh water and salt water. Diatoms are part of a larger group known as the stramenopiles.

A member of the Alveolata, a large protist clade. The alveolates, which are characterized by a system of sacs beneath the cell membrane, include ciliates, dinoflagellates, and apicomplexans.

A member of a protist group that includes photosynthetic forms in which two flagella project through armor-like plates; abundant in oceans; can reproduce rapidly, causing "red tides." Dinoflagellates are part of a larger group known as the alveolates.

An amoebozoan protist that uses a characteristic streaming mode of locomotion by extending a cellular projection called a *pseudopod*. Also known as *lobose amoebas*.

A member of a protist group characterized by two nuclei and multiple flagella. Diplomonads, which include disease-causing parasites such as *Giardia*, are part of a larger group known as the excavates.

A member of the Amoebozoa, a protist clade. The amoebozoans, which generally lack shells and move by extending pseudopods, include the lobose amoebas and the slime molds.

A member of a protist group characterized by one or more whiplike flagella, which are used for locomotion, and by a photoreceptor, which detects light; are photosynthetic. Euglenids are part of a larger group known as euglenozoans.

A member of the protist clade Apicomplexa, which includes mostly parasitic, single-celled eukaryotes such as *Plasmodium*, which causes malaria in humans. Apicomplexans are part of a larger group known as the alveolates.

A member of the Euglenozoa, a protist clade. The euglenozoans, which are characterized by mitochondrial membranes that appear under the microscope to be shaped like a stack of disks, include the euglenids and the kinetoplastids.

A type of organism consisting of individual amoeboid cells that can aggregate to form a sluglike mass, which in turn forms a fruiting body. Cellular slime molds are members of the protist clade Amoebozoa.

excavate	protozoan (plural, protozoa)
foraminiferan	pseudoplasmodium
kinetoplastid	pseudopod
parabasalid	radiolarian
phytoplankton	rhizarian
plasmodium	stramenopile
protist	water mold

A nonphotosynthetic, single-celled protist.

A member of the Excavata, a protist clade. The excavates, which generally lack mitochondria, include the diplomonads and the parabasalids.

An aggregation of individual amoeboid cells that form a sluglike mass.

A member of a protist group characterized by pseudopods and elaborate calcium carbonate shells. Foraminiferans are generally aquatic (largely marine) and are part of a larger group known as rhizarians.

An extension of the plasma membrane by which certain cells, such as amoebas, locomote and engulf prey.

A member of a protist group characterized by distinctively structured mitochondria. Kinetoplastids are mostly flagellated and include parasitic forms such as *Trypanosoma*, which causes sleeping sickness. Kinetoplastids are part of a larger group known as euglenozoans.

A member of a protist group characterized by pseudopods and typically elaborate silica shells. Radiolarians are largely aquatic (mostly marine) and are part of a larger group known as rhizarians.

A member of a protist group characterized by mutualistic or parasitic relationships with the animal species inside which they live. Parabasalids are part of a larger group known as the excavates.

A member of Rhizaria, a protist clade. Rhizarians, which use thin pseudopods to move and capture prey and which often have hard shells, include the foraminiferans and the radiolarians.

Photosynthetic protists that are abundant in marine and freshwater environments.

A member of Stramenopila, a large protist clade. Stramenopiles, which are characterized by hair-like projections on their flagella, include the water molds, the diatoms, and the brown algae.

A sluglike mass of cytoplasm containing thousands of nuclei that are not confined within individual cells.

A member of a protist group that includes species with filamentous shapes that give them a superficially fungus-like appearance. Water molds, which include species that cause economically important plant diseases, are part of a larger group known as the stramenopiles.

A eukaryotic organism that is not a plant, animal, or fungus. The term encompasses a diverse array of organisms and does not represent a monophyletic group.

SELF TEST

1. The kingdom of protists is BEST described as
   a. a closely related evolutionary taxon.
   b. eukaryotes that are not plants, animals, or fungi.
   c. the single-celled eukaryotes.
   d. a clearly defined clade.

2. Which of the following correctly describes a protozoan?
   a. prokaryotic
   b. heterotrophic
   c. photosynthetic
   d. cell wall with peptidoglycan

3. Which of the following structures in a single-celled organism functions like a stomach in a multicelled organism?
   a. vacuole
   b. pseudopod
   c. eyespot
   d. oral groove

4. Which of the following organelles evolved first, and why?
   a. mitochondria, because all living organisms require oxygen
   b. mitochondria, because almost all eukaryotes have mitochondria
   c. chloroplasts, because all living organisms require energy
   d. chloroplasts, because more eukaryotes have chloroplasts than mitochondria

5. Pseudopods, finger-like extensions of the plasma membrane, are most often found in protists that
   a. photosynthesize, such as many of the single-celled green algae.
   b. ingest their food, such as the way that the predatory *Amoeba proteus* does.
   c. absorb their food, such as the way that downy mildew does when it infests grapes.
   d. acquire chemical energy from volcanic fluids, such as at deep-sea hot vents.

6. The red algae, brown algae, dinoflagellates, diatoms, and green algae
   a. are all photosynthetic.
   b. are all strictly single celled.
   c. all have flagella for mobility.
   d. together, constitute all of the members of a clade.

7. Which group of protists includes neither predators nor parasites?
   a. ciliates
   b. parabasalids
   c. diatoms
   d. amoebas
   e. kinetoplastids

8. Which of the following is NOT a mode of reproduction used by the protists?
   a. asexual reproduction
   b. sexual reproduction
   c. mitosis
   d. binary fission

9. Which group of protists does NOT include species that infect and directly harm humans?
   a. amoebas
   b. water molds
   c. diplomonads
   d. parabasalids
   e. apicomplexans

10. Which group of protists does not have mitochondria?
    a. stramenopiles
    b. euglenozoans
    c. alveolates
    d. excavates

11. When hikers drink untreated water from streams, they may suffer severe diarrhea, dehydration, nausea, vomiting, and cramps, caused by *Giardia*, which is a parasitic
    a. kinetoplastid.
    b. euglenid.
    c. parabasalid.
    d. diplomonad.

12. *Trypanosoma* is a parasitic kinetoplastid that causes a potentially fatal human disease known as
    a. African sleeping sickness.
    b. hiker's sickness.
    c. dysentery.
    d. malaria.

13. Which protists use flagella for locomotion?
    a. cellular slime molds
    b. ciliates
    c. diatoms
    d. euglenids
    e. red algae

14. Consider the location of the eyespot of *Euglena* in the figure below. The eyespot can detect whether it is being exposed to direct light or whether it is in the shade (including self-shading by the *Euglena* cell). Draw a light source at some location near the cell, placing it where the eyespot will be cast into the greatest shade.

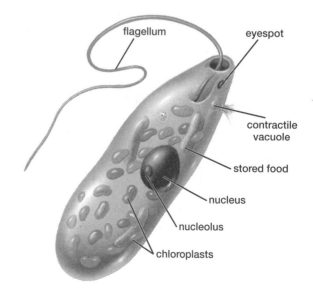

15. The devastating potato famine in Ireland was caused by a water mold from which of the following groups?
    a. stramenopiles
    b. euglenozoans
    c. alveolates
    d. excavates

16. The following tests are performed on an unidentified organism, with the results as shown. How should you classify the organism?

    Chemical A glows when it binds to a plasma membrane. Result = glowing.
    Chemical B turns blue in the presence of chloroplasts. Result = blue color develops.
    Chemical C fizzes when it binds to a nuclear membrane. Result = fizzing.
    Chemical D produces a bad odor when the organism is multicellular. Result = no bad odor.
    a. cyanobacteria   c. ciliates
    b. red algae   d. diatoms

17. Why do red algae, green algae, and brown algae exhibit different colors?
    a. They all have green chlorophyll pigments, but in addition, the red and brown algae also have other pigments that mask the green color.
    b. Only the green algae have chlorophyll, whereas the red and brown algae use entirely different sets of chemicals to absorb light energy.
    c. They all have chlorophyll, but the red algae also have other pigments that are specialized to absorb the bright light in shallow equatorial tide pools, and the brown algae are usually infested with brown-colored, parasitic bacteria.
    d. Although all three types of algae possess green chlorophyll, the chloroplasts of the brown algae are surrounded by an extra brown-colored membrane, and the red algae possess other pigments that are especially effective at absorbing red light.

18. The towering life-forms in the picture at right belong to the subcategory of the _____, in the Kingdom _____.
    a. flowering marine plants such as seagrasses; Plantae
    b. red algae; Protista
    c. brown algae; Protista
    d. filamentous cyanobacteria; Bacteria
    e. green algae; Protista

19. Which of the following protists are photosynthetic and bioluminescent?
    a. euglenids   c. water molds
    b. green algae   d. dinoflagellates

20. Malaria has had a huge impact on the human population. It is caused by a parasitic protist called *Plasmodium* that spends part of its life cycle in mosquitoes before moving into humans. Unfortunately, drug-resistant forms of this parasite are spreading across Africa. These parasitic protists are
    a. apicomplexans   c. parabasalids
    b. dinoflagellates   d. diplomonads

21. What phytoplankton group can reproduce so prodigiously that it can cause "red tides," killing large numbers of fish by clogging their gills?
    a. red algae   c. diatoms
    b. brown algae   d. dinoflagellates

22. Which of the following categories includes complex, single-celled predators that are capable of swallowing prey nearly their own size?
    a. apicomplexans   c. radiolarians
    b. ciliates   d. diplomonads

Kelp forest

23. Single-celled organisms that predominantly live inside of hard shells, but that also move and capture food by using thin finger-like extensions of their cells, are called
    a. amoebas.   c. diatoms.
    b. rhizarians.   d. apicomplexans.

24. Which of the following consume dead organic materials and bacteria in their ecosystems?
    a. slime molds
    b. sporozoans/apicomplexans
    c. cyanobacteria
    d. viruses
    e. green algae

25. What is the essential difference between a plasmodium and a pseudoplasmodium?
    a. The cells in the plasmodium are always haploid, whereas the cells in the pseudoplasmodium are always diploid.
    b. A plasmodium is formed from many individual cells that fuse and then lose their plasma membranes, whereas a pseudoplasmodium always exists as a multicellular colony and never as distinctly separate cells.
    c. The plasmodium contains one giant nucleus, whereas the pseudoplasmodium contains many normal nuclei.
    d. The plasmodium contains many nuclei but is not divided into separate cells, whereas the pseudoplasmodium is divided into distinct cells.

26. Which of the following algae dominate in deep, clear tropical waters, where their pigments absorb the deeply penetrating blue-green light and transfer this light energy to chlorophyll, where it is used in photosynthesis?
    a. plants   c. brown algae
    b. green algae   d. red algae

27. Besides photosynthesis, what is another important ecological function of the red algae?
    a. Because the tissues of many species are impregnated with calcium carbonate, they help in the formation of reefs.
    b. Most species are able to chemically "fix" phosphorus from its gaseous state, in the atmosphere, into the phosphate form essential to all life-forms.
    c. The single-celled species are essential symbiotic organisms with many marine animals, such as corals.
    d. They form large underwater "forests" of tall seaweeds, in the cool waters along the California coast.

28. Which of the following is NOT a type of green algae?
    a. *Volvox*   c. *Porphyra*
    b. *Spyrogyra*   d. *Ulva*

29. Which of the following shares a common ancestor with plants and is most like the earliest plants?
    a. cyanobacteria   d. amoebas
    b. slime molds   e. viroids
    c. green algae

30. Which of the following is NOT a reason that the invasive green algae *Caulerpa taxifolia* has been able to spread so well in marine environments such as in the Mediterranean?
    a. Fish and invertebrate animals in the new habitats do not eat them.
    b. A single cell of the algae can form thousands of new, distinct cells in less than a month, each cell complete with its own nucleus, mitochondria, chloroplasts, and cell membrane.
    c. If a fragment of the algae is broken off, it quickly seals off the breach.
    d. Although the algae were originally native to the tropics, captive breeding produced a form that could tolerate cooler, temperate waters.

ANSWER KEY

1. b
2. b
3. a
4. b
5. b
6. a
7. c
8. d
9. b
10. d
11. d
12. a
13. d
14. See figure below
15. a

16. d
17. a
18. c
19. d
20. a
21. d
22. b
23. b
24. a
25. d
26. d
27. a
28. c
29. c
30. b

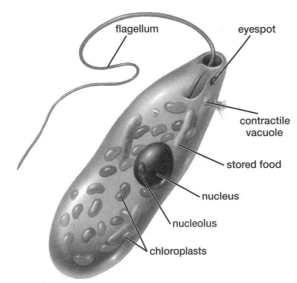

flagellum   eyespot

contractile vacuole

stored food

nucleus

nucleolus

chloroplasts

LIGHT SOURCE

CHAPTER 21 THE DIVERSITY OF PLANTS

OUTLINE

Section 21.1 What Are the Key Features of Plants?

In most terrestrial parts of the world, plants are the most conspicuous organisms. Like many protists and prokaryotes, they carry out photosynthesis.

Plants have alternating multicellular haploid (**gametophyte**) and diploid (**sporophyte**) generations, with embryos that are dependent on the parent plant for nutrients (**Figure 21-1**). The presence of multicelled, dependent embryos distinguishes plants from photosynthetic protists and prokaryotes.

Section 21.2 How Do Plants Affect Other Organisms?

Plants are crucial components of terrestrial ecosystems. They capture light energy, storing it in carbohydrate that becomes food for heterotrophs. Plants also release oxygen gas to the atmosphere, and build and stabilize soil (**Figure 21-2**). They also retain water within ecosystems, limiting the effects of destructive flooding.

Plants provide humans with most of their food. They also provide essential fuel and building materials, as well as important drugs. Another benefit of plants is that they provide humans with important emotional support, such as the pleasure that we draw from smelling **flowers**, tending to gardens, and seeing tall trees.

Section 21.3 What Is the Evolutionary Origin of Plants?

It is likely that green algae gave rise to the first plants, which were probably similar to modern multicellular green algal forms such as stoneworts (**Figure 21-3**).

Because most green algae live in freshwater habitats, it is likely that their early evolutionary history took place there. Life in water had many advantages for the ancestors of land plants, and the move to land posed many challenges.

Section 21.4 How Have Plants Adapted to Life on Land?

Plant bodies possess a number of features that allow them to resist the effects of gravity and desiccation. These features include (1) the development of roots (or rootlike structures) for water and nutrient absorption from the soil; (2) a waxy **cuticle** and **stomata**, which minimize water loss from the plant; (3) conducting vessels that transport water and nutrients throughout the plant body; and (4) **lignin** molecules that impregnate the plant and stiffen its tissues.

Plants are well adapted for reproduction in a terrestrial environment. Their adaptive features include **seeds** that provide protection and nutrients to developing embryos, and windborne **pollen**, which allows sperm to be easily transferred among plants. Flowering plants also possess flowers that attract animals that can act as more effective pollinators, as well as **fruits** that are eaten by animals, which then disperse the fruits' seeds.

Section 21.5 What Are the Major Groups of Plants?

Plants are composed of two major groups: **nonvascular plants** (also known as bryophytes) and **vascular plants** (Table 21-1).

Nonvascular plants include the hornworts, liverworts, and mosses (**Figure 21-5**). These are small land plants that lack conducting structures, and require water to disperse sperm for reproduction (**Figure 21-6**). For these reasons, most nonvascular plants live in moist environments.

Vascular plants possess lignin-infused conducting vessels that transport water and nutrients throughout the plant and support its body (allowing the plants to grow larger). Vascular plants consist of two groups: seedless vascular plants and seed plants.

Seedless vascular plants include the club mosses, horsetails, and ferns (**Figure 21-7**). As with the nonvascular plants, seedless vascular plant sperm require water for dispersal of sperm and fertilization (**Figure 21-8**).

Seed plants produce pollen that allows for windborne sperm dispersal. They also produce seeds that protect and nourish the developing embryo inside (**Figure 21-9**). Seed plants consist of two groups: **gymnosperms** and **angiosperms**.

Gymnosperms are nonflowering seed plants (**Figure 21-10**). Representative gymnosperm groups include (1) ginkgos, (2) cycads, (3) gnetophytes, and (4) **conifers**, whose seeds develop in cones (**Figure 21-11**).

Angiosperms are flowering seed plants (**Figure 21-12**) and are currently the dominant form of plant. Angiosperms utilize flowers to attract pollinators, fruits to encourage seed dispersal by animals (**Figure 21-13**), and broad leaves to capture more light for photosynthesis.

FLASH CARDS

To use the flash cards, tear the page from the book and cut along the dashed lines. The key term appears on one side of the flash card, and its definition appears on the opposite side.

alternation of generations	flower
angiosperm	fruit
antheridium (plural, antheridia)	gametophyte
archegonium (plural, archegonia)	gymnosperm
conifer	lignin
cuticle	nonvascular plant

The reproductive structure of an angiosperm plant.

A life cycle, typical of plants, in which a diploid sporophyte (spore-producing) generation alternates with a haploid gametophyte (gamete-producing) generation.

In flowering plants, the ripened ovary (plus, in some cases, other parts of the flower), which contains the seeds.

A flowering vascular plant.

The multicellular haploid stage in the life cycle of plants.

A structure in which male sex cells are produced; found in nonvascular plants and certain seedless vascular plants.

A nonflowering seed plant, such as a conifer, gnetophyte, cycad, or gingko.

A structure in which female sex cells are produced; found in nonvascular plants and certain seedless vascular plants.

A hard material that is embedded in the cell walls of vascular plants and that provides support in terrestrial species; an early and important adaptation to terrestrial life.

A member of a group of nonflowering vascular plants whose members reproduce by means of seeds formed inside cones; retains its leaves throughout the year.

A plant that lacks lignin and well-developed conducting vessels. Nonvascular plants include mosses, hornworts, and liverworts.

A waxy or fatty coating on the exposed surfaces of the epidermal cells of many land plants that aids in the retention of water.

ovule

sporophyte

pollen

stoma (plural, stomata)

seed

vascular plant

The multicellular diploid stage in the life cycle of a plant; produces haploid, asexual spores through meiosis.

A structure within the ovary of a flower, inside which the female gametophyte develops; after fertilization, develops into the seed.

An adjustable opening in the epidermis of a leaf, surrounded by a pair of guard cells, that regulates the diffusion of carbon dioxide and water into and out of the leaf.

The male gametophyte of a seed plant; also called a *pollen grain*.

A plant that has conducting vessels for transporting liquids; also called a *tracheophyte*.

The reproductive structure of a seed plant; protected by a seed coat; contains an embryonic plant and a supply of food for it.

SELF TEST

1. All plants produce
   a. spores.
   b. seeds.
   c. pollen.
   d. swimming sperm.
   e. fruits.

2. In the life cycle of plants, gametophytes are the
   a. haploid generation, which produces spores.
   b. diploid generation, which produces spores.
   c. haploid generation, which produces gametes.
   d. diploid generation, which produces gametes.

3. Examine the general plant life cycle diagram below, showing alternation of generations. Label each arrow that indicates a process in which mitosis is occurring.

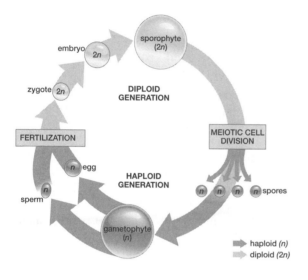

4. Which of the following human foods is NOT mainly based, either directly or indirectly, on the productivity of members of the plant kingdom?
   a. a can of tuna
   b. a fried chicken drumstick
   c. a pork chop
   d. a slice of pepperoni pizza
   e. a peanut butter and jelly sandwich

5. The Pacific Northwest of the United States receives heavy rain in the winter, and little rain in the summer. Imagine that we examine two forested, western Oregon watersheds (drainage areas), A and B, each of which drains into a single creek, A or B, respectively. As of 6 years ago, the watersheds were the same size, they had the same climate and were at the same elevation, and the forests were virtually identical. Neither watershed had experienced fires, logging, or other major disturbances for 300 years. Creeks A and B had essentially equal flow rates of water, of essentially the same quality. Five years ago, we logged and removed all of the trees in watershed A, building many forest roads and tearing up some of the soil as we did so. If we now study the two creeks at the present time, which of the following statements is most likely to be INCORRECT?
   a. The January water flow rate in creek A is now higher than in creek B.
   b. The July water flow rate in creek A is now lower than in creek B.
   c. The July temperature of the stream water in creek A is now lower than in creek B.
   d. The January turbidity (cloudiness of the water, due to suspended silt) of the water in creek A is now higher than in creek B.

6. The hypothesis that plants evolved from green algae ancestors is supported by the fact that both plants and green algae
   a. store food as glycogen.
   b. have cell walls made of chitin.
   c. use the same type of chlorophyll and accessory pigments during photosynthesis.
   d. have true roots, stems, and leaves and complex reproductive structures.

7. The rigors of the terrestrial environment led to many adaptations among terrestrial plants. Which of the following is NOT a necessary adaptation to dry land?
   a. conducting vessels
   b. cuticle and stomata
   c. lignin
   d. roots or rootlike structures
   e. separate gametophyte stage

8. Photosynthesis stops during very hot and dry weather because
   a. the stomata close, which cuts off the plant's supply of the carbon dioxide needed for photosynthesis.
   b. plants do not make lignin when it is hot and dry.
   c. most plants lack vascular tissue that would enable them to absorb water from the soil.
   d. the waxy cuticle melts in hot weather.

9. Plants are much more complicated than their green algae ancestors, with organs such as roots, stems, and leaves, as well as lignin to stiffen their structures, cuticles to cover their leaves, and internal vascular (conducting) tissues to transport water and nutrients. Which of the following is NOT a valid reason for the high structural complexity of land plants?
   a. Roots must be located below ground, but cannot perform photosynthesis there, and yet they still need the sugar that is produced above ground in the leaves.
   b. Leaves need water, but cannot absorb it directly from the air because they are covered with a waxy cuticle.

c. Leaves need mineral nutrients, but these are located below ground in the soil.

d. Roots need a source of carbon dioxide in order to gain energy by respiring sugar, but CO_2 is only available above ground in the atmosphere.

e. Plants that become rigid and erect, due to the stiffening provided by lignin, have a competitive advantage over other plants that have less lignin, in the competition for limited light energy.

10. Which of the following is NOT an example of the sporophyte stage of the alternation of generations?
  a. oak tree
  b. green moss
  c. pine tree
  d. tomato plant

11. Seed plants produce male gametophytes known as
  a. fruit.
  b. flower.
  c. pollen.
  d. seed.
  e. sporangium.

12. What is the reproductive structure of nonvascular plants and of seedless vascular plants that encloses eggs and protects them from drying out?
  a. antheridium
  b. archegonium
  c. gametophyte
  d. rhizoid
  e. sporangium

13. When we examine the sporophyte of a tall redwood tree, the largest geometric dimension of the sporophyte is closest in size to
  a. 1 millimeter.
  b. 1 centimeter.
  c. 1 meter.
  d. 100 meters.

14. The sperm of conifers
  a. swim to an egg.
  b. are carried in a pollen grain that has tiny wings.
  c. are transported to an egg by a bee.
  d. are triploid.
  e. are found in the ovary of a flower.

15. The relatively small size of the nonvascular plants is probably due to
  a. the absence of vascular tissue.
  b. the dependence on water for reproduction.
  c. their habitat.
  d. the lack of true leaves.
  e. a unique but conservative pattern of reproduction.

16. Interpret the figure below: To which other group (or groups), of the plants shown here, are the ferns most closely related?
  a. the liverworts
  b. the mosses
  c. the gymnosperms
  d. the angiosperms
  e. the liverworts and mosses, equally
  f. the gymnosperms and angiosperms, equally

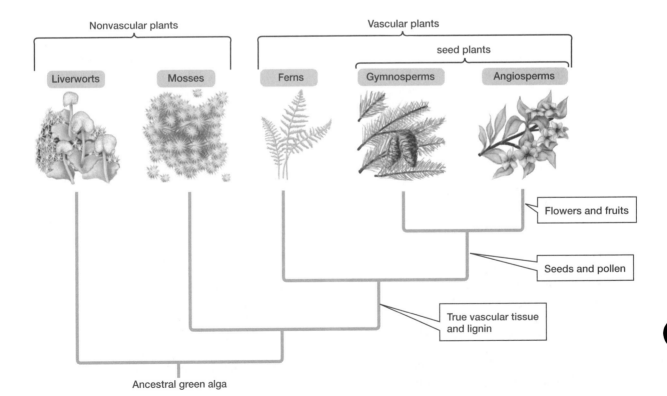

17. In the angiosperms, which major adaptation may be most vulnerable to herbivore attack, especially by insects?
    a. broad leaves
    b. fruits
    c. flowers
    d. roots

18. Even though they are vascular plants, and thus moderately advanced terrestrially, the ferns have not solved all of the problems of terrestrial life. How is this so?
    a. Their zygotes remain unprotected from desiccation.
    b. They are broad leafed.
    c. They do not possess a gametophyte stage.
    d. They lack advanced vascular tissue.
    e. Typically, they lack erect stems.

19. Which of the following is BEST adapted to a dry habitat?
    a. moss
    b. ferns
    c. horsetails
    d. crabgrass

20. Which of the following is true concerning the life cycle of a moss?
    a. Gametophytes are nutritionally dependent on sporophytes.
    b. Emerging gametophytes arise from germinated spores.
    c. The zygote is protected within specialized tissues of the sporophyte.
    d. Spores are produced within reproductive bodies called antheridia.

21. What is true about the ginkgos?
    a. Their lineage has a long fossil history, dating back more than 200 million years.
    b. They are a widespread and diverse group of gymnosperms, with various species growing in the wild on all continents except Africa and Antarctica.
    c. They are one of the most recent lineages of plants to develop, evolving much more recently than roses, lilies, sunflowers, and orchids.
    d. Like conifers and many other gymnosperms, each individual tree is bisexual, producing both male and female reproductive structures.
    e. Ginkgo trees have needle leaves, similar to pine trees and other conifers.

22. In a flower, the fertilized ovule develops into a(n) _____, whereas the ovary develops into a(n) _____.
    a. sporophyte; seed
    b. seed; fruit
    c. fruit; embryo
    d. embryo; sporophyte

23. Which of the following is a characteristic of ferns?
    a. The gametophyte is the dominant generation.
    b. They are vascular plants, capable of internal transport of water and nutrients.
    c. They produce seeds on frond leaflets.
    d. They produce nonmotile (nonswimming) sperm.

24. The anther in a complete flower (one that includes both male and female reproductive structures) has the same reproductive function as the _____ in the nonvascular plants.
    a. archegonium
    b. sporangium
    c. antheridium
    d. indusium

25. What is the correct statement?
    a. One seed can contain many fruits.
    b. Fruits form in both the gymnosperms and the angiosperms.
    c. Some plants "intend" for their fruits to be consumed.
    d. Fruits are always sweet, as with oranges and apples.
    e. Potatoes, beets, and lima beans are all examples of fruits.

26. The ovule is found in _____ and _____.
    a. seedless vascular plants; angiosperms
    b. nonvascular plants; conifers
    c. conifers; seedless vascular plants
    d. conifers; angiosperms

27. Rust-colored clusters found on the underside of fern fronds are
    a. gametophytes.
    b. sporangia.
    c. archegonia.
    d. antheridia.

28. The archegonium produces
    a. an egg cell.
    b. a seed.
    c. sperm cells.
    d. a spore cell.

29. In the flowering plants, the gametophyte
    a. develops in the flower.
    b. lives independently of the sporophyte.
    c. is the dominant generation.
    d. develops into the plant's fruit.

30. With the stinking corpse lily (*Rafflesia*), what is the most likely means by which one flower fertilizes another?
    a. Hummingbirds are attracted to the beautiful red flowers and carry pollen between them.
    b. Flies are attracted by the rotting-meat smell and carry pollen from one stinking flower to another.

c. The heavy rainfall in the tropical rain forest physically washes the swimming sperm from one flower to another.

d. Wind carries the pollen, which is typically the case for flowers that have unpleasant smells and repel all potential animal pollinators.

e. Honeybees are the agents of pollination here, just like they are for almost every species of flowering plant.

ANSWER KEY

1. a
2. c
3.

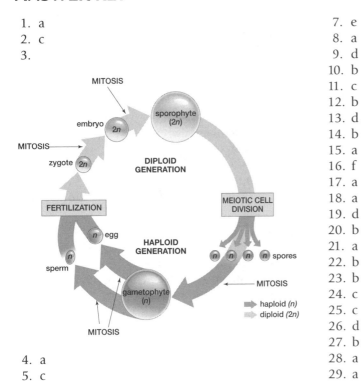

4. a
5. c
6. c

7. e
8. a
9. d
10. b
11. c
12. b
13. d
14. b
15. a
16. f
17. a
18. a
19. d
20. b
21. a
22. b
23. b
24. c
25. c
26. d
27. b
28. a
29. a
30. b

CHAPTER 22 THE DIVERSITY OF FUNGI

OUTLINE

Section 22.1 What Are the Key Features of Fungi?

Fungal bodies are composed of filamentous threads (**hyphae**) that form an interwoven mass (**mycelium**). Hyphal threads are either multicellular or multinucleated (**Figure 22-1**). Fungal cells are surrounded by a chitin-reinforced cell wall and are typically haploid.

Fungi secrete enzymes and break down nutrients outside their bodies. They can function as decomposers, parasites, or predators (**Figure 22-2**) and can also form mutualistic associations with other organisms.

Fungi propagate by haploid **spores**, which can be distributed over a wide area (**Figure 22-3**). Fungi can reproduce asexually (by mitotic spore production) or sexually (by meiotic spore production).

Section 22.2 What Are the Major Groups of Fungi?

Chytrid fungi (**Figure 22-5**) live in water, which is required for reproduction because they utilize swimming spores. The oldest known fossil fungi are chytrids.

Zygomycetes live in soil or on decaying plant or animal material. They can produce spores asexually. They can also reproduce sexually via the fusion of hyphae from individuals with different mating types, eventually forming stalked **sporangia** that release haploid spores (**Figure 22-6**).

Glomeromycetes live in close contact with plant roots, penetrating root cells (**Figure 22-7**). They form mutually beneficial mycorrhizal relationships with most plant species and are only known to reproduce asexually.

Basidiomycetes typically reproduce sexually and produce club-shaped reproductive structures called **basidia** (**Figure 22-8**). They form the common mushrooms, as well as puffballs and shelf fungi (**Figure 22-9**).

Ascomycetes reproduce asexually and sexually, but during sexual reproduction, they form a saclike case called an **ascus** (**Figure 22-11**). They include molds and yeasts and can live in decaying forest vegetation (**Figure 22-12**). They also provide the fungal species in most **lichens**.

Section 22.3 How Do Fungi Interact with Other Species?

Lichens are symbiotic associations between a fungal species and a photosynthetic species of green algae or cyanobacteria (**Figure 22-13**). This arrangement protects the photosynthetic cells from harsh environments while the fungus receives sugars made by the photosynthetic cells. Lichens can take on many physical forms (**Figure 22-14**).

Mycorrhizae are symbiotic associations between fungi and plant roots (**Figure 22-15**). This arrangement allows plants to absorb more water and minerals from the soil, whereas the fungi receive sugars from the plant. Most plant species form beneficial mycorrhizal relationships with fungi.

Endophytes are symbiotic associations with fungi and plant leaves and stems. This arrangement can protect plant structures from herbivores because of distasteful substances produced by the fungi.

Fungi are the most effective decomposers of dead wood. Saprophytes are fungi that decompose dead organisms, thus acting as essential recyclers for Earth.

Section 22.4 How Do Fungi Affect Humans?

Fungi affect humans by attacking useful plants (**Figure 22-16**). Some plants that are or have been damaged by fungi include grain crops, American elms, and American chestnuts. However, some fungi attack insect and other arthropod crop pests, and can be used as nontoxic alternatives to pesticides (**Figure 22-17**).

Fungi also cause human diseases, including skin diseases such as ringworm and athlete's foot. Other fungi can cause vaginal and lung infections.

Some fungi produce toxins, such as the carcinogenic aflatoxins produced by *Aspergillus* molds, as well as the poisons produced by *Claviceps purpurea* when it infests rye plants. Some species of mushrooms are also deadly poisonous if eaten (**Figure E22-1**).

Some fungi, such as *Penicillium* molds, produce important antibiotics that fight bacterial infections (**Figure 22-19**). Other fungi are delicious to eat directly (**Figure 22-20**), or else are used to produce foods such as wine, beer, bread, and certain cheeses.

FLASH CARDS

To use the flash cards, tear the page from the book and cut along the dashed lines. The key term appears on one side of the flash card, and its definition appears on the opposite side.

ascomycete	glomeromycete
ascus (plural, asci)	hypha (plural, hyphae)
basidiomycete	lichen
basidium (plural, basidia)	mycelium (plural, mycelia)
chytrid	mycorrhiza (plural, mycorrhizae)
club fungus	sac fungus

A member of the fungus phylum Glomeromycota, which includes species that form mycorrhizal associations with plant roots and that form bush-shaped branching structures inside plant cells.

A member of the fungus phylum Ascomycota, whose members form sexual spores in a saclike case known as an *ascus*.

A threadlike structure that consists of elongated cells, typically with many haploid nuclei; many hyphae make up the fungal body.

A saclike case in which sexual spores are formed by members of the fungus phylum Ascomycota.

A symbiotic association between an alga or cyanobacterium and a fungus, resulting in a composite organism.

A member of the fungus phylum Basidiomycota, which includes species that produce sexual spores in club-shaped cells known as *basidia*.

The body of a fungus, consisting of a mass of hyphae.

A diploid cell, typically club-shaped, formed by members of the fungus phylum Basidiomycota; produces basidiospores by meiosis.

A symbiotic association between a fungus and the roots of a land plant that facilitates mineral extraction and absorption.

A member of the fungus phylum Chytridomycota, which includes species with flagellated swimming spores.

A member of the fungus phylum Ascomycota, whose members form spores in a saclike case called an *ascus*.

A fungus of the division Basidiomycota, whose members (which include mushrooms, puffballs, and shelf fungi) reproduce by means of basidiospores.

septum (plural, septa)

spore

sporangium (plural, sporangia)

zygomycete

In plants and fungi, a haploid cell capable of developing into an adult without fusing with another cell (without fertilization). In bacteria and some other organisms, a stage of the life cycle that is resistant to extreme environmental conditions.

A partition that separates the fungal hypha into individual cells; pores in septa allow the transfer of materials between cells.

A fungus of the phylum Zygomycota, which includes the species that cause fruit rot and bread mold.

A structure in which spores are produced.

SELF TEST

1. If you examined the structure of the body of a typical fungus, you would find a mass of
   a. round cells called coccae, which form a long chain of cells connected by chitin.
   b. square cells called hexae, which are clumped together, forming the structures we observe on rotting organic material.
   c. flattened cells called squamae, which are stacked on top of each other, forming a mass called a mycelium.
   d. threadlike cells called hyphae that are only one cell thick.

2. A haploid asexual spore is formed by a haploid mycelium via
   a. fertilization.
   b. meiosis.
   c. mitosis.
   d. pollination.

3. Both fungi and animals
   a. photosynthesize.
   b. form embryos.
   c. interact to form lichens.
   d. fix nitrogen.
   e. are heterotrophic.

4. Which of the following characteristics is typical of fungi?
   a. autotrophism
   b. possession of cell walls of cellulose
   c. cytoplasmic connections between the cells
   d. diploid nature
   e. sexual production of eggs and sperm

5. The fungal spores resulting from sexual reproduction are produced after
   a. the mycelia of two compatible mating types fuse.
   b. the septa between hyphal cells break down and allow the nuclei to fuse.
   c. a male mycelium encounters a mature female mycelium.
   d. the mycelium encounters a favorable environment for growth.

6. One predominant value of fungi, in ecosystem functioning, is BEST described by which of the following statements? Consider the effects of fungi alone, rather than the effects of the mutualistic partners of certain fungi.
   a. Fungi "fix" nitrogen gas (N_2) from the atmosphere, converting it into a mineral form that can be used by plants.
   b. Fungi decompose complex organic molecules, thereby releasing nutrients into ecosystems.
   c. Fungi trap the energy from sunlight, and store it in the form of carbohydrates.
   d. Various fungal infections, taken together, are the fourth largest worldwide cause of deaths among humans.

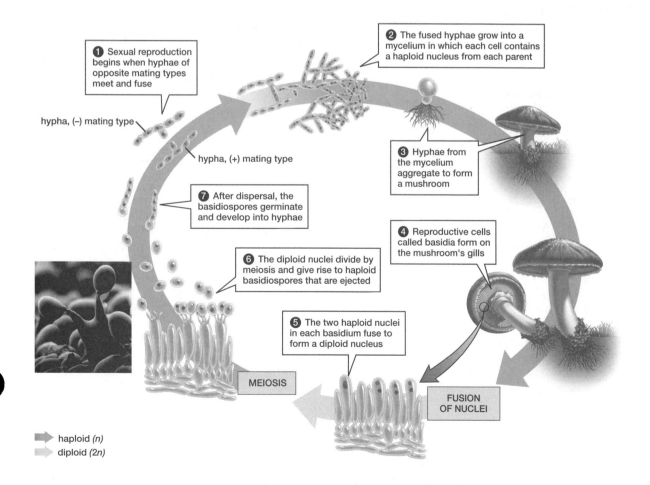

1 Sexual reproduction begins when hyphae of opposite mating types meet and fuse

2 The fused hyphae grow into a mycelium in which each cell contains a haploid nucleus from each parent

3 Hyphae from the mycelium aggregate to form a mushroom

4 Reproductive cells called basidia form on the mushroom's gills

5 The two haploid nuclei in each basidium fuse to form a diploid nucleus

6 The diploid nuclei divide by meiosis and give rise to haploid basidiospores that are ejected

7 After dispersal, the basidiospores germinate and develop into hyphae

hypha, (–) mating type

hypha, (+) mating type

MEIOSIS

FUSION OF NUCLEI

haploid (n)
diploid (2n)

7. Examine the figure on page 265, noting how the basidia are arranged on the gills of a mushroom, and how these basidia then produce and release spores. A mushroom normally grows so that its gills are oriented vertically. Why?
   a. If the mushroom is tilted, the gills will collapse under their own weight.
   b. Vertically oriented gills are best able to capture the energy from sunlight.
   c. This orientation allows spores to most easily fall free from the mushroom.
   d. This orientation prevents the gills from being eaten by small invertebrates such as beetles and worms, before they have had a chance to release their mature spores.

8. Which of the following is NOT associated with the ascomycetes?
   a. most fungal diseases that directly infect people
   b. production of bread and beer
   c. nearly all species of lichens
   d. sexual reproductive structures that make eight spores at a time
   e. most species form mycorrhizal relationships with plant roots

9. Which of the following is NOT associated with the mostly aquatic chytrids?
   a. youngest of all fungal phyla
   b. cell walls of chitin
   c. eukaryotic cells
   d. haploid body
   e. flagellated spores

10. One of the more important characteristics used in the classification of fungi is their form of
    a. asexual reproduction.
    b. sexual reproduction.
    c. mycelium.
    d. septa.

11. Soft fruit rot and black bread mold belong to which division of fungi?
    a. Ascomycota
    b. Basidiomycota
    c. Glomeromycota
    d. Zygomycota

12. A zygospore undergoes _____ to produce haploid spores.
    a. germination
    b. meiosis
    c. fertilization
    d. mitosis

13. The mushrooms on the pizza you might have eaten last night belong to the phylum
    a. Zygomycota.
    b. Basidiomycota.
    c. Glomeromycota.
    d. Ascomycota.

14. In the basidiomycete life cycle, after a diploid cell is first formed, how many times does it then divide by mitosis, before meiosis produces the haploid spores?
    a. 0
    b. 1
    c. 2
    d. 3
    e. 4

15. During sexual reproduction in the Basidiomycota and the Ascomycota, _____ produces _____ spores, which give rise to a new mycelium.
    a. meiosis; diploid
    b. meiosis; haploid
    c. mitosis; diploid
    d. mitosis; haploid

16. The clublike structure producing the spores of typical mushrooms is called a(n)
    a. ascus.
    b. basidium.
    c. conidia.
    d. sporangia.

17. Which of the following associations is NOT correct?
    a. zygomycetes–zygospores
    b. basidiomycetes–large fruiting bodies
    c. glomeromycetes–absence of sexual reproduction
    d. chytrids–lightweight, windblown sexual spores
    e. ascomycetes–asci

18. The spores of chytrids are unique because they are
    a. lightweight, windblown structures.
    b. flagellated.
    c. found in club-shaped reproductive structures called basidia.
    d. especially well adapted to dispersal in dry environments.

19. Some fungi seldom or never reproduce sexually. Suppose that instead of being able to inspect a distinctive sexual reproductive structure, such as that of a particular mushroom, you must instead classify a fungus based on its hyphae. Which of the following would be the easiest to distinguish from the other three, based on the overall configuration of its hyphae?
    a. a chytrid
    b. a glomeromycete
    c. a basidiomycete
    d. a zygomycete

20. Examine the figure below, showing the algal (autotrophic) and fungal (heterotrophic) components of a typical lichen. Note the relative proportions of each, as shown in the diagram. Consider that the sugars produced in lichen photosynthesis are needed to (1) support respiration and maintenance of life, in both types of organisms, and (2) construct new cells and allow growth, of both types of organisms. Consider, also, that photosynthetic organisms in general need water if they are going to carry out photosynthesis, and that even if they can survive desiccation (which is the case with lichens and mosses), photosynthetic organisms still become inactive if they dry out. With these facts in mind, which of these organisms do you think should grow the slowest?
    a. a brown algae kelp, growing near the sea surface in the cool Pacific waters off the coast of California
    b. a 5-year-old pine tree sapling, growing in a site in Alabama that is well supplied with water, sunlight, and soil nutrients
    c. a typical lichen, growing attached to a tree branch in a moist, low-elevation forest in western Oregon
    d. a corn plant, growing from seed in a farm field in Iowa

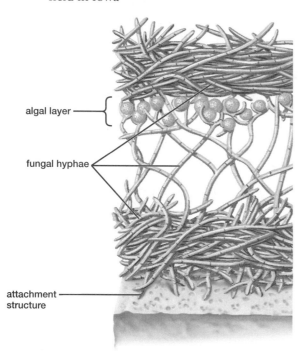

algal layer

fungal hyphae

attachment structure

21. A test reveals that a lichen contains a prokaryotic symbiont. Which kind of organism is this prokaryotic symbiont?
    a. fungus
    b. heterotrophic bacterium
    c. moss
    d. cyanobacterium
    e. virus

22. Which of the following is the most important role fungi have in their ecosystems?
    a. recyclers of nutrients such as carbon and nitrogen from dead animal and plant bodies
    b. parasites on crops such as corn, tobacco, and apples
    c. symbionts with algae or cyanobacteria
    d. pathogens that cause diseases such as ringworm, athlete's foot, and yeast infections
    e. edible fungi such as truffles, morels, and mushrooms

23. What beneficial agricultural role do fungi play?
    a. Fungal pathogens act as fungal pesticides to protect numerous crop species from various insect species.
    b. The application of rusts and smuts changes the coloration of crop leaves and confuses potential insect pests.
    c. The American elm and American chestnut have benefited from application of ascomycete fungi.
    d. Plants that have chytrid fungi attached to their roots grow faster than other plants that are not attached to chytrid fungi.

24. Which of the following is NOT true about mycorrhizal relationships?
    a. Most, but not all plants with roots, form such relationships with certain soil fungi.
    b. Each partner in a mycorrhizal relationship gains an important resource from the other partner, albeit at a small cost to itself.
    c. Some plants do not perform photosynthesis, and a mycorrhizal soil fungus would gain nothing by attaching to the roots of such a plant.
    d. Mycorrhizae only developed long after plants and fungi had separately become well established in the continental regions of the planet.

25. Consider the action of yeast in producing ethanol (see both Chapters 8 and 22). In which of the following situations should the greatest concentration of ethanol be produced?
    a. Grape juice and yeast are combined, and placed into a sealed container for several months.
    b. Dry yeast is mixed with 1 cup of water, 4 cups of whole-wheat bread flour (predominantly starch, with very little sugar), 1 teaspoon of salt, and 1 tablespoon of sugar, and the entire mix is kneaded and then allowed to rise for 2 hours.

c. Malted barley and hops are combined with yeast and water, and the whole mix is placed into an open-top vat and stirred constantly and kept warm for several weeks.

d. Yeast and olive oil are mixed, bottled, and carefully stored in a dark, room temperature environment for at least 1 year.

26. If a person accidentally consumes the deadly mushroom *Amanita virosa* (destroying angel), what is the most likely way for the person to die?

a. The person's limbs will become gangrenous and fall off.

b. The person will inevitably develop untreatable, deadly cancer.

c. Liver failure and subsequent death are almost certain to happen.

d. Blood vessels in the brain rapidly deteriorate, leading to a fatal stroke.

e. Within hours of eating the mushroom, the person will have a massive heart attack.

27. Which of the following is a fungal disease?

a. athlete's foot

b. botulism

c. Lyme disease

d. malaria

28. If yeasts are responsible for the alcohol in wine and beer, why don't we get a little tipsy from eating bread?

a. Alcohol is not produced by yeast inoculated into bread dough.

b. Fermentation is impaired by the nature of the dough.

c. Baking the bread evaporates the alcohol produced by the fermenting yeasts.

d. Yeasts are actually not used in breads, but an ascomycete mold is.

29. Which of the following economic problems or diseases is NOT caused by a fungus?

a. corn smut

b. Dutch elm disease

c. histoplasmosis

d. mad cow disease

e. ringworm

30. With the extensive growth of an *Armillaria ostoyae* mycelium in eastern Oregon, the fungus mainly gains its nutrition by

a. forming mutualistic relationships with green algae, in which the algae provide the fungus with sugar via photosynthesis, while the fungus shelters the algae from the harsh environment.

b. harmlessly decomposing the abundant dead organic material in the soil.

c. forming mutually beneficial mycorrhizal relationships with most trees in the area.

d. infesting tree roots, decomposing the living tissue, and killing the trees.

e. poisoning careless mushroom hunters and then decomposing their remains.

ANSWER KEY

1. d
2. c
3. e
4. c
5. a
6. b
7. c
8. e
9. a
10. b
11. d
12. b
13. b
14. a
15. b
16. b
17. d
18. b
19. c
20. c
21. d
22. a
23. a
24. d
25. a
26. c
27. a
28. c
29. d
30. d

CHAPTER 23 ANIMAL DIVERSITY I:
INVERTEBRATES

OUTLINE

Section 23.1 What Are the Key Features of Animals?

Animals are multicellular, motile heterotrophs that respond to external environmental stimuli and normally undergo sexual reproduction. Their cells lack a cell wall.

None of these features is, in itself, unique to animals, but as a collective set of features they do distinguish animals from other categories of organisms.

Section 23.2 Which Anatomical Features Mark Branch Points on the Animal Evolutionary Tree?

The anatomical features that mark animal evolutionary branch points are summarized in **Figure 23-1**.

The earliest branch point is the development of **tissues**, which is expressed in all animals except sponges.

Animals with tissues can be divided into those that have **radial symmetry** and those that have **bilateral symmetry** (**Figure 23-2**). Radially symmetric animals have two embryonic germ layers, whereas bilaterally symmetric animals have three.

Bilaterally symmetrical animals usually express **cephalization** and some form of body cavity (**Figure 23-3**). Bilaterally symmetrical animals can be divided into two main groups based on patterns of embryonic development: **protostomes** and **deuterostomes**.

Protostome animals develop a body cavity within the space between the body wall and digestive cavity. They include the annelids, arthropods, and mollusks.

Deuterostome animals develop a body cavity as an outgrowth of the digestive cavity. They include the echinoderms and chordates.

Section 23.3 What Are the Major Animal Phyla?

As a matter of convenience, biologists usually categorize animals as being either **vertebrates** (having a backbone or vertebral column), such as mammals and fish, or **invertebrates** (lacking a backbone). Most phyla and species of animals are invertebrates.

Animals of the phylum Porifera (sponges, **Figure 23-4**) are aquatic filter feeders that lack tissues. Their bodies are generally asymmetrical and are composed of three major cell types (**Figure 23-5**). Their bodies are stiffened by spicules, small units of calcium carbonate, silica, or protein.

Animals of the phylum Cnidaria (sea jellies, sea anemones, corals, and hydrozoans, **Figure 23-6**) possess tissues, radial symmetry, stinging cnidocytes (**Figure 23-8**), a simple nervous system, and a central gastrovascular cavity that allows extracellular digestion of food. Most cnidarians alternate between polyp and medusa forms during their life cycles (**Figure 23-7**).

Animals of the phylum Ctenophora (comb jellies) superficially resemble sea jellies. Most are small, all move by beating at the water with rows or combs of cilia, and they capture prey with sticky tentacles.

Animals of the phylum Platyhelminthes (flatworms, **Figure 23-9**) have bilateral symmetry and cephalization but lack a body cavity. They possess a more advanced nervous system than the cnidarians and can be either free living or **parasites** (**Figure 23-10**).

Animals of the phylum Annelida (segmented worms) have a **closed circulatory system**, a well-developed excretory system, and a compartmentalized digestive system. They have a true **coelom** and **segmentation** (**Figures 23-11** and **23-12**). Annelids consist of the oligochaetes, polychaetes, and leeches.

Animals of the phylum Mollusca (snails, clams, and squid) are soft bodied and usually protect themselves with some form of shell (**Figure 23-13**). They have gills and annelid-like nervous systems. Most have **open circulatory systems**. Mollusks consist of the gastropods (**Figure 23-14**), bivalves (**Figure 23-15**), and cephalopods (**Figure 23-16**).

Animals of the phylum Arthropoda are the most successful animals on earth. They have **exoskeletons** (**Figure 23-17**) that they periodically **molt** (**Figure 23-18**), jointed appendages, well-developed nervous systems (including **compound eyes**, **Figure 23-20**), and specialized respiratory structures that have allowed them to invade nearly every type of aquatic and terrestrial habitat. Some arthropods (the insects) can fly. Arthropods consist of the insects (**Figure 23-21**), arachnids (**Figure 23-22**), myriapods (**Figure 23-23**), and crustaceans (**Figure 23-24**).

Animals of the phylum Nematoda (roundworms, **Figure 23-25**) are found nearly everywhere. They lack circulatory and respiratory systems, relying, like the flatworms, on diffusion for gas and nutrient exchange. They also possess a **pseudocoelom** and have both free-living and parasitic forms (**Figure 23-26**).

Animals of the phylum Echinodermata (sea cucumbers, sea urchins, and sea stars, **Figure 23-27**) are bilaterally symmetrical as **larvae** and radially symmetrical as adults. They possess a nonliving **endoskeleton** that projects through the skin, as well as a water-vascular system (**Figure 23-28**).

Animals of the phylum Chordata include all vertebrates, as well as the invertebrate tunicates and lancelets.

FLASH CARDS

To use the flash cards, tear the page from the book and cut along the dashed lines. The key term appears on one side of the flash card, and its definition appears on the opposite side.

bilateral symmetry	ectoderm
budding	endoderm
cephalization	endoskeleton
closed circulatory system	exoskeleton
coelom	ganglion (plural, ganglia)
compound eye	hemocoel
deuterostome	hermaphroditic

The outermost embryonic tissue layer, which gives rise to structures such as hair, the epidermis of the skin, and the nervous system.

A body plan in which only a single plane through the central axis will divide the body into mirror-image halves.

The innermost embryonic tissue layer, which gives rise to structures such as the lining of the digestive and respiratory tracts.

Asexual reproduction by the growth of a miniature copy, or bud, of the adult animal on the body of the parent. The bud breaks off to begin independent existence.

A rigid internal skeleton with flexible joints that allow for movement.

Concentration of sensory organs and nervous tissue in the anterior (head) portion of the body.

A rigid external skeleton that supports the body, protects the internal organs, and has flexible joints that allow for movement.

A circulatory system in which the blood is always confined within the heart and vessels.

A cluster of neurons.

In animals, a space or cavity, lined with tissue derived from mesoderm, that separates the body wall from the inner organs.

A blood cavity within the bodies of certain invertebrates in which blood bathes tissues directly; part of an open circulatory system.

A type of eye, found in arthropods, that is composed of numerous independent subunits called *ommatidia*. Each ommatidium apparently contributes a piece of a mosaic-like image perceived by the animal.

Possessing both male and female sexual organs. Some hermaphroditic animals can fertilize themselves; others must exchange sex cells with a mate.

An animal with a mode of embryonic development in which the coelom is derived from outpocketings of the gut; characteristic of echinoderms and chordates.

hydrostatic skeleton

parasite

invertebrate

protostome

larva (plural, larvae)

pseudocoelom

mesoderm

pupa (plural, pupae)

metamorphosis

radial symmetry

molt

segmentation

nerve cord

tissue

open circulatory system

vertebrate

An organism that lives in or on, and feeds on, a larger organism called a *host*, weakening it.

In invertebrate animals, a body structure in which fluid-filled compartments provide support and rigid structures against which muscles can exert force.

An animal with a mode of embryonic development in which the coelom is derived from splits in the mesoderm; characteristic of arthropods, annelids, and mollusks.

An animal that lacks a vertebral column.

In animals, a "false coelom," that is, a space or cavity, partially but not fully lined with tissue derived from mesoderm, that separates the body wall from the inner organs; found in roundworms.

An immature form of an animal that subsequently undergoes metamorphosis into its adult form; includes the caterpillars of moths and butterflies, the maggots of flies, and the tadpoles of frogs and toads.

A developmental stage in some insect species in which the organism stops moving and feeding and may be encased in a cocoon; occurs between the larval and the adult phases.

The middle embryonic tissue layer, lying between the endoderm and ectoderm, and normally the last to develop; gives rise to structures such as muscle and skeleton.

A body plan in which any plane along a central axis will divide the body into approximately mirror-image halves. Cnidarians and many adult echinoderms exhibit radial symmetry.

In animals with indirect development, a radical change in body form from larva to sexually mature adult, as seen in amphibians (e.g., tadpole to frog) and insects (e.g., caterpillar to butterfly).

An animal body plan in which the body is divided into repeated, typically similar units.

To shed an external body covering, such as an exoskeleton, skin, feathers, or fur.

A group of (normally similar) cells that together carry out a specific function, for example, muscle; may include extracellular material produced by its cells.

A paired neural structure, present in most animals, that conducts nervous signals to and from the ganglia; in chordates, the nerve cord lies along the dorsal side of the body and is also called the *spinal cord*.

An animal that has a vertebral column.

A type of circulatory system found in some invertebrates, such as arthropods and mollusks, that includes an open space (the hemocoel) in which blood directly bathes body tissues.

SELF TEST

1. Which of the following is NOT a characteristic of most animals?
   a. heterotrophism
   b. ability to move
   c. unicellularity
   d. sexual reproduction

2. Which of the following characteristics is NOT shared by plants and animals?
   a. multicellularity
   b. eukaryotic nature
   c. sexual reproduction
   d. heterotrophic nature

3. How does radial symmetry differ from bilateral symmetry?
   a. A radially symmetrical animal has dorsal and ventral surfaces, whereas bilaterally symmetrical animals do not.
   b. Radially symmetrical animals can be divided into symmetrical halves with many planes through the central axis, whereas bilaterally symmetrical animals can be divided into equal halves by only one specific plane through a central axis.
   c. Radially symmetrical animals possess three embryonic germ layers, whereas bilaterally symmetrical animals possess only two.
   d. Most radially symmetrical animals are active, free-moving organisms throughout their lives, whereas bilaterally symmetrical animals are not.

4. An animal with cephalization will have
   a. two embryonic germ layers.
   b. a complete, tubular digestive tract, with a mouth at one end and an anus at the other.
   c. sensory cells/organs and nerve cells clustered at the anterior end of the animal.
   d. a completely lined fluid-filled body cavity.

5. Which of the following is NOT a function associated with a body cavity?
   a. support
   b. digestion
   c. protection of internal organs
   d. allowing internal organs to operate independently of the body wall

6. Species of the animal phylum _____ are deuterostomes.
   a. Annelida
   b. Arthropoda
   c. Chordata
   d. Mollusca
   e. Nematoda

7. All of the following animal phyla have coeloms EXCEPT
   a. arthropods.
   b. nematodes.
   c. echinoderms.
   d. mollusks.
   e. annelids.

8. In an animal such as a human, which portion of the body is produced by the ectoderm?
   a. nerve
   b. muscle
   c. artery
   d. bone
   e. liver

9. Astronauts experience a weightless environment. Consider that humans have a coelomate body cavity arrangement, similar to that shown for the earthworm in the diagram below. Why do the internal organs of astronauts stay in their proper locations, even in the absence of weight? Label the diagram, drawing an arrow to the structure(s) that keep the organs of astronauts in place.

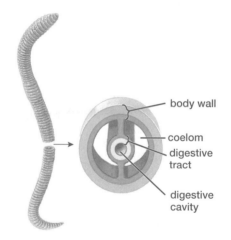

10. The vast majority of animals
    a. are vertebrates.
    b. lack tissues.
    c. are producers.
    d. lack a backbone.

11. A digestive cavity with a single opening is the characteristic digestive system of animals in the _____ phylum.
    a. Arthropoda
    b. Nematoda
    c. Mollusca
    d. Platyhelminthes
    e. Porifera

12. Which of the following is NOT associated with sponges?
    a. epithelial cells
    b. endoderm and ectoderm
    c. asexual reproduction
    d. collar cells
    e. active larvae

13. What is the excretory structure in annelid worms?
    a. anus
    b. kidney
    c. crop
    d. nephridium

14. What are the two functions of the gastrovascular cavity of a cnidarian?
    a. digestion and prey capture
    b. respiration and sexual reproduction
    c. digestion and distribution of nutrients
    d. movement and response to threats
    e. filtering of blood and removal of waste

15. Animals in the _____ phyla are segmented.
    a. Annelida
    b. Arthropoda
    c. Echinodermata
    d. Mollusca
    e. both a and b
    f. both a and d

16. Which of the following do you expect to be associated with, or characteristic of, a parasitic flatworm such as a tapeworm?
    a. hooks and suckers
    b. eyespots
    c. a respiratory system
    d. cilia
    e. a complex digestive system

17. A radula is a
    a. flexible supportive rod on the dorsal surface of chordates.
    b. stinging cell used by sea anemones to capture prey.
    c. gas-exchange structure in insects.
    d. spiny ribbon of tissue used for feeding in snails.
    e. locomotory structure of sea stars.

18. Annelids (segmented worms) have many structures comparable to those observed in vertebrates (such as humans). Which of the following is a vertebrate structure to which there is nothing comparable in an annelid?
    a. heart
    b. kidney
    c. lung
    d. brain

19. Which of the following is NOT a mollusk?
    a. barnacle
    b. clam
    c. octopus
    d. slug
    e. snail

20. Animals in the _____ phylum have collar cells.
    a. Porifera
    b. Cnidaria
    c. Annelida
    d. Mollusca
    e. Chordata

21. The success of the cephalopods as predators is supported by the presence of all of the following EXCEPT
    a. beaklike jaws.
    b. a complex eye.
    c. a large and complex brain.
    d. tentacles.
    e. a single large and prominent foot.

22. Which of the following animals molts its exoskeleton, allowing the animal to grow larger?
    a. blue crab
    b. bat sea star
    c. scallop
    d. sea urchin
    e. Venus clam

23. Besides being predators of fish and other marine animals, what is another way that some cnidarians obtain their nutrition?
    a. Some cnidarians are parasites that attach to larger marine animals and slowly drain nutrients from them.
    b. Some of the cnidarians are scavengers, detecting and moving toward dead animals that have sunk to the sea bottom, upon which they then feed.
    c. Some cnidarians are hosts to beneficial, photosynthetic dinoflagellates that live inside of their tissues and produce sugar that they share with their hosts.
    d. Some especially large cnidarians filter single-celled organisms out of the water, straining tons of water daily to obtain only a few grams of food.

24. What is one way in which the sea jellies and comb jellies differ?
    a. Only the sea jellies have stinging cells.
    b. Only the comb jellies have tentacles.
    c. Only the sea jellies are predators.
    d. Only the comb jellies have bodies as large as 1 meter across.

25. Examine the diagram of the five-armed sea star, and consider its radial symmetry. Imagine that the sea star is resting on a flat and level surface, and its arms are pointing straight out from the center, and equally spaced from each other. In this case, exactly how many vertical planes may be passed through the sea star in ways that will divide it into mirror-image halves?
    a. an infinite number
    b. 10
    c. 5
    d. 2
    e. 1

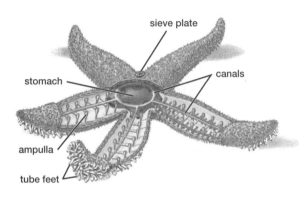

26. What is one way in which humans may become infected with both parasitic tapeworms and *Trichinella* roundworms?
    a. by swimming in water that is contaminated with larval tapeworms and roundworms
    b. by having sexual contact with someone who is already infested with tapeworms and roundworms
    c. by eating a salad with contaminated lettuce that has not been thoroughly washed
    d. by eating undercooked pork that is infested with cysts of tapeworms and roundworms

27. Which is considered an example of a social insect?
    a. a beetle
    b. a butterfly
    c. a housefly
    d. an ant
    e. a cricket

28. Which of the following statements is NOT true about the myriapods?
    a. Despite what is implied by their names, centipedes actually have more legs than millipedes.
    b. All species live on land.
    c. Centipedes have one pair of legs per body segment, whereas millipedes have two pairs per segment.
    d. Centipedes are predators, with poisonous claws.
    e. Myriapods have eyes that are of very poor quality, when compared to the eyes of a dragonfly.

29. Which of the following statements is NOT true about the arachnids?
    a. Scorpions and spiders have eight legs.
    b. The strongest spider silk has about the same strength as cotton thread.
    c. Spiders liquefy prey animals before eating them.
    d. Unlike the insects, arachnids do not have compound eyes.
    e. Not all spiders make webs.

30. Regarding giant squids (*Architeuthis*), of which aspect of their lives do we have the best understanding?
    a. how long they live
    b. how they grasp their prey
    c. whether they are more solitary or social
    d. their mating behavior
    e. what they prefer to eat

ANSWER KEY

1. c
2. d
3. b
4. c
5. b
6. c
7. b
8. a
9.

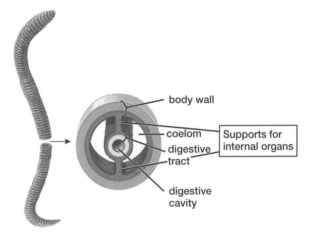

10. d
11. d
12. b
13. d
14. c
15. e
16. a
17. d
18. c
19. a
20. a
21. e
22. a
23. c
24. a
25. c
26. d
27. d
28. a
29. b
30. b

CHAPTER 24 ANIMAL DIVERSITY II: VERTEBRATES

OUTLINE

Section 24.1 What Are the Key Features of Chordates?

Animals of the phylum Chordata have the following features at some point in their development: (1) a **notochord**; (2) a dorsal, hollow **nerve cord**; (3) **pharyngeal gill slits**; and (4) a **post-anal tail**.

Section 24.2 Which Clades Make Up the Chordates?

Invertebrate chordates include the tunicates, lancelets (**Figure 24-3**), and hagfishes.

Lancelets are small chordates that resemble fish and have all four defining chordate features as adults. They feed by filtering particles from the seawater. Tunicates have the four defining chordate features as mobile larvae, but most become sessile and take on a basket-like shape as adults.

Hagfish (class Myxini, **Figure 24-4**) are **craniates**, having a cartilaginous skull, but lacking vertebrae. Thus, they are not vertebrates. They reside on the ocean floor and produce copious amounts of slime as a defense against predators.

Vertebrate chordates have a **vertebral column** composed of either **cartilage** or bone, which is part of their living endoskeleton. All vertebrates have a skull composed of either cartilage or bone.

Section 24.3 What Are the Major Groups of Vertebrates?

Lampreys (class Petromyzontiformes) are jawless vertebrates that often parasitize fish (**Figure 24-5**).

Jawed fishes can grasp, tear, and crush food. They consist of cartilaginous fishes, ray-finned fish, and lobe-finned fish. Cartilaginous fish (sharks, skates, and rays, class Chondrichthyes) are marine predators that lack bony skeletons (**Figure 24-6**). Ray-finned fish (class Actinopterygii) are the most diverse vertebrates (**Figure 24-7**) and have webbed fins supported by spines. Lobe-finned fish have fleshy fins made of bone and muscle. Some lobe-finned fish (such as lungfish) have both gills and primitive lungs and are the most closely related to land-dwelling vertebrates (**Figure 24-8**).

Amphibians (class Amphibia) have limbs and simple lungs, allowing for a more terrestrial existence (**Figure 24-9**). Most amphibians need to live near damp environments to keep their permeable skin moist and to facilitate external fertilization and the development of aquatic eggs and larvae.

Reptiles (class Reptilia) have well-developed lungs; scaly, waterproof skin; and shelled, **amniotic eggs** that cause them to be well adapted for a terrestrial lifestyle (**Figures 24-10** and **24-11**). Reptile groups include lizards, snakes, crocodilians, turtles, and birds. Extinct reptiles also include the dinosaurs.

Birds are a group of terrestrial reptiles that have endothermic bodies modified for flight (**Figures 24-12** and **24-13**). These modifications include feathers, hollow bones, well-developed respiratory systems, and well-developed circulatory systems. Birds are warm blooded (endothermic) and have very high metabolic rates.

Mammals (class Mammalia) are endotherms that have hair and well-developed nervous systems and use **mammary glands** to provide milk to their young. Mammal groups include the **monotremes** (egg-laying mammals such as platypuses, **Figure 24-14**), **marsupials** (pouched mammals such as wallabies, **Figure 24-15**), and **placental** mammals (which retain young within the uterus for a very long time, such as cheetahs, **Figure 24-16**). Marsupials are the dominant native mammals of Australia, whereas placental mammals dominate in other parts of the world.

FLASH CARDS

To use the flash cards, tear the page from the book and cut along the dashed lines. The key term appears on one side of the flash card, and its definition appears on the opposite side.

amnion	monotreme
amniotic egg	nerve cord
cartilage	notochord
craniate	pharyngeal gill slit
mammary gland	placenta
marsupial	placental

A member of the clade Monotremata, which includes mammals that lay eggs; platypuses and spiny anteaters are monotremes.

One of the embryonic membranes of reptiles (including birds) and mammals; encloses a fluid-filled cavity that envelops the embryo.

A paired neural structure, present in most animals, that conducts nervous signals to and from the ganglia; in chordates, the nerve cord lies along the dorsal side of the body and is also called the *spinal cord*.

The egg of reptiles, including birds; contains a membrane, the amnion, that surrounds the embryo, enclosing it in a watery environment and allowing the egg to be laid on dry land.

A stiff but somewhat flexible, supportive rod that extends along the head-to-tail axis and is found in all members of the phylum Chordata at some stage of development.

A form of connective tissue that forms portions of the skeleton; consists of chondrocytes and their extracellular secretion of collagen; resembles flexible bone.

One of a series of openings, located just posterior to the mouth, that connects the throat to the outside environment; present (as some stage of life) in all chordates.

An animal that has a skull.

In mammals, a structure formed by a complex interweaving of the uterine lining and the embryonic membranes, especially the chorion; functions in gas, nutrient, and waste exchange between embryonic and maternal circulatory systems and secretes hormones.

A milk-producing gland used by female mammals to nourish their young.

Referring to a member of a group of mammal species in which the placenta is especially complex. All mammals except marsupials and monotremes are placental mammals.

A member of the clade Marsupialia, which includes mammals whose young are born at an extremely immature stage and undergo further development in a pouch, where they remain attached to a mammary gland; kangaroos, opossums, and koalas are marsupials.

post-anal tail

vertebrate

vertebral column

An animal that has a vertebral column.

A tail that extends beyond the anus; found in all chordates at some stage of development.

A column of serially arranged skeletal units (the vertebrae) that enclose the nerve cord in vertebrates; the backbone.

SELF TEST

1. Species within the animal phylum (or phyla) _____ are deuterostomes.
   a. Annelida
   b. Arthropoda
   c. Chordata
   d. Echinodermata
   e. both a and b
   f. both c and d

2. All members of the phylum Chordata, whether human or lancelet, share certain key features. Which of the following traits is NOT characteristic of all chordates?
   a. dorsal, hollow nerve cord
   b. notochord
   c. pharyngeal gill slits
   d. tail
   e. bony endoskeleton

3. The only chordate feature present in adult humans is the
   a. post-anal tail.
   b. dorsal, hollow nerve cord.
   c. pharyngeal gill slits.
   d. notochord.

4. Invertebrate chordates differ from vertebrate chordates in their lack of
   a. pharyngeal gill slits.
   b. a post-anal tail.
   c. a backbone.
   d. a dorsal, hollow nerve cord.

5. Study the chordate evolutionary tree below. It shows that the first chordates to possess jaws were early members of the
   a. lancelets (Cephalochordata).
   b. cartilaginous fishes (Chondrichthyes).
   c. ray-finned fishes (Actinopterygii).
   d. coelacanths (Actinistia).

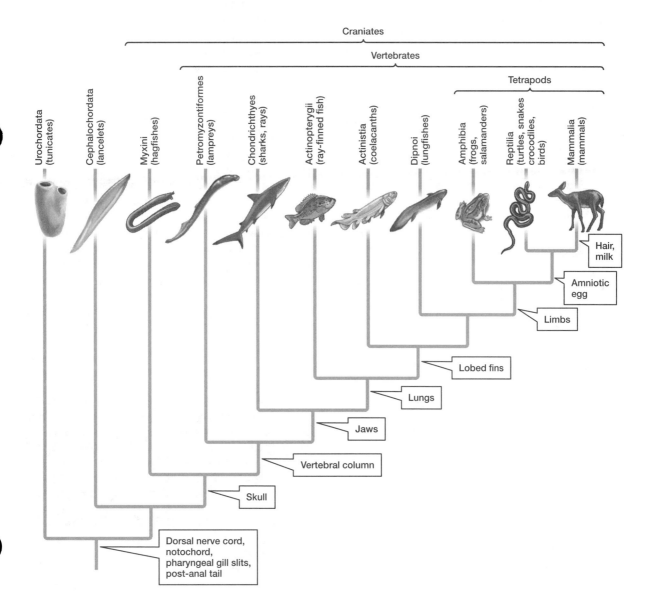

6. What of the following statements is true about craniates?
   a. They all have a bony skull.
   b. They all have a vertebral column.
   c. Some do not have a brain.
   d. Hagfishes represent the most primitive craniates.

7. What is one way to distinguish between the lancelets and the tunicates?
   a. Lancelets display the defining four chordate features in adulthood, whereas tunicates display the same features in their larval stage but lose some of them in adulthood.
   b. Adult lancelets are sessile (firmly attached to surfaces such as rocks), whereas adult tunicates remain highly active and mobile.
   c. Adult lancelets lack the notochord that their larval stage possesses, whereas adult tunicates lack the pharyngeal slits that their larval stage possesses.
   d. Lancelets have a cartilaginous skeleton as adults, whereas tunicates have such a skeleton in their larval stage, but lose it in their adult stage.

8. What of the following statements about vertebrates is NOT true?
   a. Vertebrae may consist of either bone or cartilage.
   b. The backbone provides protection for nervous tissue.
   c. The greatest number of vertebrate species possess three-chambered hearts.
   d. Some of the earliest vertebrates were large, armored fish.
   e. Lampreys represent the most primitive vertebrates.

9. Which vertebrate group is the most diverse but is often overlooked because of humans' habitat bias?
   a. bony fish
   b. jawless fish
   c. mammals
   d. birds

10. Cartilaginous fish are characterized by
    a. a three-chambered heart.
    b. poorly developed lungs.
    c. a skeleton formed entirely of cartilage.
    d. milk-producing mammary glands.

11. Which of these does not belong to the same phylum as the other four?
    a. caecelian
    b. condor
    c. cephalopod
    d. coelacanth
    e. chondrichthyes

12. What defines, or distinguishes, a mammal from other vertebrates?
    a. its hairless exterior
    b. its primitive, simple brain
    c. milk-producing glands
    d. the fact that most mammals complete the great majority of their development outside the uterus

13. Amphibians are most like
    a. mosses.
    b. flowering plants.
    c. conifers.
    d. ferns.

14. A long period of uterine development and gas, nutrient, and waste exchange between the mother and embryo are characteristic of
    a. all mammals.
    b. birds.
    c. marsupials.
    d. placental mammals.
    e. monotremes.

15. Reptiles are well adapted to living in drier habitats because of their
    a. hollow bones.
    b. production of a shelled amniotic egg.
    c. two-chambered heart.
    d. moist skin used as a supplemental respiratory organ.
    e. external fertilization.

16. Study the image of the crocodile hatching from its egg. What would NOT be a difference between this image and a similar image of a salamander hatching from its egg?
    a. The salamander egg would be gelatinous, unlike the hard-shelled, amniotic egg of the crocodile.
    b. The salamander egg would be underwater, unlike the crocodile egg.
    c. The salamander would have moist, gas-permeable skin, unlike the tough and scaly skin of the crocodile.
    d. The salamander would not be a tetrapod, whereas the crocodile is a tetrapod.

17. Lampreys have all of the following features EXCEPT
    a. a skull.
    b. fleshy, muscular fins.
    c. vertebrae.
    d. three embryonic tissue layers.
    e. a dorsal nerve cord.

18. Which of the following is a group of invertebrates?
    a. Reptilia
    b. Amphibia
    c. Mammalia
    d. Aves
    e. Echinodermata

19. Which of the following characteristics are shared by both arthropods and mammals?
    a. a well-developed nervous system
    b. a closed circulatory system
    c. an internal skeleton
    d. compound eyes

20. Class Chondrichthyes includes
    a. whales.
    b. lampreys.
    c. all fish.
    d. frogs.
    e. sharks.

21. The terrestrial vertebrates that may be indicators of environmental degradation are
    a. amphibians.   c. lancelets.
    b. bony fishes.   d. reptiles.

22. What substance mostly fills the cranial space of a coelacanth?
    a. bone   c. fat
    b. brain   d. blood

23. The great size and mobility of vertebrates is associated with
    a. four-chambered hearts.
    b. lungs used for respiration.
    c. lightweight endoskeletons.
    d. uterine development of offspring.
    e. increased brain size and complexity.

24. Which of the following statements is NOT true about lungfishes and lizards?
    a. Both are able to breathe air.
    b. Both lay their eggs on land.
    c. Both have muscular, bony appendages.
    d. Both are ectotherms.

25. Which of the following has a ventral nerve cord?
    a. earthworm
    b. shark

    c. coelacanth
    d. frog
    e. hummingbird

26. The ability of birds to fly is facilitated by all of the following, EXCEPT their
    a. ectothermic nature.
    b. lungs supplemented by air sacs.
    c. external development in a shelled egg.
    d. hollow bones.
    e. four-chambered heart.

27. An animal's ability to live successfully on land is increased by
    a. external fertilization.
    b. a two-chambered heart.
    c. moist skin used for gas exchange.
    d. gills for breathing.
    e. development in a shelled egg.

28. The high body temperature of birds and mammals is due to
    a. lots of energy lost as heat during metabolism.
    b. the presence of sweat, scent, and sebaceous glands.
    c. the fur that covers and insulates them.
    d. behaviors such as basking in the sun or seeking shade.
    e. the exchange of gases and nutrients via the placenta.

29. Reptilian embryos will not dry out in a desert habitat because
    a. reptiles produce lots of defensive slime.
    b. reptiles' eggs are protected by a jelly-like coating.
    c. reptiles' placentas facilitate exchanges between mother and embryo.
    d. reptiles produce shelled amniotic eggs.
    e. reptiles are warm blooded.

30. Many fish live in the deep ocean and have balloon-like swim bladders. What would happen to such a fish if it was caught in the net of a trawler and rapidly raised to the surface?
    a. It would swell up.
    b. It would shrink.
    c. It would stay about the same size.
    d. There is no way to tell from the information given.

ANSWER KEY

1. f		16. d	
2. e		17. b	
3. b		18. e	
4. c		19. a	
5. b		20. e	
6. d		21. a	
7. a		22. c	
8. c		23. c	
9. a		24. b	
10. c		25. a	
11. c		26. a	
12. c		27. e	
13. a		28. a	
14. d		29. d	
15. b		30. a	

CHAPTER 25 ANIMAL BEHAVIOR

OUTLINE

Section 25.1 How Do Innate and Learned Behaviors Differ?

Innate behaviors can be performed without prior experience (**Figure 25-1**), whereas **learned behaviors** are modified by experience. Learned behaviors include **habituation**, **trial-and-error learning**, and **insight learning** (**Figures 25-2** and **25-3**).

Learned behaviors can modify innate behaviors to make them more appropriate for an animal's specific situation (**Figure 25-5**).

Some forms of learning (e.g., **imprinting**) occur only within innate constraints.

Section 25.2 How Do Animals Communicate?

An animal communicates by producing signals that alter another animal's behavior in a beneficial way for the animal producing the signal.

Visual **communication** is effective over short distances, whereas communication by sound is effective over long distances (**Figures 25-8, 25-9, 25-10,** and **25-11**). Both methods convey information rapidly.

Chemical communication (such as by **pheromones**) is also effective over long distances, but it persists longer and is difficult to vary in intensity (**Figure 25-12**).

Communication by touch is a method of forming social bonds among members of an animal group and can be used by some animals for sexual communication (**Figure 25-13**).

Section 25.3 How Do Animals Compete for Resources?

Aggressive behavior or **aggression** (e.g., visual and vocal displays or ritualized combat) helps secure resources from members of the same species (**Figures 25-14** and **25-15**). These displays rarely result in serious injury.

Dominance hierarchies are used to manage aggressive interactions, which can disrupt other important tasks (**Figure 25-16**). Dominant animals generally have access to the most resources and/or mates.

Territoriality minimizes aggressive encounters when animals manage and defend boundaries of specific areas that contain resources (**Figure 25-18**).

Section 25.4 How Do Animals Find Mates?

For reproduction to occur, potential mates must first recognize each other as members of the same species and opposite sex and also recognize that they are sexually receptive.

Finding mates is accomplished by acoustic mating signals, visual mating signals (**Figures 25-21, 25-22, and 25-23**), and chemical mating signals (**Figure 25-24**).

Section 25.5 Why Do Animals Play?

Many animals play, either alone or in social groups (**Figure 25-25**). Play usually utilizes behaviors that are essential in the day-to-day life of the organism (feeding, hunting, defense), and animals will discontinue play in order to find food or mates. Because individuals can be distracted by play, it can be very dangerous.

Play does have evolutionary value because (1) it can provide practice of important behaviors and (2) it reinforces and teaches behaviors that are important as adults.

Section 25.6 What Kinds of Societies Do Animals Form?

Living in a group has both advantages and disadvantages, with the degree of sociality varying among species.

Some species, particularly insects and mammals, form complex societies that may sacrifice individuals for the good of the group (**altruism, Figure 25-26**).

Honeybees and naked mole rats have formed particularly complex societies whose members follow specific roles throughout their lives (**Figures 25-27** and **25-28**).

Section 25.7 Can Biology Explain Human Behavior?

Studies on newborn infants suggest strong correlations between some behaviors and physiology (e.g., suckling and the sounds of a mother's voice, **Figure 25-30**).

Cross-cultural behavioral comparisons suggest an innate human signaling system, as demonstrated, for example, by facial expressions.

Studies on identical twins can reveal the genetic basis for human behavior, especially in twins who were separated at birth.

FLASH CARDS

To use the flash cards, tear the page from the book and cut along the dashed lines. The key term appears on one side of the flash card, and its definition appears on the opposite side.

aggression	habituation
altruism	imprinting
behavior	innate
communication	insight learning
dominance hierarchy	kin selection
ethology	operant conditioning

A type of simple learning characterized by a decline in response to a repeated stimulus.

Antagonistic behavior, normally among members of the same species, often resulting from competition for resources.

A type of learning in which an animal acquires a particular type of information during a specific sensitive phase of development.

A behavior that benefits other individuals while reducing the fitness of the individual that performs the behavior.

Inborn; instinctive; an innate behavior is performed correctly the first time it is attempted.

Any observable activity of a living animal.

A type of learning in which a problem is solved by understanding the relationships among the components of the problem rather than through trial and error.

The act of producing a signal that causes a receiver, normally another animal of the same species, to change its behavior in a way that is, on average, beneficial to both signaler and receiver.

A type of natural selection that favors traits that enhance the survival or reproduction of an individual's relatives, even if the traits reduce the fitness of the individuals bearing them.

A social structure that arises when the animals in a social group establish individual ranks that determine access to resources; ranks are usually established through aggressive interactions.

A laboratory training procedure in which an animal learns to perform a response (such as pressing a lever) through reward or punishment.

The study of animal behavior.

pheromone

trial-and-error learning

territoriality

waggle dance

A type of learning in which behavior is modified in response to the positive or negative consequences of an action.

A chemical produced by an organism that alters the behavior or physiological state of another member of the same species.

A symbolic form of communication used by honeybee foragers to communicate the location of a food source to their hive mates.

The defense of an area in which important resources are located.

SELF TEST

1. Innate behavior is
   a. any observable response to external or internal stimuli.
   b. a behavior performed reasonably completely the first time.
   c. behavior that is changed on the basis of experience.
   d. a change in the speed of random movements.
   e. a directed movement toward or away from a stimulus.

2. Learned behavior is
   a. any observable response to external or internal stimuli.
   b. a behavior performed reasonably completely for the first time.
   c. behavior that is changed on the basis of experience.
   d. a change in the speed of random movements.
   e. a directed movement toward or away from a stimulus.

3. Which of the following is NOT a type of innate behavior?
   a. movements
   b. fixed action patterns
   c. feeding
   d. habituation

4. The ability to change a behavior as a result of new experiences (i.e., learning) is most closely associated with which of the following behaviors?
   a. instincts   c. reflexes
   b. habituation   d. all of the above

5. Fish hatchery workers know that if they want artificially spawned salmon to return to a specific stream when they reach adulthood, they have to raise fry (baby salmon) in water from that stream during a certain critical time in the baby salmon's early development. This is a clear demonstration of
   a. operant conditioning.
   b. trial-and-error learning.
   c. habituation.
   d. imprinting.
   e. none of the above.

6. Imprinting
   a. takes place only in a sensitive period of the organism.
   b. is preprogrammed.
   c. takes place only in birds.
   d. is a method for maintaining territories for feeding.

7. In reference to a bee's waggle dance described in the figure below, if Bee A completes a dance on a horizontal surface and circles at a slow rate, and Bee B completes a dance on a vertical surface at a slow rate, then you could say that
   a. the food source for Bee A is farther than the food source for Bee B.
   b. the distance to the food source may be the same, but the food source of Bee A is harder to get to.
   c. the distance may be the same, but the two bees are describing different locations for the food sources.
   d. distance cannot be determined, and the direction could be the same if the angle of Bee B is the same as that of Bee A.

If the dance is performed on a vertical wall inside the hive, the angle (from vertical) of the waggle run represents the angle between the sun and the food source

up

40°

40°

The rate of circling communicates the distance to the food source

If the dance is performed on a horizontal surface outside, the waggle run is aimed at the food source

8. Pheromones are
   a. behaviors that utilize fixed action patterns.
   b. chemicals that are given off by all organisms except humans.
   c. a way for organisms to communicate.
   d. chemicals that signal a female to begin nesting behaviors.

9. Social leaf-cutter ants lay down pheromone trails to lead other members of their colony to a rich food source they have found. This behavior is a classic example of
   a. communication.
   b. habituation.
   c. insight learning.
   d. social hierarchy.

10. Territoriality
    a. is a behavior in which an animal shares resources with another of its species.
    b. increases the aggressive encounters an individual is likely to have.
    c. usually leads to a reduction in fitness.
    d. involves the active defense of an area containing important resources.

11. Trimming the feathers on the left side of a peacock's tail but leaving the right side unaltered would MOST likely _____ the male's chance of mating.
    a. increase
    b. decrease
    c. have no effect on

12. Which of the following is UNLIKELY to be a consequence of sociality?
    a. conservation of energy
    b. ease of finding mates
    c. increased access to limited resources such as food or nest sites
    d. an increase in foraging efficiency
    e. an increased ability to deter predators

13. An individual benefits from a social grouping with other animals through increased
    a. ability to detect, repel, or confuse predators.
    b. hunting efficiency.
    c. likelihood of finding mates.
    d. efficiency resulting from division of labor.
    e. all of the above.

14. Which of the following is an advantage of using identical twins in studies on human behavior?
    a. You get twice as much information from twins compared to nontwins.
    b. Twins get along well.
    c. Genetic influences on behavior can be factored out.
    d. They allow you to factor out environmental influences on behavior.

15. Which of the following is NOT a feature of play?
    a. It occurs more often in young animals than in adults.
    b. It may borrow movements from other behaviors.
    c. It uses considerable energy.
    d. It always has a clear, immediate function.
    e. It is potentially dangerous.

16. Grasshopper mice (genus Onychomys) are insectivores. Among their prey are beetles that, when threatened, elevate their rear ends and eject a spray of acetic acid into the faces of potential predators. Adult grasshopper mice avoid this by plunging the beetle's rear end into the ground and biting off its head. Juvenile grasshopper mice, when first exposed to these beetles, perform the "jam and bite" behavior but do not always get the right end of the beetle up! Which of the following is the best explanation for this feeding behavior?
    a. The "jam and bite" behavior is innate but modified by learning.
    b. The behavior is a fixed action pattern.
    c. The behavior is completely learned.

17. Scientists think that play occurs in order to
    a. keep the young occupied and out of danger from predators.
    b. develop bonds among closely related individuals.
    c. prevent disease by exposure to antibodies carried by closely related individuals.
    d. practice behaviors that will be needed as an adult for survival and feeding.

18. Females choosing males based on ornamentation are selecting partially based on
    a. innate responses to the ornamentation of the male.
    b. the male being at the right place at the right time.
    c. the pheromones associated with the ornamentation or nest building of the male.
    d. symmetry in the ornamentation or nest building of the male.

19. Which of the following is NOT likely to be a behavior that can be explained in biological terms?
    a. similar behaviors in identical twins
    b. responses to pheromones among members of the same species
    c. avoidance of mating with closely related individuals
    d. all of the above
    e. none of the above

20. The behavior of the frog in the figure at right is an example of which type of behavior?
    a. trial-and-error learning
    b. innate behavior
    c. imprinting
    d. mimicry

21. The hybrid offspring produced by mating a member of the eastern European population of blackcaps (a bird) to a member of the western European population of the same species follow a migration route intermediate between that of their parents. This occurs even if the offspring are raised in isolation. From this information, we can reasonably conclude that migration in this species
    a. uses the stars to determine direction.
    b. is a purely learned behavior.
    c. is controlled entirely by a single gene pair.
    d. is imprinted on the birds when they are very young.
    e. has a genetic component.

22. An example of a dominance hierarchy would be
    a. two male deer vying for the right to mate with the females in a herd.
    b. a female black bear fighting an intruder.
    c. ground squirrels calling to other members of the colony when a predator approaches.
    d. a pack of wolves in which only the top female mates and produces young.

23. A disadvantage of male frogs using loud calls to attract a female for mating would be that
    a. it will also attract predators.
    b. females may not be receptive at the times the males are calling.
    c. females may not be able to find the male if only auditory and no visual cues are used.
    d. females must learn the call of the potential mate before she will be receptive to his call.

24. Which of the following types of communication would be LEAST appropriate in the dark confines of a beehive?
    a. chemical messages such as pheromones
    b. sound signals such as buzzing
    c. touch
    d. visual signals

25. The basis for viewer interest in the popular television show *Survivor*, as well as its spinoffs, stems from our fascination with which of the following behaviors?
    a. territoriality
    b. dominance hierarchy
    c. aggression
    d. all of the above

❶ A naive toad is presented with a bee.

❷ While trying to eat the bee, the toad is stung painfully on the tongue.

❸ Presented with a harmless robber fly, which resembles a bee, the toad cringes.

❹ The toad is presented with a dragonfly.

❺ The toad immediately eats the dragonfly, demonstrating that the learned aversion is specific to bees and insects resembling bees.

26. Which of the following is an important function of mating behaviors?
    a. to communicate species identity
    b. to communicate gender
    c. to communicate sexual receptivity
    d. to defuse aggressive responses
    e. all of the above

27. Which of the following conditions would be MOST likely to force animals to associate into loosely organized social groupings?
    a. when weather conditions are extreme
    b. when prey numbers are low but predator numbers are high
    c. when predator numbers decrease but prey numbers remain high
    d. both a and b
    e. all of the above

28. Small birds will often mob large predatory birds such as owls and hawks. This behavior involves the smaller birds gathering or flying near the larger bird, calling loudly, and even striking the larger bird and may result in the larger bird leaving the area. This behavior is risky for the smaller birds but is commonly seen. From this description, we can reasonably conclude that

a. birds are not intelligent enough to know what is good for them.
b. there must be some advantage of this behavior, such as protecting their chicks, that makes the risks worthwhile.
c. these birds must have learned the behavior by operant conditioning.
d. this behavior is clearly maladaptive and should be eliminated by natural selection.

29. Which of the following would NOT be an example of aggressive behavior?
    a. pheromones released on a trail by ants
    b. the baring of fangs by a wolf
    c. the clashing of antlers of male deer trying to win the rights to mate with a female
    d. the raising of quills on the back of a porcupine

30. Territories are important when
    a. there is a limited amount of a particular resource.
    b. there are more males than females in a habitat.
    c. animals rely on imprinting for survival of young.
    d. animals must migrate long distances for food or nesting sites.

ANSWER KEY

1. b
2. c
3. d
4. b
5. d
6. a
7. c
8. b
9. a
10. d
11. b
12. 2
13. e
14. c
15. d
16. a
17. d
18. d
19. e
20. a
21. e
22. f
23. a
24. d
25. d
26. e
27. d
28. b
29. a
30. a

Chapter 26 Population Growth and Regulation

OUTLINE

Section 26.1 How Does Population Size Change?

A **population** is a group of organisms of the same species that live within the same ecosystem. Populations grow when the number of births and immigrants exceeds that of deaths and emigrants. Populations decline when the reverse is true.

The size of a stable population is regulated by (1) the maximum rate a population can increase under ideal conditions (**biotic potential**) and (2) the abiotic and biotic limits on population growth (**environmental resistance**).

Unchecked, the biotic potential of a population will result in **exponential growth** (**Figure 26-2**). Biotic potential is influenced by (1) an organism's earliest reproductive age, (2) the frequency of reproduction, (3) the number of offspring produced per reproductive event, (4) an organism's reproductive life span, and (5) the **death rate** under ideal conditions (**Figure 26-3**).

Section 26.2 How Is Population Growth Regulated?

A population undergoing exponential growth will either stabilize or undergo **boom-and-bust cycles** due to the influence of environmental resistance (**Figures 26-4** and **26-5**). Exponential growth occurs most commonly when organisms invade new habitats with abundant resources, which is commonly observed with **invasive species**.

Environmental resistance restrains population growth by increasing the death rate and decreasing the birth rate, resulting in **logistic population growth** (**Figures 26-6, 26-8,** and **26-9**).

The formula for logistic growth includes a variable for **carrying capacity (K)** (**Figure 26-6a**), which is defined as the maximum sustainable population size for an ecosystem. If K is exceeded, population size may (1) oscillate around K, (2) crash and stabilize at a lower K, or (3) reach zero, resulting in the elimination of that population (**Figure 26-6b**).

Density-independent factors, such as climate and weather, limit populations independently of population density.

Density-dependent factors, such as predation, parasitism, and competition, limit populations more effectively as population density increases (**Figures 26-9 to 26-12**).

Section 26.3 How Are Populations Distributed in Space and Time?

Populations exhibit three different spatial distributions: (1) **clumped**, (2) **uniform**, and (3) **random** (**Figure 26-14**). Members of a population with a clumped distribution live in groups. Members of a population with a uniform distribution maintain relatively constant distances between individuals. Members of a population with a random distribution do not form social groups and have resources available throughout their habitat.

Populations show differences in the likelihood of survival (survivorship) at different ages, which can be represented as **survivorship curves** (**Figure 26-15**). Survivorship curves for **late-loss populations** are convex and represent populations in which individuals produce few offspring with low juvenile death rates. Survivorship curves for **constant-loss populations** have relatively constant slopes and represent populations in which individuals have an equal chance of dying at any time during their life span. Survivorship curves that are concave and represent **early-loss populations** where individuals in the population produce large numbers of offspring that have a low chance of survival.

Section 26.4 How Is the Human Population Changing?

The human population growth rate is currently following a J-shaped exponential model (**Figure 26-16**), primarily due to high birth rates and the technical, cultural, agricultural, and industrial-medical revolutions that have occurred throughout human history.

Age structure diagrams show human age groups for males and females in different populations. Expanding populations (as in Mexico) are represented by pyramidal age structure diagrams (**Figure 26-19a**). Stable populations (as in Sweden) are represented by column-shaped age structure diagrams (**Figure 26-19b**). Shrinking populations (as in Italy) are represented by age structure diagrams that are constricted at the base (**Figure 26-19c**).

Most humans live in **developing countries** with growing populations (**Figure 26-18**). Although the birth rates of many of these populations have declined, they are still going through **demographic transition**, so their populations continue to grow (**Figure 26-17**).

The population of the United States is the fastest growing of all developed nations, primarily due to high **immigration** rates and birth rates (**Figure 26-22**). This has significant environmental implications because the average U.S. citizen uses five times the energy as a citizen of other countries.

FLASH CARDS

To use the flash cards, tear the page from the book and cut along the dashed lines. The key term appears on one side of the flash card, and its definition appears on the opposite side.

age structure	community
biosphere	competition
biotic potential	constant-loss population
birth rate	contest competition
boom-and-bust cycle	death rate
carrying capacity (*K*)	demographic transition
clumped distribution	demography

All the interacting populations within an ecosystem.

The distribution of males and females in a population according to age groups.

Interaction among individuals who attempt to utilize a resource (for example, food or space) that is limited relative to the demand for it.

That part of Earth inhabited by living organisms; includes both living and nonliving components.

A population characterized by a relatively constant death rate; constant-loss populations have a roughly linear survivorship curve.

The maximum rate at which a population is able to increase, assuming ideal conditions that allow a maximum birth rate and minimum death rate.

A mechanism for resolving intraspecific competition by using social or chemical interactions to limit access of some individuals to important limited resources, such as food, mates, or territories.

The number of births per individual in a specified unit of time, such as a year.

The number of deaths per individual in a specified unit of time, such as a year.

A population cycle characterized by rapid exponential growth followed by a sudden massive die-off; seen in seasonal species, such as many insects living in temperate climates, and in some populations of small rodents, such as lemmings.

A change in population dynamic in which a fairly stable population with both high birth rates and high death rates experiences rapid growth as death rates decline, and then returns to a stable (although much larger) population as birth rates decline.

The maximum population size that an ecosystem can support for a long period of time without damaging the ecosystem; determined primarily by the availability of space, nutrients, water, and light.

The study of the changes in human numbers over time, grouped by world regions, age, sex, educational levels, and other variables.

The distribution characteristic of populations in which individuals are clustered into groups; the groups may be social or based on the need for a localized resource.

density-dependent	ecosystem
density-independent	emigration
developed country	environmental resistance
developing country	exponential growth
early-loss population	growth rate
ecological footprint	host
ecology	immigration

All the organisms and their nonliving environment within a defined area.	Referring to any factor, such as predation, that limits population size to an increasing extent as the population density increases.
Migration of individuals out of an area.	Referring to any factor, such floods or fires, that limits a population's size regardless of its density.
Any factor that tends to counteract biotic potential, limiting population growth and the resulting population size.	A country (including Australia, New Zealand, Japan, and most countries in North America and Europe) that benefits from a relatively high average standard of living, with access to modern technology and medical care.
A continuously accelerating increase in population size; population growth is a function of the current population size; this type of growth generates a curve shaped like the letter "J."	A country (including most in Central and South America, Asia, and Africa) that has a relatively low average living standard and limited access to modern technology and medical care.
A measure of the change in population size per individual per unit of time.	A population characterized by a high birth rate, a high death rate among juveniles, and lower death rates among adults; early-loss populations have a concave survivorship curve.
The prey organism on or in which a parasite lives; the host is harmed by the relationship.	The area of productive land needed to produce the resources used and absorb the wastes (including carbon dioxide) generated by an individual person, or by an average person of a region or the world, using current technologies.
Migration of individuals into an area.	The study of the interrelationships of organisms with each other and with their nonliving environment.

interspecific competition	population
intraspecific competition	population cycle
invasive species	predator
J-curve	prey
late-loss population	random distribution
logistic population growth	replacement-level fertility (RLF)
parasite	scramble competition

All the members of a particular species within an ecosystem, found in the same time and place and actually or potentially interbreeding.	Competition among individuals of different species.
Regularly recurring, cyclic changes in population size.	Competition among individuals of the same species.
An organism that eats other organisms.	Organisms with a high biotic potential that are introduced (deliberately or accidentally) into ecosystems where they did not evolve, and where they encounter little environmental resistance and tend to displace native species.
Organisms that are eaten, and often killed, by another organism (a predator).	The J-shaped growth curve of an exponentially growing population in which increasing numbers of individuals join the population during each succeeding time period.
The distribution characteristic of populations in which the probability of finding an individual is equal in all parts of an area.	A population in which most individuals survive into adulthood; late-loss populations have a convex survivorship curve.
The average birth rate at which a reproducing population exactly replaces itself during its lifetime.	Population growth characterized by an early exponential growth phase, followed by slower growth as the population approaches its carrying capacity, and finally reaching a stable population at the carrying capacity of the environment; this type of growth generates a curve shaped like a stretched-out letter "S."
A free-for-all scramble for limited resources among individuals of the same species.	An organism that lives in or on a larger organism (its host), harming the host but usually not killing it immediately.

S-curve

survivorship table

survivorship curve

uniform distribution

A data table that groups organisms born at the same time and tracks them throughout their life span, recording how many continue to survive in each succeeding year (or other unit of time). Various parameters such as sex may be used in the groupings. Human life tables may include many other parameters (such as socioeconomic status) used by demographers.

The S-shaped growth curve produced by logistic population growth, usually describing a population of organisms introduced into a new area; consists of an initial period of exponential growth, followed by a decreasing growth rate, and finally, relative stability around a growth rate of zero.

The distribution characteristic of a population with a relatively regular spacing of individuals, commonly as a result of territorial behavior.

The curve that results when the number of individuals of each age in a population is graphed against their age, usually expressed as a percentage of their maximum life span.

SELF TEST

1. Which of the following would NOT be a suitable research project for an ecologist?
   a. the effect of parasitic worms on death rates of people in tropical Africa
   b. the recovery of the forest community around Mount St. Helens following its eruption in 1980
   c. adaptations of deep-sea fish that allow them to live under such high pressures
   d. how predation by feral cats on ground-nesting birds on an island affects the numbers of each species
   e. how the three-dimensional structure of the active site of an enzyme involved in converting glucose to glycogen in the liver affects its activity

2. Which of the following would define a population?
   a. the largemouth bass of Lake Michigan
   b. all species of Pacific salmon that inhabit the north Pacific Ocean
   c. the grizzlies, elk, pronghorn, and bison of Yellowstone National Park
   d. all organisms found in Big Bend National Park

3. The rate at which a population reproduces and grows under ideal or optimal conditions is known as the
   a. replacement-level fertility.
   b. biotic potential.
   c. carrying capacity.
   d. environmental resistance.
   e. survivorship curve.

4. The biotic potential of a species depends on the
   a. age at which the organism first reproduces.
   b. chance of survival to the age of reproduction.
   c. frequency with which reproduction occurs.
   d. average number of offspring produced each time.
   e. length of the reproductive life span of the organism.
   f. all of the above.

5. Which of the following examples illustrates the principle of exponential growth?
   a. aphids whose population numbers decrease consistently from year to year
   b. spiders whose population numbers increase one year but decrease in other years
   c. spider mites whose population numbers double every 2 weeks for the course of the summer
   d. purseweb tarantulas whose population numbers remain essentially unchanged over time

6. Exponential growth requires that
   a. there is no mortality.
   b. there are no density-independent limits.
   c. the birth rate consistently exceeds the death rate.
   d. a species reproduce very quickly.
   e. the species is an exotic invader in an ecosystem.

7. Based on what you know about the factors that influence biotic potential, which organism below would have a higher biotic potential?
   a. elephant   c. robin
   b. cat   d. mouse

8. If the elk population were drastically reduced, perhaps due to disease, which prediction could one make about the effects on the wolf population that preys on the elk?
   a. The wolf population will crash, because there are no other food sources for the wolves but elk.
   b. The wolf population will increase because there will be more dead or weakened prey to eat.
   c. The wolf population will be unaffected because it will change its food source to something else.
   d. The wolves will decline as well but will rebound if the deer/elk population increases.
   e. The wolf population will achieve carrying capacity and maintain an S-shaped curve.

9. Carrying capacity is
   a. the total number of organisms of a species that an environment can support.
   b. the population size always reached by a particular species.
   c. a measurement of the total resources in the environment.
   d. a measurement of the total resources in an environment that are used by one population.
   e. the maximum biotic potential of a species assuming no limitation of resources.

10. Fish that swim together in schools represents what type of species distribution pattern?
    a. clumped   c. uniform
    b. random   d. scrambled

11. Once established, exotic species introduced into a new area often increase dramatically in numbers for quite some time. This increase is believed to occur because
    a. they lack many of the natural enemies (predators, parasites, etc.) that would keep their numbers in check in their native range.
    b. exotic species always have very high population growth rates.

c. there are many vacant niches in most ecosystems that exotic species can readily fill.

d. exotic species always have very high carrying capacities.

e. all of the above.

12. A clan of spotted hyenas in Ngorongoro Crater (Africa) gobbling down the carcass of a zebra from which they have chased a lone lioness exhibits _____ among themselves.
   a. contest competition
   b. scramble competition
   c. intraspecific competition
   d. interspecific competition
   e. both b and c

13. Although the death rate exceeds the birth rate, the population in North America and the United States in particular continues to expand. What factor might account for this continued increase?
   a. an unstable economy
   b. failed birth control measures
   c. immigration
   d. none of the above

14. Humans have been able to expand the carrying capacity over the course of recorded history
   a. through advances in technology and medicine.
   b. by co-opting the resources of other species.
   c. by exploiting renewable resources faster than they can be replaced and nonrenewable resources that cannot be replaced.
   d. all of the above.

15. Which example below represents intraspecific competition?
   a. bison, pronghorn, and cattle competing for grass on the open plains
   b. two male deer fighting for the right to mate with a female
   c. four species of orioles segregating a tree into different areas, each species feeding at a different location on the tree
   d. the bees, butterflies, and moths that pollinate the crops in an agricultural field

16. Based on what you see in the graph at the top of the next column, what could you say about how mortality (death) affects population growth?
   a. Death rate has no effect because the curves have a J shape.
   b. The higher the death rate, the less likely the population will reach carrying capacity.
   c. If the death rate is higher, the growth rate is lower.
   d. The higher death rates keep the population from growing exponentially.

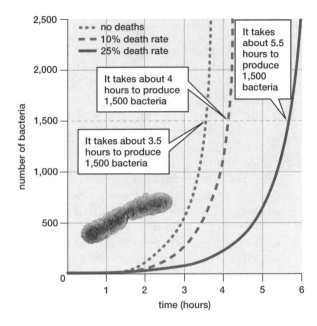

17. If the number of births in a population is greater than the number of deaths, which of the following statements is correct?
   a. The population is increasing in size.
   b. The population is decreasing in size.
   c. The value for the rate of growth (r) is positive.
   d. The value for the rate of growth (r) is negative.
   e. both a and c.

18. Assuming that the birth rate and death rate for a population were equal, what would happen to population numbers if emigration exceeded immigration?
   a. Population numbers would decrease.
   b. Population numbers would increase.
   c. Population numbers would remain stable.

19. An ecologist studying a population of ostriches records 1,000 individuals at the start of a 1-year study. Over the course of the year, he records 100 births and 60 deaths. What is the growth rate (r) for this population?
   a. 0.1
   b. 0.04
   c. 0.4
   d. 40
   e. 100

20. A population of pocket gophers has a growth rate (r) of 0.2 per year. If there are initially 100 individuals, how many pocket gophers do you expect at the end of the first and second years?
   a. 120, 144
   b. 120, 140
   c. 20, 4
   d. 120.2, 120.4
   e. 200, 400

21. Growth rate implies a change over time. If a village in Asia has a population of 750 people and 10 children were born in the village this past year, what is the birth rate for this village?
    a. 760
    b. 13%
    c. 75
    d. 0.013
    e. less than 1%

22. Which of the following factors increases populations of organisms?
    a. birth
    b. death
    c. immigration
    d. emigration
    e. both a and c

23. Environmental resistance may limit the size of populations by
    a. increasing both birth rates and death rates.
    b. decreasing both birth rates and death rates.
    c. increasing death rates and/or decreasing birth rates.
    d. decreasing death rates and/or increasing birth rates.

24. As the size of a snowshoe hare population rises, the number of deaths resulting from predation by lynx often also rises because
    a. the number of encounters between predators and prey will increase when there are more prey around.
    b. some of the lynx will switch from other prey such as grouse to snowshoe hares as the latter's numbers increase.
    c. the increased food available to lynx will eventually increase the numbers of lynx by increasing their birth rate.
    d. all of the above.

25. Based on the figure below, what would be your estimate of the carrying capacity of barnacles on this rocky seashore?
    a. about 80 barnacles per square centimeter
    b. about 60 barnacles per square centimeter
    c. about 75 barnacles per square centimeter
    d. about 50 barnacles per square centimeter

26. Of the following animals, which would represent an organism with a late-loss survivorship?
    a. elephant
    b. robin
    c. honeybee
    d. salmon

27. Which of the following factors is LEAST likely to influence population size in a density-dependent way?
    a. predation
    b. competition
    c. emigration
    d. climate and weather
    e. parasitism and disease

28. If a country has more individuals in the prereproductive category than all other categories, what will happen to the population in relation to size in 20 to 30 years?
    a. The population size will remain the same.
    b. The population will increase in size.
    c. The population will decrease in size.
    d. There is not enough information given to evaluate the future size of the population.

29. Populations are regulated independently of density by
    a. disease.
    b. competition.
    c. increased number of parasites.
    d. floods.

30. Given your knowledge of population dynamics, how could the crash of the population of reindeer on St. Paul Island have been avoided? (Review Figure 26-7 in your text if necessary.)
    a. The population could have been kept stable by increasing the amount of land area for the deer to graze.
    b. The population could have been managed at a lower size by decreasing the birth rate of females.
    c. The population could have been kept at a lower size by introducing hunting or a natural predator to increase mortality rates.
    d. Nothing could have altered the population size of the deer herd. The biotic potential of the species would have forced the population to crash.

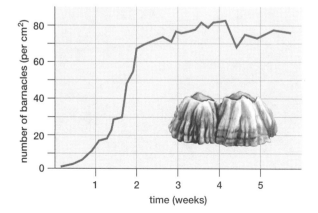

ANSWER KEY

1. e
2. a
3. b
4. f
5. c
6. c
7. d
8. d
9. a
10. a
11. a
12. e
13. c
14. d
15. b

16. c
17. e
18. a
19. b
20. a
21. d
22. e
23. c
24. d
25. c
26. a
27. d
28. b
29. d
30. c

CHAPTER 27 COMMUNITY INTERACTIONS

OUTLINE

FLASH CARDS

To use the flash cards, tear the page from the book and cut along the dashed lines. The key term appears on one side of the flash card, and its definition appears on the opposite side.

abiotic	coevolution
aggressive mimicry	community
biome	competition
biotic	competitive exclusion principle
camouflage	disturbance
carnivore	ecological niche
climax community	herbivore

The evolution of adaptations in two species due to their extensive interactions with one another, such that each species acts as a major force of natural selection on the other.

Nonliving; the abiotic portion of an ecosystem includes soil, rock, water, and the atmosphere.

All the interacting populations within an ecosystem.

The evolution of a predatory organism to resemble a harmless animal or a part of the environment, thus gaining access to prey.

Interaction among individuals who attempt to utilize a resource (for example, food or space) that is limited relative to the demand for that resource.

A terrestrial ecosystem that occupies an extensive geographic area and is characterized by a specific type of plant community; for example, deserts.

The concept that no two species can simultaneously and continuously occupy the same ecological niche.

Living.

Any event that disrupts an ecosystem by altering its community, its abiotic structure, or both; disturbance precedes succession.

Coloration and/or shape that renders an organism inconspicuous in its environment.

The role of a particular species within an ecosystem, including all aspects of its interaction with the living and nonliving environments.

Literally, "meat eater"; a predatory organism that feeds on herbivores or on other carnivores; a secondary (or higher) consumer.

Literally, "plant-eater"; an organism that feeds directly and exclusively on producers; a primary consumer.

A diverse and relatively stable community that forms the endpoint of succession.

host	mimicry
interspecific competition	mutualism
intertidal zone	parasite
intraspecific competition	pioneer
invasive species	predator
keystone species	primary succession

The situation in which a species has evolved to resemble something else, typically another type of organism.

The prey organism on or in which a parasite lives; the host is harmed by the relationship.

A symbiotic relationship in which both participating species benefit.

Competition among individuals of different species.

An organism that lives in or on a larger organism (its host), harming the host but usually not killing it immediately.

An area of the ocean shore that is alternately covered and uncovered by the tides.

An organism that is among the first to colonize an unoccupied habitat in the first stages of succession.

Competition among individuals of the same species.

An organism that eats other organisms.

Organisms that are introduced (deliberately or accidentally) into ecosystems where they did not evolve, and where they encounter little environmental resistance and tend to displace native species.

Succession that occurs in an environment, such as bare rock, in which no trace of a previous community is present.

A species whose influence on community structure is greater than its abundance would suggest.

resource partitioning

subclimax

secondary succession

succession

startle coloration

warning coloration

A community in which succession is stopped before the climax community is reached; it is maintained by regular disturbance—for example, a tallgrass prairie maintained by periodic fires.

The coexistence of two species with similar requirements, each occupying a smaller niche than either would if it were by itself; a means of minimizing the species' competitive interactions.

A structural change in a community and its nonliving environment over time. During succession, species replace one another in a somewhat predictable manner until a stable, self-sustaining climax community is reached.

Succession that occurs after an existing community is disturbed—for example, after a forest fire; secondary succession is much more rapid than primary succession.

Bright coloration that warns predators that the potential prey is distasteful or even poisonous.

A form of mimicry in which a color pattern (in many cases resembling large eyes) can be displayed suddenly by a prey organism when approached by a predator.

SELF TEST

1. Which of the following organisms is an example of an herbivore?
   a. bear
   b. coyote
   c. bison
   d. rattlesnake

2. Following applications of insecticides to agricultural fields to control pest insects, sometimes the crops suffer more insect damage than if no pesticides were applied. Knowing what you know about interactions between species and the factors that regulate population sizes, what is the most likely explanation for this?
   a. The insecticides probably killed off most of the predators that previously kept the size of the plant-eating insect population from exploding. The insects then increased in number and did more damage to the plants.
   b. The insecticides probably killed off most of the competitors that previously kept the size of the plant-eating insect population from exploding. The insects then increased in number and did more damage to the plants.
   c. The insecticides probably weakened the defenses of the plants, making them more vulnerable to attack by the surviving plant-eating insects.
   d. The insecticides probably inhibited the growth of mycorrhizae and nitrogen-fixing bacteria on which the plants rely for nutrients. The plants were therefore weakened and more prone to attack by insects.

3. No two species can occupy the same ecological niche in the same place at the same time. This statement is known as
   a. mutualism.
   b. a premating isolating mechanism.
   c. Darwin's theory of natural selection.
   d. the Hardy–Weinberg principle.
   e. the competitive exclusion principle.

4. If an island population of fruit bats is competing with other nocturnal mammals, such as the sugar glider, for the same resources of fruit, pollen, nectar, and insects, which concept best describes the splitting of this similar niche of food resources so that both species may coexist with these limited resources?
   a. resource partitioning
   b. competitive exclusion principle
   c. adaptation
   d. mimicry
   e. mutualism

5. Over time a predator adapts to blend in to the environment in order to catch a prey species that has adapted a strategy of alerting its relatives to the presence of a predator. This relationship can be explained as
   a. coevolutionary.
   b. mimicry.
   c. adaptionary.
   d. a predator/prey cycle.

6. What is the function of aggressive mimicry?
   a. to hide a prey from a predator
   b. to warn a predator that a prey is dangerous
   c. to warn a predator that a prey is distasteful
   d. to keep prey from recognizing a predator
   e. to startle a prey when it sees a predator

7. The rattles of rattlesnakes would be similar in function to which of the following?
   a. camouflage
   b. warning coloration
   c. startle coloration
   d. aggressive mimicry
   e. both b and c

8. Brownheaded cowbirds lay their eggs in the nest of other songbird species, forcing them to raise brownheaded cowbird offspring rather than their own. This relationship could be considered to be
   a. a mutualistic relationship.
   b. a parasitic relationship.
   c. a predator/prey relationship.
   d. a coevolutionary relationship.

9. Honeyguides are African birds that excitedly lead the way to a bee's nest, and ratels are the honey- and bee-eating mammals that open up and scatter the contents of the bee's nests, allowing both the ratels and the honeyguides to feed on the contents. This relationship is an example of
   a. predation.   d. mutualism.
   b. competition.   e. commensalism.
   c. parasitism.

10. *Trypanosoma* is a protozoan (single-celled organism) that lives and reproduces for an extended period in the blood of a mammalian host (e.g., a human, native antelope, or introduced cattle). Newly introduced cattle generally die from this infection if they are not treated, whereas the native antelope or cattle that have been exposed to this protozoan for several generations are less severely affected. This relationship is an example of
   a. predation.   d. mutualism.
   b. competition.   e. commensalism.
   c. parasitism.

11. The lynx is a predator that feeds on hare. If the two populations tend to cycle with each other, what would you expect to see happen to the lynx population if the hare population increased significantly?
   a. Nothing; since the lynx has many prey to choose from, it is not affected by population numbers of hare.
   b. It should go up since there are more resources to feed more offspring so that more lynx offspring survive.
   c. It should go down since there will be more hare for lynx to compete over as food resources.
   d. Nothing; since the hare resource is not limited, there would be no competition to reduce predator numbers.

12. In the intertidal zone, diverse assemblages of many invertebrate species and algae exist attached to the rocks. If an oil spill occurred that directly affected only a species of starfish that is a keystone predator in this system, totally eliminating it, what do you predict would happen to the community?
   a. No significant change in the structure of the community would be likely to occur.
   b. The community is likely to become less diverse, increasingly dominated by a few species that are good competitors for space.
   c. The community is likely to become more diverse, because strong and weak competitors can then coexist.

13. Why does a climax community tend to persist in an area?
   a. It is made up of large, hard-to-move organisms.
   b. The species in it do not alter their environment significantly.
   c. It is able to change the weather.
   d. The species in it are able to kill off competitors.

14. Boulders are often overturned by wave action during storms along the lower reaches of the rocky intertidal shore. Along the coast of Southern California, there is a fairly predictable sequence of species that colonize the bare rock as spores and then replace each other. This starts with the green algae *Ulva*, which is replaced after a year or so by various red algae, and so on. This is an example of
   a. primary succession.
   b. secondary succession.
   c. aggressive mimicry.
   d. a keystone species.

15. Plant and animal communities in a biome are regulated by
   a. climate.
   b. succession.
   c. the biological interactions of the organisms.
   d. competition for limited resources.

16. What must be true in order for competition to take place?
   a. There should be resource partitioning between the individuals involved.
   b. The individuals must be from two different species.
   c. There must be a limited resource.
   d. There has to be a struggle for a resource.

17. Corals are polyps of coelenterates (cnidarians) that contain numerous algae in their tissues. The algae contain photopigments that give the corals their color. When stressed by high temperatures, water pollution, or similar shocks, many corals expel their algae and turn white, a process called "coral bleaching." This appears to help the coelenterates survive the initial shock, but if they do not recover their native algae quickly, they soon die. Similarly, the algae cannot live for long outside the coelenterates' bodies. Based on this information, the relationship between the two organisms is most likely
   a. mutualism.
   b. interspecific competition.
   c. commensalism.
   d. parasitism.
   e. predation.

18. In a farm pond location, a landowner decided to introduce bluegill (*Lepomis macrochirus*) because she liked to fish and preferred to eat bluegill. The species of fish already living in the pond were sunfish (*Lepomis humilis*). After introduction of the bluegill, which is very similar to sunfish in habitat and food preferences, the landowner discovered several years later that there were not as many sunfish in the pond; the sunfish numbers had diminished over time. This is an example of
   a. intraspecific competition.
   b. interspecific competition.
   c. the competitive exclusion principle.
   d. both a and c.
   e. both b and c.

19. Based on what you know of predators and prey, what could be said about the individuals in the figure at the top of the facing page?
   a. They are predators lying in wait to ambush their prey.
   b. They are prey species that utilize warning coloration to avoid being eaten.
   c. The spots are a form of mimicry to mimic the eyes of the prey.
   d. The eyespots are used to startle the predator into looking elsewhere for a meal.

20. Within a year of the abandonment of agriculture on a plot of prairie, the previously bare soil is overrun with annual weeds. Light, carbon dioxide, and mineral nutrients are readily available, but soil moisture is limiting. Still, some species appear to coexist very close to one another. Upon closer examination, two such species, smartweed and bristly foxtail, are observed to have very different root systems and ways of managing water. Smartweed has a deep taproot, extending about a meter beneath the surface, tapping (literally) into a continuous deep-water supply. Bristly foxtail has a much shallower and spreading fibrous root system, reaching less than 20 cm down. However, the latter plant is able to tolerate periods of drought and to take up water rapidly after a rain. This example is a clear case of
    a. resource partitioning.
    b. competitive exclusion.
    c. intraspecific competition.
    d. commensalism.
    e. a keystone species.

21. If you were a predator and came upon the frog shown below, you would
    a. leave the frog alone since it is warning by its color that it is poisonous.
    b. leave the frog alone since it is a predator as well, and I don't want to be eaten.
    c. eat the frog since it is my preferred prey.
    d. leave the frog alone since the color tells me that it is not the right prey species for me.

22. Fireflies use their amazing bioluminescence to find mates of the right species. Males fly around and emit a series of flashes in a pattern that differs from all other species. A female, perched nearby, may respond with her own species-specific signal. The male immediately flies toward her and, after a few more exchanges, it's wedded bliss! Well, not always. Females of a few species have learned to imitate other species' signals and, when their males arrive, they eat them. These femme fatales are exhibiting a form of
    a. aggressive mimicry.
    b. startle coloration.
    c. camouflage.
    d. warning coloration.
    e. chemical warfare.

23. In many ways, parasitism and predation are similar types of interactions between species. Which of the following is NOT true about their differences?
    a. Parasites are usually much smaller than their hosts, but predators are usually larger than their prey.
    b. Parasites are usually much more numerous than their hosts, but predators are usually less numerous than their prey.
    c. Parasites usually do not kill their hosts immediately, but predators usually do kill their prey immediately.
    d. Parasites usually have no effect on their hosts, but predators usually harm their prey.

24. On many coral reefs in the Pacific, large fish bearing parasites will visit "cleaning stations" where species of small fish, known as cleaner wrasses, remove parasites and loose scales from the larger fish. The cleaner fish may even enter the mouth and gills of the larger fish to clean parasites from the soft tissues. The cleaner fish are recognized by their coloration, black with a bright blue or yellow stripe, and by a little "dance" that invites the larger fish in to be cleaned. Also on the reef, however, is a small fish known as the saber-tooth blenny. Looking and acting very much like the cleaner wrasse, the blenny also attracts the larger fish, but instead of

cleaning away parasites, the blenny bites small bits of flesh from the larger fish. The interactions among the larger fish, the cleaner wrasse, and the saber-tooth blenny represent all of the following EXCEPT

a. mutualism.
b. parasitism.
c. aggressive mimicry.
d. warning coloration.

25. What is the term for a situation in which one organism benefits from its close association with a second species, but the second species is harmed in the process?

a. commensalism
b. mutualism
c. parasitism
d. both a and b

26. In the rocky intertidal zone, along the coast of the state of Washington, space on the substrate (rocks) is a critical resource for sessile (attached) organisms, such as algae, barnacles, and mussels, and also for other organisms that graze the algae or prey on the sessile animals. For example, limpets graze the algae, and three species of barnacle dominate the algae for space. Among them they form a dominance hierarchy, with the larger *Semibalanus*-dominating *Balanus*, which in turn dominates the smaller *Chthamalus*. The mussel, *Mytilus*, is at the top of this hierarchy. A large thaid snail feeds on *Balanus* and *Chthamalus* but cannot eat the larger *Semibalanus* or the mussel. The starfish, *Pisaster*, however, prefers to feed on mussels and the large barnacle *Semibalanus*. Thus, if the starfish is present in this community, the community will be a diverse assemblage of nearly all of the species described here; but if the starfish is absent, the dominant

mussel will take over most of the available space. In this example we see

a. predation.
b. intraspecific competition.
c. interspecific competition.
d. a keystone species.
e. all of the above.

27. Invasive species

a. are usually the first to colonize after a disturbance.
b. outcompete the other species in an area.
c. drive the interactions of species in a community.
d. are not subject to resource partitioning because they are not native to the area.

28. Plants and animals invade a region recently scoured clean by a retreating glacier and over time are replaced by other species. This statement describes

a. a climax community.
b. primary succession.
c. secondary succession.
d. all of the above.

29. The "weeds" that begin growing in a plowed field would be considered

a. climax species.   c. colonizers.
b. pioneer species   d. herbivores.

30. Organisms in a successional community help cause the changes that result in their own replacement because

a. they become old and unfit for the environment.
b. they emigrate to other areas to make room for new species.
c. they change the physical environment in ways that favor competitors.

ANSWER KEY

1. c	16. c
2. a	17. a
3. e	18. e
4. a	19. d
5. a	20. a
6. d	21. a
7. e	22. a
8. b	23. d
9. d	24. d
10. c	25. c
11. b	26. e
12. b	27. b
13. b	28. b
14. a	29. b
15. a	30. c

Chapter 28 How Do Ecosystems Work?

OUTLINE

Section 28.1 How Do Energy and Nutrients Move Through Ecosystems?

Energy moves through ecosystems in a continuous one-way flow, originating as sunlight. **Nutrients** are constantly recycled within and among ecosystems (**Figure 28-2**).

Section 28.2 How Does Energy Flow Through Ecosystems?

Energy enters communities through photosynthesis (**Figure 28-2**). The energy that photosynthetic organisms store and make available to other members of the community is the **net primary productivity** (**Figure 28-3**).

Energy is passed within communities from one **trophic level** to another as a series of feeding relationships. The first trophic level consists of the **autotrophs** (**producers**), which capture energy from sunlight. The **primary consumers**, or **herbivores**, feed on the producers and form the second level. **Carnivores** (**secondary consumers**) feed on the primary consumers and form the third trophic level. Sometimes carnivores eat other carnivores (**tertiary consumers**) and form the fourth trophic level. **Omnivores** consume plants and animals and occupy multiple trophic levels.

A linear representation of feeding relationships is called a **food chain** (**Figure 28-4**). Typically, food chains interconnect, forming **food webs** (**Figure 28-5**).

Detritus feeders and **decomposers** digest dead bodies and decaying organic matter. They release the energy stored within them and also recycle nutrients back into ecosystems.

Only about 10% of available energy is transferred from one trophic level to the next one above it, which is very inefficient. This energy relationship is represented by an energy pyramid (**Figure 28-6**).

Toxic chemicals accumulate in the higher trophic levels. This process is referred to as **biological magnification**. Mercury, lead, and DDT are common chemicals that accumulate in the tissues of top carnivores.

Section 28.3 How Do Nutrients Cycle Within and Among Ecosystems?

A **nutrient cycle** depicts the movement of a particular nutrient from the **reservoir** through the biotic portion of the ecosystem and back to the reservoir.

In the **hydrologic cycle**, water in the oceans evaporates and enters the atmosphere. This water returns to Earth as precipitation, which flows into lakes, **aquifers** (underground reservoirs), and rivers, which flow into the oceans. Water is absorbed by plants and animals and enters the biotic community (**Figure 28-7**).

In the **carbon cycle**, atmospheric CO_2 enters producers through photosynthesis. It then enters the biotic community through the food web, and is released by the biotic community back into the atmosphere as a result of cellular respiration (**Figure 28-8**). Carbon is also released to the atmosphere by fire and the burning of **fossil fuels**.

In the **nitrogen cycle**, atmospheric nitrogen gas is converted, during **nitrogen fixation**, by bacteria (and human industrial activity) into ammonia and nitrate, which are used by plants. When herbivores ingest plants, nitrogen enters the biotic component of the ecosystem. Nitrogen returns to the soil during excretion by organisms and when organisms die. **Denitrifying bacteria** then convert ammonia to nitrogen and release nitrogen back to the atmosphere (**Figure 28-9**).

In the **phosphorus cycle**, phosphate in rocks dissolves in rainwater and is absorbed by producers and then passes through food webs. Phosphate returns to the soil and water via excretion and the action of decomposers. Some phosphate is carried to the oceans, where it is deposited in sediments (**Figure 28-10**).

Section 28.4 What Happens When Humans Disrupt Nutrient Cycles?

Human use of nitrogen and phosphorous fertilizers has led to phytoplankton blooms in lakes and oceans due to fertilizer that runs off of farm fields and enters streams and rivers. This leads to fish kills in lakes and dead zones in the oceans.

Human consumption of fossil fuels and industrial processes cause the release of sulfur dioxide and nitrogen oxides into the atmosphere. These compounds are converted to sulfuric acid, which falls to Earth as **acid deposition** (acid rain). Acidification of lake and forest ecosystems has reduced their ability to sustain life (**Figure 28-13**).

Elevated atmospheric CO_2, a **greenhouse gas** produced by the burning of fossil fuels, is correlated with increased global temperatures (**global warming, Figures 28-14, 28-15, 28-16, and 28-17**).

Global warming is causing ancient ice to melt (**Figure 28-17**) and is predicted to cause extreme weather conditions, which has an influence on seasonal wildlife activity.

FLASH CARDS

To use the flash cards, tear the page from the book and cut along the dashed lines. The key term appears on one side of the flash card, and its definition appears on the opposite side.

acid deposition	carbon cycle
aquifer	carnivore
autotroph	consumer
biodegradable	decomposer
biogeochemical cycle	deforestation
biological magnification	denitrifying bacteria
biomass	detritus feeder

The steps by which carbon moves from its reservoirs in the atmosphere and oceans through producers and into higher trophic levels, and then back to its reservoirs.

The deposition of nitric or sulfuric acid, either in rain (acid rain) or in the form of dry particles, as a result of the production of nitrogen oxides or sulfur dioxide through burning, primarily of fossil fuels.

Literally, "meat eater"; a predatory organism that feeds on herbivores or on other carnivores; a secondary (or higher) consumer.

An underground deposit of fresh water, often used as a source for irrigation.

An organism that eats other organisms; a heterotroph.

Literally, "self-feeder"; normally, a photosynthetic organism; a producer.

An organism, usually a fungus or bacterium, that digests organic material by secreting digestive enzymes into the environment, in the process liberating nutrients into the environment.

Able to be broken down into harmless substances by decomposers.

The excessive cutting of forests. In recent years, deforestation has occurred primarily in rain forests in the Tropics, to clear space for agriculture.

The pathways of a specific nutrient (such as carbon, nitrogen, phosphorus, or water) through the living and nonliving portions of an ecosystem; also called a *nutrient cycle*.

Bacteria that break down nitrates, releasing nitrogen gas to the atmosphere.

The increasing accumulation of a toxic substance in progressively higher trophic levels.

One of a diverse group of organisms, ranging from worms to vultures, that live off the wastes and dead remains of other organisms.

The total weight of all living material within a defined area.

energy pyramid	greenhouse gas
estuary	herbivore
food chain	heterotroph
food web	hydrologic cycle
fossil fuel	legume
global warming	macronutrient
greenhouse effect	micronutrient

A gas, such as carbon dioxide or methane, that traps sunlight energy in a planet's atmosphere as heat; a gas that participates in the greenhouse effect.

A graphical representation of the energy contained in succeeding trophic levels, with maximum energy at the base (primary producers) and steadily diminishing amounts at higher levels.

Literally, "plant-eater"; an organism that feeds directly and exclusively on producers; a primary consumer.

A wetland formed where a river meets the ocean; the salinity is quite variable, but lower than that of sea water and higher than that of fresh water.

Literally, "other-feeder"; an organism that eats other organisms; a consumer.

A linear feeding relationship in a community, using a single representative from each of the trophic levels.

The pathway that water takes as it travels from its major reservoir, the oceans, through the atmosphere to reservoirs in freshwater lakes, rivers, and groundwater, and back into the oceans. The hydrologic cycle is driven by solar energy. Nearly all water remains as water throughout the cycle (rather than being used in the synthesis of new molecules).

A representation of the complex feeding relationships (in terms of interacting food chains) within a community, including many organisms at various trophic levels, with many of the consumers occupying more than one level simultaneously.

A member of a family of plants characterized by root swellings in which nitrogen-fixing bacteria are housed; includes peas, soybeans, lupines, alfalfa, and clover.

A fuel such as coal, oil, and natural gas, derived from the remains of ancient organisms.

A nutrient required by an organism in relatively large quantities.

A gradual rise in global atmospheric temperature as a result of an amplification of the natural greenhouse effect due to human activities.

A nutrient required by an organism in relatively small quantities.

The process in which certain gases such as carbon dioxide and methane trap sunlight energy in a planet's atmosphere as heat; the glass in a greenhouse does the same. The result, global warming, is being enhanced by the production of these gases by humans.

net primary production	primary consumer
nitrogen cycle	producer
nitrogen fixation	reservoir
nutrient	secondary consumer
nutrient cycle	tertiary consumer
omnivore	transpiration
phosphorus cycle	trophic level
phytoplankton	zooplankton

An organism that feeds on producers; an herbivore.

The energy stored in the autotrophs of an ecosystem over a given time period that is available to other members of the community.

A photosynthetic organism; an autotroph.

The process by which nitrogen moves from its primary reservoir of nitrogen gas in the atmosphere via nitrogen-fixing bacteria to reservoirs in soil and water, through producers and into higher trophic levels, and then back to its reservoirs.

The major source and storage site of a nutrient in an ecosystem, normally in the abiotic portion.

The process that combines atmospheric nitrogen with hydrogen to form ammonium (NH_4^+).

An organism that feeds on primary consumers; a carnivore.

A substance acquired from the environment and needed for the survival, growth, and development of an organism.

A carnivore that feeds on other carnivores (secondary consumers).

The pathways of a specific nutrient (such as carbon, nitrogen, phosphorus, or water) through the living and nonliving portions of an ecosystem; also called a *biogeochemical cycle*.

The evaporation of water through stomata, chiefly in leaves.

An organism that consumes both plants and animals.

Literally, "feeding level"; the categories of organisms in a community, and the position of an organism in a food chain, defined by the organism's source of energy; includes producers, primary consumers, secondary consumers, and so on.

The process by which phosphorus moves from its primary reservoir—phosphate-rich rock—to reservoirs of phosphate in soil and water, through producers and into higher trophic levels, and then back to its reservoirs.

Nonphotosynthetic protists that are abundant in marine and freshwater environments.

Photosynthetic protists that are abundant in marine and freshwater environments.

SELF TEST

1. An ecologist studying a plot of ground in the tundra excludes all herbivores from her study area and estimates the plant biomass at 530 grams per square meter. She comes back to the same plot one year later and estimates that the biomass has increased to 670 grams per square meter. The difference in these two values, 140 grams, represents the _____ for that year.
   a. biological magnification
   b. food chain length
   c. food web complexity
   d. net primary productivity
   e. trophic transfer efficiency

2. Of the energy found in organisms, what percentage is passed from one trophic level to the next higher trophic level?
   a. 0.01%   c. 10%
   b. 1%   d. 100%

3. Earthworms are
   a. detritus feeders.
   b. herbivores.
   c. primary consumers.
   d. producers.
   e. secondary consumers.

4. The base of the energy pyramid represents
   a. producers.
   b. decomposers.
   c. primary consumers.
   d. secondary consumers.
   e. tertiary consumers.

5. Which of the following are heterotrophic?
   a. producers
   b. decomposers
   c. primary consumers
   d. secondary consumers
   e. b, c, and d

6. Of the following trophic levels, which would support the fewest organisms?
   a. producer
   b. decomposer
   c. primary consumer
   d. secondary consumer
   e. tertiary consumer

7. What might be a possible result if the hawk in this figure were to become extinct?
   a. Nothing, the hawk is a top predator, so the food web would be unchanged.
   b. The badger would have more food to eat since it wouldn't be competing with the hawk.
   c. The organisms that the hawk feeds on would have higher population numbers after a few years.
   d. The other bird species will show an increase in population numbers since there would be an open niche in the bird community.

8. If primary productivity is related to diversity of organisms, which biome in the figure above would have the highest biodiversity?
   a. grassland
   b. temperate deciduous forest
   c. tundra
   d. tropical rain forest

9. In the carbon cycle, carbon is returned to the atmosphere by
   a. photosynthesis.
   b. evaporation of water.
   c. burning of fossil fuels.
   d. respiration of plants and animals.
   e. both c and d.

10. For which of the following nutrients is rock a major reservoir?
    a. water
    b. oxygen
    c. carbon
    d. nitrogen
    e. phosphorus

11. How is nitrogen released back to the atmosphere once it has been incorporated into the body of an organism?
    a. through nitrogen fixation
    b. through symbiotic association with a legume
    c. by decomposers and denitrifying bacteria
    d. both a and b

12. Which of the following is a major contributor to the problem of acid deposition?
    a. oxygen
    b. carbon dioxide
    c. sulfur dioxide

    d. nitrogen oxides
    e. both c and d

13. So far, global warming has been documented to be causing
    a. changes in precipitation patterns on land, with some areas subjected to more severe and frequent droughts, and other areas more frequent and severe floods.
    b. melting of ice sheets and retreat of glaciers at unprecedented rates.
    c. shifts in the distribution and abundance of a number of plant and animal species.
    d. shifts in the timing of spring events, which are occurring much earlier than previously.
    e. all of the above.
    f. none of the above; to date, global warming has had no effects on any of these events.

14. How does using wood derived from trees as a source of fuel for cooking affect the carbon cycle?
    a. It decreases the uptake of CO_2 from the atmosphere because fewer trees carry on photosynthesis.
    b. It has no effect on the carbon cycle.
    c. It increases the release of CO_2 into the atmosphere as the wood is burned.
    d. Both a and c.

15. Which of the organisms below would be most likely to experience issues related with biological magnification?
    a. phytoplankton
    b. bald eagle
    c. field mouse
    d. rattlesnake

16. Which of the chemicals below would be most likely to accumulate in the tissues of organisms in food webs?
    a. lead
    b. glucose
    c. phosphorus
    d. nitrogen

17. A spider that feeds on an aphid that in turn feeds on germinating blades of wheat would belong to which consumer category?
    a. herbivore
    b. primary consumer
    c. secondary consumer
    d. tertiary consumer

18. Which of the following animals are omnivores?
    a. deer
    b. wolves
    c. hyenas
    d. black bears
    e. fungi

19. What are the primary consumers of the ocean environment?
    a. phytoplankton
    b. zooplankton
    c. small fish such as anchovies
    d. big fish such as tuna
    e. jellyfish

20. Which of the following organisms is considered to be a detritus feeder or decomposer?
    a. bacteria
    b. worms
    c. fungi
    d. all of the above

21. What is the base of the food pyramid or the food chain in a marine ecosystem?
    a. plankton
    b. zooplankton
    c. phytoplankton
    d. small fish
    e. coral

22. Which biogeochemical cycle is the only one that does not have an atmospheric component?
    a. hydrologic cycle
    b. carbon cycle
    c. nitrogen cycle
    d. phosphorus cycle

23. Which biogeochemical cycle includes aquifers?
    a. hydrologic
    b. carbon
    c. nitrogen
    d. phosphorus

24. Which trophic level has members that require uptake of ammonia, a complex form of nitrogen?
    a. producers
    b. herbivores
    c. carnivores
    d. decomposers

25. In ecosystems, elements such as carbon and nitrogen
    a. are neither created nor destroyed, but may change molecular form as they pass from organism to organism and between abiotic and biotic components.
    b. are produced by the sun, travel to Earth, and pass briefly through ecosystems, being degraded in the process, and are ultimately lost to space.
    c. play no role whatsoever.

26. Dead zones are a result of
    a. sulfur and nitrogen oxides put in the atmosphere due to industrial pollution.
    b. greenhouse gases killing vegetation in tropical forests.
    c. accumulation of mercury in the food webs of coastal waters.
    d. nitrogen and phosphorus that get into water supplies and end up in the ocean.

27. Why is acid rain, or acid deposition, considered harmful?
    a. Moisture in the air becomes acidified and then falls on plants and the soil below, harming them.
    b. Acid rain leeches essential nutrients out of the soil (e.g., potassium and calcium) and kills decomposers in the soil.
    c. Acid rain makes it difficult for osmosis to take place in the roots because there are too many solutes.
    d. Both a and b are reasons acid rain is harmful.

28. Which of the following greenhouse gases does phytoplankton in the marine environment require for the process of photosynthesis, thereby assisting in the regulation of Earth's atmospheric composition?
    a. oxygen
    b. hydrogen
    c. carbon monoxide
    d. carbon dioxide
    e. sulfur dioxide

29. Which of the following is NOT a cause of the increase in the amount of greenhouse gases?
    a. agriculture
    b. burning fossil fuels
    c. mining
    d. photosynthesis

30. If you were to remove wolves from a nature preserve, you could expect
  a. fewer coyotes because they are competitors of wolves.
  b. an increase in the number of deer and elk, which would lead to a decrease in habitat diversity.
  c. a decrease in the number of deer because there would be an increase in elk due to lack of predation and the two species compete for food resources.
  d. nothing to change because the wolves are only a small part of the overall food web of the nature preserve.

ANSWER KEY

1. d
2. c
3. a
4. a
5. e
6. e
7. c
8. d
9. e
10. e
11. c
12. e
13. e
14. d
15. b

16. a
17. c
18. d
19. b
20. d
21. c
22. d
23. a
24. a
25. a
26. d
27. d
28. d
29. d
30. b

Chapter 29 Earth's Diverse Ecosystems

OUTLINE

Section 29.1 What Factors Influence Earth's Climate?

The **climate** for a region is determined by the amount of sunlight and water and by the range of temperatures over decades or centuries. **Weather**, on the other hand, is short term and refers to the conditions in a local area.

The curvature and tilt of Earth influence climate because they affect the amount of light that strikes different latitudes at different times of the year (**Figure 29-1**).

Uneven heating of Earth's surface generates air currents that produce broad climactic regions such as rain forests and deserts (**Figure 29-2**). Earth's rotation, winds, and solar heating of the oceans generate circular patterns of ocean currents (**gyres**) that moderate nearshore climates (**Figure 29-3**).

As elevations increase, temperatures decrease, creating biomes similar to those of higher latitudes (**Figure 29-4**). Rainfall patterns are also modified by air movement around mountains. It will rain on the near side of the mountain, but as air moves down the mountain it absorbs moisture, creating a local dry area of low precipitation (a **rain shadow**, **Figure 29-5**).

Section 29.2 What Conditions Does Life Require?

Life requires nutrients, energy, liquid water, and organism-appropriate temperatures. These resources are not distributed evenly throughout Earth's surface, which limits the types of organisms that can live in different areas.

Section 29.3 How Is Life on Land Distributed?

Temperature and liquid water are the crucial limiting factors for terrestrial ecosystems. Regions with similar climates (**biomes**) have similar vegetation as determined by temperature and water availability (**Figure 29-7**).

Tropical rain forests are equatorial, warm, and wet. They are dominated by broadleaf evergreen trees, with animal life being primarily arboreal (**Figure 29-9**). This biome has the highest level of **biodiversity**, and most nutrients are tied up in the vegetation rather than the soil.

Tropical deciduous forests are slightly less equatorial than tropical rain forests and have pronounced wet and dry seasons. Plants shed their leaves during the dry season to minimize water loss.

Savannahs are extensive drought-resistant grasslands (no more than 12 inches of annual rainfall) with pronounced wet and dry seasons (**Figure 29-11**).

Deserts are hot and dry, with less than 10 to 20 inches of annual rainfall (**Figure 29-13**). Desert plants are extremely drought resistant, and desert animals use behavioral and physiological mechanisms to conserve water and thermoregulate.

A **chaparral** is a desert-like area (up to 30 inches of annual rainfall), whose climate is moderated by its proximity to the coastline (**Figure 29-16**). Chaparral vegetation consists of small trees and bushes.

Grasslands are located in the center of continents and receive no more than 30 inches of annual rainfall. They are composed almost entirely of continuous grass cover with few to no trees (**Figures 29-17** and **29-18**).

Temperate deciduous forests retain enough moisture (up to 60 inches of annual rainfall) to support trees, but seasonal freezing temperatures cause deciduous plants to drop their leaves in the fall to conserve water (**Figure 29-20**).

Temperate rain forests are seasonal forests that experience moderate temperatures and an abundance of rainfall (as much as 160 inches annually). The trees of these forests are typically conifers (**Figure 29-21**).

Taiga has a harsher climate than temperate deciduous forest, with long, cold winters and short growing seasons. Taiga consists almost entirely of evergreen coniferous trees (**Figure 29-22**).

Tundra is a treeless region bordering the Arctic Ocean that is a frozen desert (up to 10 inches of annual precipitation). **Permafrost** prevents tree growth and bushes are stunted in size, but perennial flowers and lichens exist (**Figure 29-24**).

Section 29.4 How Is Life in Water Distributed?

Energy and nutrients are the crucial limiting factors for aquatic ecosystems. Nutrients are located either in bottom sediments or along the shore, where they are washed in from the surrounding land.

Freshwater lakes are diverse ecosystems that are divided into **littoral, limnetic,** and **profundal zones** (life zones that correspond to specific depths) (**Figure 29-25**). Lakes can be classified based on poor nutrient availability (**oligotrophic**) or rich nutrient availability (**eutrophic, Figure 29-26**).

Streams begin at a source where water is provided by rain or snow. Source water is clear, oxygenated, and nutrient poor. Streams join at lower elevations, forming rivers that carry nutrients from water at higher elevations (**Figure 29-27**). Rivers meander through floodplains on the way to lakes or oceans.

Most oceanic life is found in shallow water (where sunlight can penetrate) near areas of **upwelling** (where nutrients are plentiful). Coastal waters contain the most abundant life and consist of **intertidal** and **nearshore zones**, each with its own particular autotrophic and heterotrophic organisms (**Figure 29-28**). **Coral reefs** occur in warm tropical waters and consist of specialized algae and corals that build reefs out of calcium carbonate skeletons. Coral reefs are the most diverse undersea ecosystems (**Figure 29-30**).

Most open ocean life is found in the **photic zone**, where light supports phytoplankton. The **aphotic zone** is supported by nutrients that drift down from the photic zone (**Figure 29-28**).

Deep ocean life is supported by excrement and dead bodies that drift down from above (**Figure 29-32**). Deep ocean life is often bioluminescent—the light is used to see, attract mates, and attract prey.

Hydrothermal vent communities are supported by chemosynthetic bacteria that act as autotrophs that use hydrogen sulfide (discharged from cracks in Earth's crust) to make energy (**Figure 29-33**). Many hydrothermal vent organisms use these chemosynthetic bacteria either as food or as symbionts to manufacture energy.

FLASH CARDS

To use the flash cards, tear the page from the book and cut along the dashed lines. The key term appears on one side of the flash card, and its definition appears on the opposite side.

aphotic zone	desert
biodiversity	desertification
biome	estuary
chaparral	eutrophic lake
chemosynthesis	grassland
climate	gyre
coral reef	hydrothermal vent community

A biome in which less than 10 to 20 inches (25 to 50 centimeters) of rain fall each year.

The region of the ocean below 200 m, where sunlight does not penetrate.

The process by which relatively dry, drought-prone regions are converted to desert as a result of drought and overuse of the land, for example, by overgrazing or cutting of trees.

The diversity of living organisms; often measured as the variety of different species in an individual ecosystem or in the entire biosphere.

A wetland formed where a river meets the ocean; the salinity is quite variable, but lower than in seawater and higher than in fresh water.

A terrestrial ecosystem that occupies an extensive geographical area and is characterized by a specific type of plant community; for example, deserts.

A lake that receives sufficiently large inputs of sediments, organic material, and inorganic nutrients from its surroundings to support dense communities, especially of plants and phytoplankton; contains murky water with poor light penetration.

A biome located in coastal regions, with very low annual rainfall; is characterized by shrubs and small trees.

A biome, located in the centers of continents, that primarily supports grasses; also called a *prairie*.

The process of oxidizing inorganic molecules, such as hydrogen sulfide, to obtain energy. Producers in hydrothermal vent communities, where light is absent, use chemosynthesis instead of photosynthesis.

A roughly circular pattern of ocean currents, formed because continents interrupt the flow of the current; rotates clockwise in the Northern Hemisphere and counterclockwise in the Southern Hemisphere.

Patterns of weather that prevail for long periods of time (from years to centuries) in a given region.

A community of unusual organisms, living in the deep ocean near hydrothermal vents, that depends on the chemosynthetic activities of sulfur bacteria.

A biome created by animals (reef-building corals) and plants in warm tropical waters.

kelp forest	ozone layer
intertidal zone	pelagic
limnetic zone	permafrost
littoral zone	photic zone
nearshore zone	phytoplankton
northern coniferous forest	plankton
oligotrophic lake	prairie
ozone hole	profundal zone

The ozone-enriched layer of the upper atmosphere (stratosphere) that filters out much of the sun's ultraviolet radiation.

A diverse ecosystem consisting of stands of tall brown algae and associated marine life. Kelp forests occur in oceans worldwide in nutrient-rich cool coastal waters.

Free swimming or floating.

An area of the ocean shore that is alternately covered by water during high tides and exposed to the air during low tides.

A permanently frozen layer of soil, usually found in tundra of the arctic or high mountains.

The part of a lake in which enough light penetrates to support photosynthesis.

The region of an ocean where light is strong enough to support photosynthesis.

The part of a lake, usually close to the shore, in which the water is shallow and plants find abundant light, anchorage, and adequate nutrients.

Photosynthetic protists that are abundant in marine and freshwater environments.

The region of coastal water that is relatively shallow but constantly submerged, and that can support large plants or seaweeds; includes bays and coastal wetlands.

Microscopic organisms that live in marine or freshwater environments; includes phytoplankton and zooplankton.

A biome with long, cold winters and only a few months of warm weather; dominated by evergreen coniferous trees; also called *taiga*.

A biome, located in the centers of continents, that primarily supports grasses; also called *grassland*.

A lake that is very low in nutrients and hence supports little phytoplankton, plant, and algal life; contains clear water with deep light penetration.

The part of a lake in which light is insufficient to support photosynthesis.

A region of severe ozone loss in the stratosphere caused by ozone-depleting chemicals; maximum ozone loss occurs from September to early October over Antarctica.

rain shadow

savanna

taiga

temperate deciduous forest

temperate rain forest

tropical deciduous forest

tropical rain forest

tropical scrub forest

tundra

upwelling

weather

wetland

zooplankton

A biome, warm all year-round, with pronounced wet and dry seasons (drier conditions than in tropical deciduous forests); characterized by short, deciduous, often thorn-bearing trees with grasses growing beneath them.

A local dry area, usually located on the downwind side of a mountain range, that blocks the prevailing moisture-bearing winds.

A biome with severe weather conditions (extreme cold and wind, and little rainfall) that cannot support trees.

A biome that is dominated by grasses and supports scattered trees; typically has a rainy season during which most of the year's precipitation falls, followed by a dry season during which virtually no precipitation occurs.

An upward flow that brings cold, nutrient-laden water from the ocean depths to the surface.

A biome with long, cold winters and only a few months of warm weather; dominated by evergreen coniferous trees; also called *northern coniferous forest*.

Short-term fluctuations in temperature, humidity, cloud cover, wind, and precipitation in a region over periods of hours to days.

A biome having cold winters and warm summers, with enough summer rainfall for trees to grow and shade out grasses; characterized by trees that drop their leaves in winter (deciduous trees), an adaptation that minimizes water loss when the soil is frozen.

A region (sometimes called a marsh, swamp, or bog) in which the soil is covered by, or saturated with, water for a significant part of the year.

A temperate biome with abundant liquid water year-round, dominated by conifers.

Nonphotosynthetic protists that are abundant in marine and freshwater environments.

A biome, warm all year-round, with pronounced wet and dry seasons; characterized by trees that shed their leaves during the dry season (deciduous trees), an adaptation that minimizes water loss.

A biome with evenly warm, evenly moist conditions year-round, dominated by broadleaf evergreen trees; the most diverse biome.

SELF TEST

1. The Great Basin Desert in Nevada and Utah is a result of
   a. a lack of drainage.
   b. overgrazing.
   c. a permanent high barometric pressure zone.
   d. poor soil.
   e. a rain shadow.

2. Which of the following statements would be true of gyres?
   a. They contribute to the moderation of coastal climates.
   b. They contribute to the warming of the Earth by reducing the amount of ozone in the atmosphere.
   c. The contribute to the El Niño/La Niña events in the Pacific Ocean.
   d. They regulate the vegetation that is unique to each biome.

3. Which of the following categories has organisms that utilize the strategy of chemosynthesis?
   a. lake
   b. coral reef
   c. estuary
   d. hydrothermal vent zones

4. Every few years, the trade winds die down, allowing warm surface waters to flow back eastward in the southern Pacific Ocean. This causes winter rains in Peru, droughts in Indonesia and South Africa, and a reduction in the anchovy harvest off the coast of Peru. This phenomenon is called
   a. El Niño.
   b. La Niña.
   c. the Gulf Stream.
   d. eutrophication.
   e. upwelling.

5. At about 30° north and south of the equator, there are very dry regions on Earth. Why does this occur?
   a. Cool air falls, is warmed, and absorbs moisture.
   b. Warm air falls and absorbs moisture.
   c. Cool air rises and water condenses.
   d. Warm air rises and water is evaporated.

6. The greatest diversity of plants and animals in terrestrial (land-based) environments, in terms of numbers of species, is found in
   a. broad-leafed (deciduous) forests.
   b. coniferous (evergreen) forests.
   c. savannas.
   d. tropical rain forests.
   e. deserts.

7. In which terrestrial biome would you expect to find the least amount of precipitation?
   a. taiga
   b. tundra
   c. desert
   d. chaparral

8. The largest expanses of undisturbed and uncut forests exist in the
   a. taiga.
   b. tropical rain forest.
   c. temperate deciduous forest.
   d. temperate rain forest.
   e. tropical deciduous forest.

9. _____ biomes are dominated by grasses but have scattered trees and thorny bushes. They have distinct wet and dry seasons. In Africa, this biome type supports huge herds of migrating large mammals.
   a. Savanna
   b. Tropical rain forest
   c. Taiga
   d. Prairie
   e. Tundra

10. Tropical rain-forest soils make poor agricultural land because
    a. there are too many nutrients in the soil that "burn" the plants.
    b. all the nutrients are tied up in the vegetation rather than in the soil.
    c. there is too much precipitation, which causes most agricultural crops to be waterlogged in the soil.
    d. there is too much shading due to the large trees and the canopy.

11. The biome associated with 30° latitude is a
    a. savanna.
    b. taiga.
    c. desert.
    d. temperate deciduous forest.

12. Large herds of migratory animals are associated with which biome?
    a. savanna
    b. temperate deciduous forest
    c. tropical rain forest
    d. chaparral

13. Extensive, shallow root systems that can quickly soak up water after infrequent rainstorms and spiny leaves and waxy coatings that reduce water loss are adaptations seen in most plants in a
    a. desert.
    b. prairie.
    c. tropical rain forest.
    d. tundra.
    e. temperate deciduous forest.

14. Damage from sediments; excess nutrients from logging, farming, and development on land; and rising water temperatures resulting from global warming are critically endangering which marine ecosystem?
    a. coral reefs
    b. intertidal zones
    c. estuaries
    d. open oceans
    e. hydrothermal vent communities

15. Where would you most likely have to go to encounter a geothermal vent that spews hydrogen sulfide that is ingested by sulfur bacteria?
    a. photic zone
    b. aphotic zone
    c. littoral zone
    d. pelagic zone

16. Rising air _____; this causes water vapor to _____.
    a. heats; evaporate
    b. heats; condense
    c. cools; evaporate
    d. cools; condense

17. Mountain ranges create deserts by
    a. lifting land up into colder, drier air.
    b. completely blocking the flow of air into desert areas, thus preventing clouds from moving in.
    c. forcing air to first rise and then fall, thus causing rain on one side of the mountains and desert on the other.
    d. causing the global wind patterns that make certain latitudes very dry.
    e. causing very steep slopes that are subject to erosion.

18. The primary driver of both weather and climate is
    a. ocean currents.
    b. the sun.
    c. the ozone layer.
    d. elevation.

19. Wetlands are important because
    a. they provide a natural water filtration system.
    b. they are nursery areas for freshwater and saltwater organisms.
    c. of all the aquatic environments, they are the most diverse.
    d. the plants associated with wetlands provide more nitrogen to the soil than other plants in the local area.

20. Which of the following factors is MOST significant for explaining why Earth experiences seasons?
    a. elevation
    b. tilt of Earth
    c. distance of Earth from the sun
    d. none of the above

21. What is the primary reason that plants from distant, but climatically similar, places commonly look the same?
    a. common ancestry
    b. adaptation to the same physical conditions
    c. adaptation to similar herbivores
    d. continental drift
    e. effects of past climate change

22. Which of the following is NOT one of the four fundamental resources required for life?
    a. nutrients
    b. energy
    c. water vapor
    d. suitable temperature

23. You would expect to find the plants in this figure in which biome?
    a. tropical deciduous forest
    b. chaparral
    c. desert
    d. savanna

24. Because of reduced evaporation resulting from fog, many coastal regions bordering deserts are characterized by small woody plants with adaptations to conserve water. What is the name of this biome?
    a. chaparral
    b. savanna
    c. steppe
    d. taiga
    e. tundra

25. The habitat on the right side of the fence in the photograph above is likely a result of
  a. desertification.
  b. overgrazing.
  c. deforestation.
  d. trawling.

26. In what terrestrial biome would you expect the climate to allow the most biological productivity?
  a. chaparral
  b. grassland
  c. northern coniferous forest
  d. temperate deciduous forest
  e. temperate rain forest
  f. tropical rain forest

27. Why are so many desert animals active only at night (nocturnal)?
  a. Reflection of light from the bright sun makes it too difficult for visual predators to hunt.
  b. The surface temperature of the ground would be too hot to walk on.

  c. Temperatures are lower and humidity is higher at night.
  d. Plants only produce oxygen at night.

28. Which biome contains soils that are extremely rich in nutrients resulting from the accumulation of dead organic matter over many centuries?
  a. prairie
  b. tropical rain forest
  c. desert
  d. taiga
  e. temperate deciduous forest

29. Atlantic cod were once phenomenally abundant in the shallow waters off the coast of New England and the Maritime Provinces of eastern Canada. In fact, it has been said that this fishery alone fed the early European colonists to North America for more than a century. In the last half-century, however, the Atlantic cod fishery has almost completely collapsed. This decline has resulted from
  a. overfishing taking far more fish than could be replaced by normal reproduction.
  b. acid rain causing the marine food web to collapse.
  c. the death of eggs and fry as the ozone hole has increased UV damage.
  d. global warming.
  e. overpopulation exceeding the carrying capacity of the system for the cod, causing massive starvation.

30. Trout, fish that require high levels of oxygen and clear water, would likely be found in which lake type listed below?
  a. oligotrophic lakes
  b. eutrophic lakes
  c. photic lakes
  d. estuaries

ANSWER KEY

1. e	16. d
2. a	17. c
3. d	18. b
4. a	19. a
5. b	20. b
6. d	21. b
7. c	22. c
8. a	23. c
9. a	24. a
10. b	25. b
11. c	26. f
12. a	27. c
13. a	28. a
14. a	29. a
15. b	30. a

CHAPTER 30 CONSERVING EARTH'S BIODIVERSITY

OUTLINE

Section 30.1 What Is Conservation Biology?

Biodiversity describes the diversity of organisms, their genes, and the ecosystems in which they live. **Conservation biology** is the discipline that tries to preserve biodiversity.

Section 30.2 Why Is Biodiversity Important?

Ecosystem biodiversity provides many beneficial services that help sustain human life (**ecosystem services, Figure 30-1**), either directly or indirectly. Direct ecosystem services include food, building materials, medicines, natural fibers and fabrics, oxygen replenishment, and fuel. Indirect ecosystem services include soil formation, erosion control, climate regulation, genetic resources, and recreation.

The field of ecological economics attempts to measure the value of ecosystem services and assess the consequences when ecosystems are damaged by human profit-making activities.

Section 30.3 Is Earth's Biodiversity Diminishing?

The background extinction rate describes the natural rate of extinction in the absence of cataclysmic events. This occurs naturally at a low rate.

Mass extinctions describe cataclysmic events in which many forms of life go extinct. Five mass extinctions have occurred previously and many biologists believe that human activities are now causing a sixth mass extinction.

The International Union for the Conservation of Nature (IUCN) has established a list that classifies at-risk species.

Section 30.4 What Are the Major Threats to Biodiversity?

Biodiversity decline is a result of the use of an increasing amount of Earth's resources to support humanity (exceeding Earth's biocapacity, **Figure 30-6**) and the direct impact of human activity.

Human activities that directly impact Earth include **habitat destruction** and **fragmentation** (**Figures 30-7** and **30-8**), **species overexploitation**, displacement of native wildlife by **invasive species** (**Figure 30-9**), pollution, and global warming (**Figure 30-10**).

Section 30.5 How Can Conservation Biology Help to Preserve Biodiversity?

Conservation biology is an integrated science that seeks to understand the impact of human activities on natural ecosystems, to preserve and restore natural communities, to reverse the escalating loss of Earth's biodiversity, and to foster the sustainable use of Earth's resources.

Conservation efforts seek to establish **core reserves** that are protected from human activity and connect the reserves with **wildlife corridors**, thus promoting functional, self-sustainable communities (**Figures 30-11** and **30-12**).

Section 30.6 Why Is Sustainability the Key to Conservation?

Sustainable development promotes long-term ecological and human well-being without compromising the future. This can be accomplished by maintaining biodiversity, recycling raw materials, and using renewable resources.

Biosphere Reserves provide models for conservation and sustainable development by maintaining biodiversity while preserving local cultural values (**Figure 30-13**).

Sustainable agriculture helps preserve natural communities by using agricultural techniques that minimize adverse impacts on the environment (**Table 30-1**).

Human population growth is unsustainable and is causing resource consumption beyond Earth's biocapacity. Ultimately, human population growth must be curtailed if Earth's resources are to support life in the long term.

FLASH CARDS

To use the flash cards, tear the page from the book and cut along the dashed lines. The key term appears on one side of the flash card, and its definition appears on the opposite side.

biocapacity	ecosystem services
biodiversity	endangered species
Biosphere Reserve	habitat fragmentation
conservation biology	keystone species
core reserve	mass extinction
critically endangered species	minimum viable population (MVP)
ecological footprint	no-till

The processes through which natural ecosystems and their living communities sustain and fulfill human life. Ecosystem services include purifying air and water, replenishing oxygen, pollinating plants, reducing flooding, providing wildlife habitat, and many more.

An estimate of the sustainable resources and waste-absorbing capacity actually available on Earth. Biocapacity calculations are subject to change as new technologies change the way people use resources.

A species that faces a high risk of extinction in the wild in the near future.

The diversity of living organisms; often measured as the variety of different species in an individual ecosystem or in the entire biosphere.

The process by which human development and activities produce patches of wildlife habitat that may not be large enough to sustain minimum viable populations.

Designated by the United Nations, a Biosphere Reserve is a region intended to maintain biodiversity and evaluate techniques for sustainable human development while maintaining local cultural values.

A species whose influence on community structure is greater than its abundance would suggest.

The application of knowledge from ecology and other areas of biology to understand and conserve biodiversity.

A relatively sudden loss of many forms of life as a result of environmental change. The fossil record reveals five mass extinctions over geologic time.

A natural area protected from most human uses that encompasses enough space to preserve all of the biodiversity of the ecosystems in that area.

The smallest isolated population that can persist indefinitely and survive likely natural events such as fires and floods.

A species that faces an extreme risk of extinction in the wild in the immediate future.

A method of growing crops that leaves the remains of harvested crops in place, with the next year's crops being planted directly in the remains of last year's crops without significant disturbance of the soil.

The area of productive land needed to produce the resources used and absorb the wastes (including carbon dioxide) generated by an individual person, or by an average person of a specific part of the world (for example, an individual country), or of the entire world, using current technologies.

overexploitation

vulnerable species

sustainable development

wildlife corridor

threatened species

A species that is likely to become endangered unless conditions that threaten its survival improve.

Hunting or harvesting natural populations at a rate that exceeds those populations' ability to replenish their numbers.

A strip of protected land linking larger areas. Wildlife corridors allow animals to move freely and safely between habitats that would otherwise be isolated by human activities.

Human activities that meet present needs for a reasonable quality of life without exceeding nature's limits and without compromising the ability of future generations to meet their needs.

All species classified as critically endangered, endangered, or vulnerable.

SELF TEST

1. Looking at the graph below, what can you say about biocapacity and footprints?
   a. Biocapacity is directly correlated with the human footprint.
   b. The current human footprint has exceeded the biocapacity of Earth.
   c. We are seeing the trend toward a declining human footprint.
   d. It would take more than two planet Earths to support the human population at our current level of consumption.

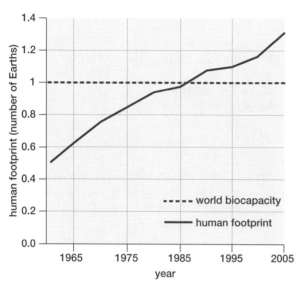

2. According to the graph below, which of the following statements is true?
   a. The current human footprint is stable.
   b. The global human footprint is equivalent to the individual human footprint.
   c. The current global human footprint is a result of human population growth.
   d. Human population growth is beginning to level off.

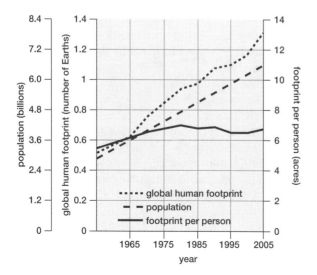

3. The genetic variability of our crop plants would be considered
   a. an indirect value of biodiversity.
   b. an economic value of biodiversity.
   c. necessary for the proper function of crops in sustainable agriculture.
   d. before the IUCN would list a species as threatened or endangered.

4. Which of the following is NOT true about extinction of species?
   a. Extinction is a natural process.
   b. Extinctions occur at a relatively low rate.
   c. Mass extinctions have occurred in the past.
   d. The current rate of extinction is equal to background rates.

5. Pollination of crop plants by insects and the input of nitrogen into the soil by bacteria associated with the roots of some plants are examples of
   a. genetic and species diversity in insects and plants.
   b. the ecosystem services provided by organisms.
   c. sustainable agricultural processes.
   d. methods used by conservation biologists to restore ecosystems.

6. Currently, extinction
   a. is higher than background levels.
   b. is due mostly to human activities.
   c. is affecting biodiversity.
   d. is explained by all of the statements above.

7. Your personal "ecological footprint" depends on
   a. the size of your house, how well it is insulated, and the climate in the region where you live.
   b. the types of transportation you use, how much you travel, and whether you travel alone or with a group.
   c. the types and amounts of foods you eat and how much you waste.
   d. how many goods you consume and the waste you produce.
   e. all of the above.

8. Clear-cutting forests, conversion of land to agricultural fields, damming of rivers, and draining of wetlands are
   a. examples of habitat fragmentation.
   b. among the primary reasons for the loss of biodiversity.
   c. listed by the IUCN as "watch list" activities.
   d. a result of the increase in human population growth and the ecological footprint.

9. A recent study estimates that at the present rate of harvest, the world's fisheries will be depleted by 2050. This is an example of
   a. pollution.   c. habitat destruction.
   b. overexploitation.   d. competition.

10. Which of the following is NOT a goal of conservation biologists?
    a. Understand the negative impact of animal habitats on human activity.
    b. Preserve and restore natural communities.
    c. Reverse the escalating loss of Earth's biodiversity that is caused by human activities.
    d. Foster sustainable use of Earth's resources.

11. Conservation biology depends on the support and expertise of
    a. people in a various fields in biology.
    b. government officials.
    c. educators.
    d. individuals.
    e. all of the above.

12. In Yellowstone Park, the removal of wolves had a wide-reaching negative effect on the level of biodiversity. Wolves are predators of elk and keep their populations in check. Without the wolves, elk eat the young aspens, which means that aspen groves could not regenerate. Aspen groves provide shelter and homes for many species of plants and birds, as well as material for beavers to build dams. The dams create marshlands that in turn create habitat for many aquatic organisms. Because the biodiversity of these communities depends on the presence of wolves, the wolves are considered to be
    a. dominant species.
    b. keystone species.
    c. domino species.
    d. predators.

13. Which of the following is NOT a feature of sustainable development?
    a. diverse communities with a richness of community interactions
    b. relatively stable populations that remain within the carrying capacity of the environment
    c. recycling and efficient use of raw materials
    d. reliance on nonrenewable sources of energy

14. The Nile perch is considered a(n)
    a. endangered species.
    b. extinct species.
    c. invasive species.
    d. keystone species.

15. No-till farming is an example of sustainable agriculture because it
    a. uses natural herbicides and pesticides to kill plant and animal pests.
    b. leaves remains of harvested plants behind to form mulch for the next year's plants.
    c. reduces soil erosion.
    d. all of the above.

16. Which of the following is an argument for protecting biodiversity?
    a. Decreasing biodiversity can result in a loss of resources that could provide humans with new drugs to cure disease or that contain genes that would produce better strains of crops.
    b. Organisms are linked in a complex web of life. Disrupting natural ecosystems may have unintended consequences; for example, insecticides have killed many of the pollinators of our agricultural crops.
    c. A hypothesis put forth by E. O. Wilson, biophilia, suggests that humans enjoy and derive benefit from natural landscapes.
    d. All of the above.

17. Which of the following is an ecosystem service provided free of charge by Earth's diverse ecosystems?
    a. soil production
    b. plant pollination
    c. water production
    d. erosion control
    e. all of the above

18. Species with several alleles for many of their genes display _____ genetic diversity and are _____ likely to be able to adapt in response to changing conditions.
    a. high; more
    b. high; less
    c. low; more
    d. low; less

19. Which of the following strategies might a conservation biologist use to manage a population of organisms threatened with extinction?
    a. Change the harvest regulations.
    b. Provide corridors for movement among disconnected habitats.
    c. Introduce another species to take its place in the environment.
    d. Promote tourism to the area so people will spend money in the local communities.

20. A hotel that caters to tourists visiting a Biosphere Reserve would likely be located in which region?
    a. The core.
    b. The transition area.
    c. The buffer zone.
    d. The hotel cannot be located anywhere within the reserve or it will impact the movements and habits of the organisms within the natural community. It must be located in the residential community outside of the reserve.

21. Which of the following would be the hardest groups in which to document extinction rates?
    a. mammals
    b. birds
    c. insects
    d. all are equally hard

22. Rerouting the Mississippi River for commercial reasons has played a role in the loss of wetlands in southern Louisiana. In addition to the loss of territory for many species, this also removed an important buffer to storm surges, which led to the disastrous flooding of New Orleans from Hurricane Katrina in 2005. This is an example of the consequences of
    a. pollution.
    b. overexploitation.
    c. habitat destruction.
    d. competition.

23. The zebra mussel was accidentally introduced into U.S. waterways in the ballast water released from ships. It reproduces much more successfully than native mussels and is outcompeting native aquatic organisms. The zebra mussel is an example of a(n)
    a. introduced species.
    b. parasite.
    c. predator.
    d. keystone species.

24. Global climate change is likely to increase the rate of extinction worldwide. The pollutant most implicated in this problem is
    a. nitrous oxides.
    b. carbon dioxide.
    c. sulfur dioxides.
    d. chlorofluorocarbon.

25. Which of the following statements would you expect to hear from a conservation biologist?
    a. Protecting the natural environment is crucial for protecting human welfare.
    b. The elaborate relationships among organisms that have evolved over millennia should be preserved within their natural environments.
    c. Other species should be protected even if they have no value to humans.
    d. All of the above.

26. Core reserves
    a. are vital stockpiles of fossil fuels being reserved for future energy use.
    b. are vital reservoirs of fresh water for future use.
    c. are areas set aside for human use and recreation.
    d. are areas that encompass entire ecosystems and their biodiversity that are protected from all but very low impact human activity.

27. Based on the burgeoning human population, today's individual core reserves are seldom large enough to meet the minimum required areas for many species. If you were in charge of core reserves in a region of the country, which of the following would you choose as the most efficient strategy for solving this problem?
    a. spearheading human-population reduction campaigns by urging couples to have only one or two children
    b. building wildlife corridors to connect reserves, thus enlarging the overall habitat range
    c. buying and developing land around the existing park to enlarge the area
    d. none of the above

28. Which of the following is NOT an example of sustainable development?
    a. the present commercial harvesting of fish
    b. harvesting trees from forests where logging and tree replanting and regrowth are balanced
    c. using wind for energy
    d. using biofuels for energy

29. Wheat can be grown sustainably if the farmer uses
    a. irrigation.
    b. more pesticides to prevent loss due to insect infestation.
    c. no-till strategies.
    d. commercial fertilizers to increase productivity.

30. Why is shade-grown coffee an example of a sustainable crop?
    a. Because it grows in the shade, it can be grown in tracts of relatively undisturbed rain forest.
    b. The trees of the remaining forest create a habitat for a large diversity of wildlife.
    c. The trees of the remaining forest reduce soil erosion.
    d. All of the above.

ANSWER KEY

1. b	16. d
2. c	17. c
3. a	18. a
4. d	19. b
5. b	20. b
6. d	21. c
7. e	22. c
8. b	23. a
9. b	24. b
10. a	25. d
11. e	26. d
12. b	27. b
13. d	28. a
14. c	29. c
15. d	30. d

CHAPTER 31 HOMEOSTASIS AND THE ORGANIZATION OF THE ANIMAL BODY

OUTLINE

Section 31.1 Homeostasis: How Do Animals Regulate Their Internal Environment?

Homeostasis describes the dynamic equilibrium within the animal body by which physiological conditions are maintained.

Animals regulate body temperature by deriving body warmth from the environment (**ectotherms**) or from metabolic activities within their own bodies (**endotherms**). Ectotherms can usually tolerate greater body temperature extremes than endotherms (**Figure 31-1**).

Homeostasis is most often maintained by **negative feedback** (**Figure 31-2**), in which a body change causes a response that counteracts the change and returns body conditions back to normal or the set point.

Sometimes **positive feedback** occurs, in which a change initiates events that intensify that change in a single direction until a desired action is reached. An example is the control of uterine contractions during childbirth.

Section 31.2 How Is the Animal Body Organized?

Cells are the building blocks of all life.

Tissues are made of cells of similar structure that perform a specific function. Animal tissues include **epithelial**, **connective**, **muscle**, and **nerve** tissue. Epithelial tissue forms membranous coverings over internal and external body surfaces and also forms **glands** (**Figure 31-4**). Connective tissue is composed of cells and extracellular matrix that usually serve to bind other tissues. Loose connective tissue can be found in the **dermis** of the skin; fibrous connective tissue includes **tendons** and **ligaments**; and specialized connective tissue includes **cartilage**, **bone**, **adipose tissue**, **lymph**, and **blood** (**Figures 31-5, 31-6, 31-7, and 31-8**). Muscle tissue has the ability to contract. **Skeletal muscle** is under voluntary control and is responsible for skeletal movements (**Figure 31-9**). **Smooth muscle** is under involuntary control and is commonly found in tubular organs such as the stomach and intestines. **Cardiac muscle** is under involuntary control, and is found only in the heart for heart muscle contraction. Nerve tissue (**Figure 31-10**) allows you to sense and respond to stimuli by the transmission of electrical stimuli. Nerve tissue is composed of cells that transmit electrical impulses (**neurons**) and cells that support neurons (**glial cells**).

Organs are made of more than one tissue type and perform complex functions. An example is the skin (**Figure 31-11**).

Organ systems are groups of two or more organs that work together for a common function. Organ systems include the integumentary, respiratory, circulatory, lymphatic/immune, digestive, urinary, nervous, endocrine, skeletal, muscular, and reproductive systems (**Table 31-1**).

FLASH CARDS

To use the flash cards, tear the page from the book and cut along the dashed lines. The key term appears on one side of the flash card, and its definition appears on the opposite side.

adipose tissue	ectotherm
blood	endocrine gland
bone	endotherm
cardiac muscle	epidermis
cartilage	epithelial tissue
connective tissue	exocrine gland
dermis	gland

An animal that obtains most of its body warmth from its environment; body temperatures of ectotherms vary with the temperature of their surroundings.

Tissue composed of fat cells.

A ductless, hormone-producing gland consisting of cells that release their secretions into the extracellular fluid from which the secretions diffuse into nearby capillaries; most endocrine glands are composed of epithelial cells.

A specialized connective tissue, consisting of a fluid (plasma) in which blood cells are suspended; carried within the circulatory system.

An animal that obtains most of its body heat from metabolic activities; body temperatures of endotherms usually remain relatively constant within a fairly wide range of environmental temperatures.

A hard, mineralized connective tissue that is a major component of the vertebrate endoskeleton; provides support and sites for muscle attachment.

In animals, specialized stratified epithelial tissue that forms the outer layer of the skin.

The specialized muscle of the heart; able to initiate its own contraction, independent of the nervous system.

A tissue type that forms membranes that cover the body surface and line body cavities, and that also gives rise to glands.

A form of connective tissue that forms portions of the skeleton; consists primarily of an extracellular matrix of collagen and the cells (chondrocytes) that secrete it.

A gland that releases its secretions into ducts that lead to the outside of the body or into a body cavity, such as the digestive or reproductive system; most exocrine glands are composed of epithelial cells.

A tissue type consisting of diverse tissues, including bone, cartilage, fat, and blood, that generally contain large amounts of extracellular material.

A cluster of cells that are specialized to secrete substances such as sweat, mucus, enzymes, or hormones.

The layer of skin beneath the epidermis; composed of connective tissue and containing blood vessels, muscles, nerve endings, and glands.

glial cell	negative feedback
hair follicle	nerve tissue
homeostasis	neuron
hormone	organ
ligament	organ system
lymph	positive feedback
muscle tissue	simple epithelium

A physiological process in which a change causes responses that tend to counteract the change and restore the original state. Negative feedback in physiological systems maintains homeostasis.

A cell of the nervous system that provides support and insulation for neurons, and regulates the composition of the extracellular fluid in the nervous system.

The tissue that makes up the brain, spinal cord, and nerves; consists of neurons and glial cells.

A cluster of specialized epithelial cells located in the dermis of mammalian skin, which produces a hair.

A single nerve cell.

The maintenance of the relatively constant internal environment that is required for the optimal functioning of cells.

A structure (such as the liver, kidney, or skin) composed of two or more distinct tissue types that function together.

A chemical that is synthesized by a group of cells, secreted, and usually carried in the bloodstream to other cells, whose activity is influenced by reception of the hormone.

Two or more organs that work together to perform a specific function; for example, the digestive system.

A tough band of dense connective tissue that joins two bones.

A physiological mechanism in which a change causes responses that tend to amplify the original change.

A pale fluid found within the lymphatic system; composed primarily of interstitial fluid and white blood cells.

A type of epithelial tissue, one cell layer thick, that lines many hollow organs such as those of the respiratory, digestive, urinary, reproductive, and circulatory systems.

Tissue composed of one of three types of contractile cells (smooth, skeletal, or cardiac).

skeletal muscle

tendon

smooth muscle

tissue

stratified epithelium

A tough band of dense connective tissue that connects a muscle to a bone.

The type of muscle that is attached to and moves the skeleton and is under the direct, normally voluntary, control of the nervous system; also called *striated muscle*.

A group of (normally similar) cells that together carry out a specific function, for example, muscle tissue; may include extracellular material produced by its cells.

The type of muscle that surrounds hollow organs, such as the digestive tract, bladder, and blood vessels; normally not under voluntary control.

A type of epithelial tissue composed of several cell layers, usually strong and waterproof; mostly found on the surface of the skin.

SELF TEST

1. The constancy of an animal body's internal environment is maintained by
   a. a single feedback mechanism.
   b. a few independent feedback systems.
   c. a coordinated, integrated network of systems.
   d. consistency in the functioning of all the organ systems.

2. Special sensory nerve endings in the skin of the hand are responsive to temperature. When an extremely hot object is encountered, nerves conduct this information to the spinal cord, which, in turn, sends a signal to skeletal muscle, causing it to contract and pull the affected part of the body away from the stimulus (often before the sensation of a burn is felt). In this scenario, the control center is
   a. the nerve endings in the skin.
   b. the spinal cord.
   c. the skeletal muscles of the hand.
   d. the nerves conducting impulses from the sensory nerves to the spinal cord.

3. Fish do not maintain whole-body temperatures different from the temperature of the water in which they live. However, many fish, if given a choice of water temperatures (say, in an experimental aquarium that offers a gradient of water temperatures), will select a narrow range of water temperatures in which to live. Thus, they exhibit a "preferred temperature" that they can maintain by controlling the amount of time they spend in water of different temperatures. Does this represent true homeostasis in the fullest sense of the term? Why or why not?
   a. Yes, because a constant body temperature is maintained.
   b. Yes, because body temperature is actively regulated such that internal physiological variables are kept within the range that cells need to function.
   c. No, because, even though the fish are maintaining relatively constant internal conditions, they are not using a feedback system in order to maintain these conditions.
   d. No, because homeostasis involves the control of a physiological variable within very narrow limits so that cells can function. The body temperature of a fish fluctuates with the temperature of the external environment.

4. How do animals regulate their physiology so that the physiological parameters (e.g., regulation of pH, body temperature, electrolyte balance) stay within narrow limits?
   a. homeostasis
   b. dynamic equilibrium

   c. negative feedback systems
   d. all of the above
   e. none of the above

5. Which of the following organisms would be considered an ectotherm?
   a. bear
   b. human
   c. fish
   d. bird

6. Which of the glands below would be considered an endocrine gland?
   a. sweat gland
   b. thyroid gland
   c. salivary gland
   d. mammary gland

7. In an unfortunate accident, a young man is struck in the leg by a bullet that tears through the femoral artery just above his right knee. The subsequent blood loss leads to a rapid drop in blood pressure. Baroreceptors in the aorta monitor blood pressure and send signals to the vasomotor center of the brain. In this case, the vasomotor center will increase its sympathetic stimulation of the blood vessels, which will cause them to constrict. This constriction leads to an increase in blood pressure. This is a description of
   a. a positive feedback system.
   b. a negative feedback system.
   c. both a and b.

8. Which tissue type would serve the function of contraction?
   a. epithelial
   b. connective
   c. muscle
   d. nervous

9. The skin contains
   a. epithelial tissue.
   b. connective tissue.
   c. nerve tissue.
   d. muscle tissue.
   e. all of the above.

10. Glands that become separated from the epithelium that produced them are called _____ glands.
    a. sebaceous
    b. sweat
    c. exocrine
    d. endocrine
    e. saliva

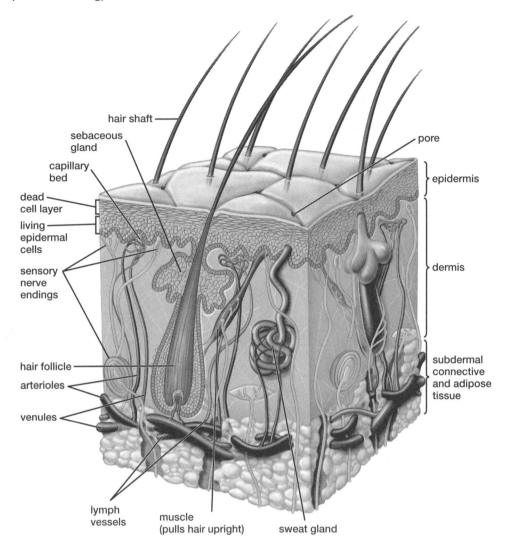

hair shaft

sebaceous gland

capillary bed

dead cell layer

living epidermal cells

sensory nerve endings

hair follicle

arterioles

venules

lymph vessels

muscle (pulls hair upright)

sweat gland

pore

epidermis

dermis

subdermal connective and adipose tissue

11. Which skin layer shown here would you expect to be thicker in cold-adapted species such as polar bears, whales, and walruses?
    a. epidermis
    b. dermis
    c. subdermal connective and adipose tissue
    d. hair

12. All of the following are examples of connective tissue EXCEPT
    a. tendons.
    b. ligaments.
    c. blood.
    d. muscle.
    e. adipose tissue.

13. Which of the following statements about muscle is true?
    a. Smooth muscle is important in locomotion.
    b. Skeletal muscle is not under conscious control.
    c. Cardiac muscle utilizes gap junctions.
    d. Smooth muscle is called voluntary muscle.
    e. Smooth muscle moves the skeleton.

14. All of the following are found in the dermis EXCEPT
    a. arteries.
    b. sensory nerve endings.
    c. hair follicles.
    d. sebaceous glands.
    e. cells packed with keratin.

15. Homeostasis is the condition in which the body maintains
    a. a low level of energy usage.
    b. a relatively constant internal environment.
    c. a body temperature without limits.
    d. a stable external environment in which to live.

16. Which of the following statements about feedback systems is true?
    a. Blood sugar levels are regulated by positive feedback systems.
    b. Negative feedback systems work to increase the original stimulus in the feedback system.
    c. Positive feedback systems are always harmful because the original stimulus runs "out of control."
    d. Negative feedback systems work to keep physiological variables within limits around a set point.

(a) Desert pupfish

(b) Ruby-throated hummingbird

(c) Collared lizard

17. In relation to body temperature, which of the organisms in this figure likely would be able to continue its normal routine during a prolonged cold snap where air temperatures are unseasonably cool?
    a. pupfish
    b. hummingbird
    c. lizard
    d. All animals would have trouble with their body temperature if it is unseasonably cold for a long period of time.

18. A friend told you he had heard that you could slow your heart rate down to a complete stop just by meditating. You tell him to evaluate his source because cardiac muscle is
    a. involuntary and not under conscious control.
    b. not striated, so it cannot be regulated by thought processes.
    c. striated and would contract all at once, so it would contract even with meditation.
    d. voluntary, but cannot be stopped since it contains gap junctions.

19. If someone had leukemia (a form of cancer of the blood cells), the individual would have a disorder associated with which tissue type?
    a. epithelial
    b. connective
    c. muscle
    d. nervous

20. Multiple sclerosis is an autoimmune disorder that destroys the myelin sheath of nerve cells. Myelin covers and protects the nerves. If an individual had this disorder, which cell type would be affected?
    a. neurons
    b. gland cells
    c. glial cells
    d. connective cells

21. When a snowball rolls down a snowy hill, it picks up snow, which causes it to roll faster. The result is that the snowball will pick up more snow and roll even faster. This scenario would best be described as
    a. a positive feedback system.
    b. a negative feedback loop.
    c. dynamic equilibrium.
    d. homeostasis.

22. The enhancement of uterine labor contractions by oxytocin would be considered
    a. a negative feedback system.
    b. dynamic equilibrium.
    c. a positive feedback system.
    d. a homeostatic mechanism to regulate the timing of childbirth.

23. Which of the following organ system(s) would help to maintain body temperature?
    a. integumentary system
    b. digestive system
    c. adipose system
    d. cardiovascular system
    e. both a and c

24. Which of the following tissues would you expect to find in the stomach?
    a. bone
    b. cartilage
    c. smooth muscle
    d. stratified squamous epithelium

25. Desert lizards rely on energy from the sun to regulate their body temperatures. Heat from the sun penetrates the skin and warms the blood, which is then circulated to the body core and other regions, warming them. When desert lizards need to cool their bodies, they move out of the sunlight until body temperatures drop. They must also, of course, be able to maintain water balance in extremely dry environments. Which of the following features would you NOT expect to find in reptilian skin?
    a. sweat glands
    b. a thick layer of heavily keratinized cells
    c. a relatively thin dermis
    d. pigment cells

26. Which of the following is NOT one of the major categories of animal tissue?
    a. endocrine tissue
    b. connective tissue

    c. epithelial tissue
    d. muscle tissue
    e. nerve tissue

27. Which of the following is NOT a type of connective tissue?
    a. fat
    b. blood
    c. bone
    d. tendons
    e. glands

28. Organs are
    a. formed of all four tissue types.
    b. formed of two or more tissues that operate independently.
    c. formed of two or more tissues that function together.
    d. formed of multiple tissues, all of which function independently for a particular goal.

29. Which of the following represents the organizational levels that make up the body of an animal from the most simple to the most complex?
    a. cell—tissue—organ—organ system
    b. organ system—organ—tissue—cell
    c. tissue—cell—organ system—organ
    d. tissue—organ—cell—organ system

30. Tendons and ligaments are which type of connective tissue?
    a. loose connective tissue
    b. fibrous connective tissue
    c. specialized connective tissue
    d. none of the above

ANSWER KEY

1. c	16. d
2. b	17. b
3. b	18. a
4. d	19. b
5. d	20. c
6. b	21. a
7. b	22. c
8. c	23. a
9. e	24. c
10. d	25. a
11. b	26. a
12. d	27. e
13. c	28. c
14. e	29. a
15. b	30. b

CHAPTER 32 CIRCULATION

OUTLINE

Section 32.1 What Are the Major Features and Functions of Circulatory Systems?

All circulatory systems are made of **blood**, **blood vessels**, and at least one **heart**.

Open circulatory systems consist of at least one heart that pumps blood through vessels into a **hemocoel**, where blood bathes internal tissues and organs (**Figure 32-1a**). Most invertebrates have this type of circulatory system.

Closed circulatory systems consist of at least one heart and blood vessels (**Figure 32-1b**). Blood never leaves the blood vessels and is efficiently transported to tissues and organs because of the higher blood pressures. All vertebrates and some invertebrates have this type of circulatory system.

The vertebrate circulatory system transports nutrients, hormones, gases, and wastes throughout the body. It also regulates temperature, forms clots to prevent blood loss, and protects the body against bacteria and viruses.

Section 32.2 How Does the Vertebrate Heart Work?

The vertebrate heart consists of **atria**, which collect blood from the body, and **ventricles**, which send blood to the body tissues.

The first vertebrate heart that evolved is seen in fish and is two chambered (one atrium and one ventricle). As fish gave rise to amphibians and reptiles, a three-chambered heart evolved (two atria and one ventricle) that allowed for partial separation between oxygenated and deoxygenated blood. As reptiles gave rise to birds and mammals, a four-chambered heart evolved (two atria and two ventricles) that allowed for complete separation of oxygenated and deoxygenated blood (**Figure 32-2**).

Veins carry blood to the heart, depositing it in the atria of the heart. When ventricles contract, blood is carried away from the heart by **arteries**. Deoxygenated blood enters the right atrium, which contracts and forces blood into the right ventricle. The right ventricle contracts and forces blood through pulmonary arteries to the lungs, where it is oxygenated. Oxygenated blood enters the left atrium through pulmonary veins. This atrium contracts and forces blood into the left ventricle. The left ventricle contracts and forces oxygenated blood through the aorta (a major artery) to the rest of the body (**Figure 32-3**).

Cardiac muscle is composed of cells joined by **intercalated discs** that allow electrical signals to spread among cells and cause interconnected heart regions to contract almost synchronously (**Figure 32-4**).

The **cardiac cycle** consists of atrial contraction, followed by ventricular contraction and a brief period of atrial and ventricular relaxation (**Figure 32-5**). **Systolic pressure** is generated during ventricular contraction, whereas **diastolic pressure** occurs during ventricular relaxation (**Figure 32-6**).

One-way heart valves maintain the direction of blood flow through the heart (**Figures 32-3 and 32-5**). The **atrioventricular valves** are found between the atria of the heart. The **semilunar valves** are found between the ventricles and the vessels.

Heart contractions are initiated by electrical impulses generated by the **sinoatrial node**. The contractions are then coordinated by the **atrioventricular node** and Purkinje fibers (**Figure 32-7**). The nervous system and hormones modify heart rate.

Section 32.3 What Is Blood?

Blood is composed of fluid **plasma** and cellular components (red blood cells, white blood cells, and **platelets**) (**Figure 32-8** and **Table 32-1**).

Plasma is composed of water, proteins, salts, nutrients, and wastes.

Red blood cells (**erythrocytes**) contain iron-containing proteins (**hemoglobin**) that carry oxygen (**Figure 32-9**). Red blood cell production occurs in the bone marrow and is influenced by the hormone erythropoietin (**Figure 32-10**).

White blood cells (**leukocytes**) come in five types and are specialized to fight infection and disease (**Figure 32-11**).

Platelets are fragments of cells (**megakaryocytes**) that stick to damaged blood vessels and release substances that result in the formation of sticky protein threads (**fibrin**) that form a **blood clot** (**Figures 32-12** and **32-13**).

Section 32.4 What Are the Types and Functions of Blood Vessels?

Once blood leaves the heart, it travels sequentially through arteries, **arterioles**, **capillaries**, **venules**, and veins (**Figure 32-14**). Veins return blood to the heart.

Arteries are vessels with thick muscular walls (**Figure 32-15**). They have elastic properties that help them to recoil during ventricular relaxation, thus maintaining blood flow.

Arterioles are small-diameter muscular vessels (**Figure 32-15**) that control the distribution of blood flow by constricting, or by controlling precapillary sphincter muscles. Arteriole diameter is controlled by the sympathetic nervous system and local factors, whereas **precapillary sphincters** are controlled by local factors.

Capillaries receive blood from arterioles. They are microscopic and have thin walls that allow for the exchange of nutrients and wastes between cells and the blood (**Figure 32-15**).

Venules collect capillary blood and deposit it into larger veins. Veins are wide and provide a low-resistance pathway for blood flow (**Figure 32-15**). Veins contain one-way valves that maintain blood flow in spite of the low blood pressure in veins (**Figure 32-17**).

Section 32.5 How Does the Lymphatic System Work with the Circulatory System?

The **lymphatic system** includes the lymphatic vessels, lymph nodes, tonsils, thymus, and spleen (**Figure 32-18**). It functions to transport small-intestine fats to the bloodstream, return excess interstitial fluid to the bloodstream, and defend the body against bacteria and viruses.

Lymph capillaries collect excess interstitial fluid (leaked from blood capillaries) through one-way valves. The fluid then enters lymph vessels that deposit it back into the bloodstream (**Figure 32-19**). This collected fluid is called **lymph**.

The tonsils, lymph nodes, and thymus house white blood cells that cleanse the lymph of bacteria and viruses that are delivered to them by the lymph vessels. The spleen performs a similar task, except it cleanses blood instead of lymph.

FLASH CARDS

To use the flash cards, tear the page from the book and cut along the dashed lines. The key term appears on one side of the flash card, and its definition appears on the opposite side.

angina	blood
arteriole	blood clotting
artery	blood vessel
atherosclerosis	capillary
atrioventricular (AV) node	cardiac cycle
atrioventricular valve	cardiac muscle
atrium (plural, atria)	closed circulatory system

A specialized connective tissue, consisting of a fluid (plasma) in which blood cells are suspended, carried within the circulatory system.

Chest pain associated with reduced blood flow to the heart muscle; caused by an obstruction of the coronary arteries.

A complex process by which platelets, the protein fibrin, and red blood cells block an irregular surface in or on the body, such as a damaged blood vessel, sealing the wound.

A small artery that empties into capillaries. Constriction of arterioles regulates blood flow to various parts of the body.

Any of several types of tubes that carry blood throughout the body.

A vessel with muscular, elastic walls that conducts blood away from the heart.

The smallest type of blood vessel, connecting arterioles with venules. Capillary walls, through which the exchange of nutrients and wastes occurs, are only one cell thick.

A disease characterized by the obstruction of arteries by cholesterol deposits and thickening of the arterial walls.

The alternation of contraction and relaxation of the heart chambers.

A specialized mass of muscle at the base of the right atrium through which the electrical activity initiated in the sinoatrial node is transmitted to the ventricles.

The specialized muscle of the heart, able to initiate its own contraction, independent of the nervous system.

A heart valve that separates each atrium from each ventricle, preventing the backflow of blood into the atria during ventricular contraction.

A type of circulatory system, found in certain worms and vertebrates, in which the blood is always confined within the heart and vessels.

A chamber of the heart that receives venous blood and passes it to a ventricle.

diastolic pressure	heart
erythrocyte	heart attack
erythropoietin	hemocoel
extracellular fluid	hemoglobin
fibrillation	hemolymph
fibrin	hypertension
fibrinogen	intercalated disc

A muscular organ responsible for pumping blood within the circulatory system throughout the body.

The blood pressure measured during relaxation of the ventricles; the lower of the two blood pressure readings.

A severe reduction or blockage of blood flow through a coronary artery, depriving some of the heart muscle of its blood supply.

A red blood cell, which contains the oxygen-binding protein hemoglobin and thus transports oxygen in the circulatory system.

A blood cavity within the bodies of certain invertebrates in which blood bathes tissues directly; part of an open circulatory system.

A hormone produced by the kidneys in response to oxygen deficiency; stimulates the production of red blood cells by the bone marrow.

The iron-containing protein that gives red blood cells their color; binds to oxygen in the lungs and releases it in the tissues.

Fluid that bathes the cells of the body; in mammals, extracellular fluid leaks from capillaries and is similar in composition to blood plasma, but lacking the large proteins found in plasma.

In animals with an open circulatory system, the fluid that is located within the hemocoel and that bathes all the body cells, therefore serving as both blood and extracellular fluid.

The rapid, uncoordinated, and ineffective contraction of heart muscle cells.

Arterial blood pressure that is chronically elevated above the normal level.

A clotting protein formed in the blood in response to a wound; binds with other fibrin molecules and provides a matrix around which a blood clot forms.

Junctions connecting individual cardiac muscle cells that serve both to attach adjacent cells to one another and to allow electrical signals to pass between cells.

The inactive form of the clotting protein fibrin. Fibrinogen is converted into fibrin by the enzyme thrombin, which is produced in response to injury.

leukocyte

plasma

lymph

platelet

lymphatic system

precapillary sphincter

megakaryocyte

semilunar valve

open circulatory system

sinoatrial (SA) node

pacemaker

spleen

plaque

stem cell

The fluid, noncellular portion of the blood.

Any of the white blood cells circulating in the blood.

A cell fragment that is formed from megakaryocytes in bone marrow; platelets lack a nucleus; they circulate in the blood and play a role in blood clotting.

A pale fluid found within the lymphatic system, composed primarily of extracellular fluid and white blood cells, especially lymphocytes.

A ring of smooth muscle between an arteriole and a capillary that regulates the flow of blood into the capillary bed.

A system consisting of lymph vessels, lymph capillaries, lymph nodes, and the thymus and spleen; helps protect the body against infection, absorbs fats, and returns excess fluid and small proteins to the blood circulatory system.

A valve located between the right ventricle of the heart and the pulmonary artery or between the left ventricle and the aorta; prevents the backflow of blood into the ventricles when they relax.

A large cell type in the bone marrow, which pinches off pieces of itself that enter the circulation as platelets.

A small mass of specialized muscle in the wall of the right atrium; generates electrical signals rhythmically and spontaneously and serves as the heart's pacemaker.

A type of circulatory system found in some invertebrates, such as arthropods and mollusks, that includes an open space (the hemocoel) in which blood directly bathes body tissues.

An organ of the lymphatic system located in the abdomen that serves several functions: (1) blood is filtered past lymphocytes and macrophages, which remove foreign particles, including bacteria; (2) lymphocytes multiply in the spleen during infection; (3) aged red blood cells are removed from the blood.

A cluster of specialized muscle cells in the upper-right atrium of the heart that produce spontaneous electrical signals at a regular rate; the sinoatrial node.

An undifferentiated cell that is capable of dividing and giving rise to one or more distinct types of differentiated cell(s).

A deposit of cholesterol and other fatty substances within the wall of an artery.

stroke

vein

systolic pressure

ventricle

thrombin

venule

In vertebrates, a large-diameter, thin-walled vessel that carries blood from venules back to the heart.

An interruption of blood flow to part of the brain caused by the rupture of an artery or the blocking of an artery by a blood clot. Loss of blood supply leads to rapid death of the area of the brain affected.

The lower muscular chamber on each side of the heart that pumps blood out through the arteries. The right ventricle sends blood to the lungs; the left ventricle pumps blood to the rest of the body.

The blood pressure measured at the peak of contraction of the ventricles; the higher of the two blood pressure readings.

A narrow vessel with thin walls that carries blood from capillaries to veins.

An enzyme produced in the blood as a result of injury to a blood vessel; catalyzes the production of fibrin, a protein that assists in blood clot formation.

SELF TEST

1. The _____ is an organism with an open circulatory system.
   a. snail
   b. sponge
   c. octopus
   d. frog
   e. human

2. The disorder represented in the figure shown here results from a round worm infection associated with
   a. arteries.
   b. veins.
   c. lymph vessels.
   d. heart valves.

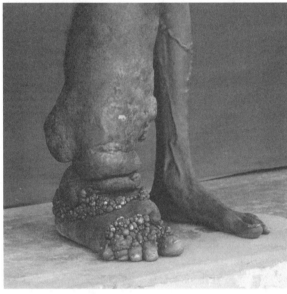

3. The area at the base of the heart in the figure shown below represents the
   a. area where oxygenated and deoxygenated blood mix.
   b. area that becomes the septum in a fish heart.
   c. region where we find the Purkinje fibers.
   d. area of capillaries in the heart.

4. Which of the following represents the highest blood pressure during ventricular relaxation?
   a. systolic pressure of 130
   b. blood pressure of 110/70
   c. diastolic pressure of 90
   d. blood pressure of 100/80

5. The evolution of the cardiovascular system in vertebrates involved several important changes. Which one of the following represents the order in which a major change occurred?
   a. a closed to an open circulatory system
   b. three-chambered to a two-chambered and then a four-chambered heart
   c. four-chambered to a three-chambered and then a two-chambered heart
   d. an open to a closed circulatory system

6. Mixing of oxygenated and deoxygenated blood within the heart occurs in the
   a. bird.
   b. frog.
   c. dog.
   d. fish.
   e. flatworm.

7. Varicose veins are a result of
   a. leaky valves in veins.
   b. poor capillary diffusion.
   c. blocked lymphatic vessels.
   d. poor skeletal muscle contraction.

8. Which statement BEST describes the role of the heart in a circulatory system?
   a. It acts as a reservoir for the storage of blood.
   b. It acts as a pump to move blood through the circulatory system.
   c. It exchanges oxygen and carbon dioxide with the outside air.
   d. It is where the blood cells are produced for the circulatory system.

9. Oxygenated blood is blood that has
   a. lost some of its oxygen after dumping it off at the lungs or gills.
   b. had its oxygen levels restored after passing through the body's tissues.
   c. lost some of its oxygen to the body's tissues.
   d. had its oxygen levels restored after picking up oxygen at the lungs.

10. Which event initiates blood clotting?
    a. contact with an irregular surface by platelets and other factors in plasma
    b. production of the enzyme thrombin
    c. conversion of fibrinogen into fibrin
    d. conversion of fibrin into fibrinogen
    e. excess flow of blood through a capillary

11. Which of the following is NOT a component of plasma?
    a. water
    b. globulins
    c. fibrinogen
    d. albumins
    e. platelets

12. A heart murmur is due to
    a. a genetic defect.
    b. a birth defect.
    c. leaky valves in the heart.
    d. low blood pressure in the heart.

13. How are capillaries specialized for the exchange of respiratory gases?
    a. Capillary walls are only one cell thick.
    b. Capillary walls are wrapped in a layer of smooth muscle.
    c. Capillaries allow the passage of fluids through spaces between the cells.
    d. Capillaries have one-way valves.
    e. Both a and c are correct.

14. The most important factor in the return of blood flow to the heart is
    a. the pumping of the heart.
    b. high pressure.
    c. skeletal muscle contraction.
    d. valves in the veins.

15. Lymph most closely resembles
    a. blood.
    b. urine.
    c. plasma.
    d. interstitial fluid.

16. We cannot say that arteries carry oxygenated blood because
    a. arterioles do not carry oxygenated blood.
    b. the pulmonary arteries carry deoxygenated blood.
    c. the aorta is the only artery that has oxygenated blood.
    d. the coronary arteries carry deoxygenated blood from the heart.

17. Which of the following is NOT an important function of the vertebrate circulatory system?
    a. transport of nutrients and respiratory gases
    b. regulation of body temperature
    c. protection of the body by circulating antibodies
    d. removal of waste products for excretion from the body
    e. defense against blood loss, through clotting

18. The wall of the left ventricle is thicker than the wall of the right ventricle in order to
    a. allow for a larger blood volume.
    b. pump blood with greater pressure.
    c. send blood to the left atrium under pressure.
    d. pump blood through a smaller valve.

19. Osmotic pressure is regulated by
    a. fibrinogen.
    b. albumin.
    c. globulin.
    d. valves in vessels.

20. Deoxygenated blood is delivered to the heart through the
    a. pulmonary artery.
    b. aorta.
    c. pulmonary veins.
    d. superior and inferior vena cavae.

21. Blood that flows through the pulmonary veins will be carried to the
    a. left ventricle.
    b. right ventricle.
    c. left atrium.
    d. right atrium.

22. The aortic semilunar valve opens when
    a. the left ventricle relaxes.
    b. the atria contract.
    c. all chambers are relaxed.
    d. the left ventricle contracts.

23. If your father was having a stent put in, he would be undergoing surgery that
    a. involves taking a portion of a leg vein and attaching it to the heart.
    b. opens a clogged artery.
    c. replaces the sinoatrial node to regulate heart rhythm.
    d. helps keep valves closed to eliminate murmurs.

24. Epinephrine is given as a treatment for shock victims because it
    a. is an anticoagulant.
    b. eliminates excess water and salts from the body, reducing blood pressure.
    c. increases heart rate and blood pressure.
    d. dilates the coronary artery carrying oxygenated blood to the heart muscle.

25. Why do people with kidney damage, particularly those who are on hemodialysis, tend to be severely anemic?
    a. The damaged kidneys are destroying red blood cells.
    b. There is a lack of erythropoietin.
    c. The bone marrow cannot get enough cholesterol to make the necessary red blood cells.
    d. Excessive urea excreted by the kidneys into urine prevents the bone marrow from making blood cells.

26. What is the determining factor that stimulates erythropoietin release from the kidney?
    a. too much carbon dioxide in the blood
    b. too little carbon dioxide in the blood
    c. too much oxygen in the blood
    d. too little oxygen in the blood

27. What would happen if a person received a drug that acted as a thrombin inhibitor?
    a. The heart rate would drop.
    b. The heart rate would increase.
    c. Anemia would result because of a decrease in the number of red blood cells.
    d. The ability to produce blood clots would be inhibited.

28. The liver is responsible for producing a variety of proteins found in plasma. When a person suffers from liver failure, a common symptom is edema (swelling) of the tissues. Why does this occur?
    a. The person is making smaller proteins that diffuse into the tissues. Water then follows these proteins into the tissues.
    b. The person is making fewer plasma proteins, but the few that remain are diffusing into the tissues. Water then follows these proteins into the tissues.
    c. The liver is overproducing plasma protein. The excess protein displaces water and forces it into the tissues.
    d. The liver is making fewer plasma proteins and a weak osmotic gradient is produced. Water that has entered the tissue begins to accumulate there.

29. During long trips, it is suggested that you stop the car occasionally to get out and walk around. This is important because
    a. sitting for long periods of time can increase blood pressure, which could cause a heart attack.
    b. sitting for long periods of time pinches off the arteries, causing blood flow to the extremities to slow.
    c. skeletal muscle contraction helps venous blood flow back to the heart so it doesn't pool and cause clots.
    d. the lowered blood pressure will cause you to lose sensation in your legs.

30. Lacteals are found in
    a. arteries.
    b. the small intestine.
    c. the spleen.
    d. heart.

ANSWER KEY

1. a	16. b
2. c	17. d
3. a	18. b
4. c	19. b
5. d	20. d
6. b	21. c
7. a	22. d
8. b	23. b
9. d	24. c
10. a	25. b
11. e	26. d
12. c	27. d
13. e	28. d
14. c	29. c
15. d	30. b

Chapter 33 RESPIRATION

OUTLINE

Section 33.1 Why Exchange Gases?

Breathing and gas exchange support cellular respiration.

When air is inhaled, O_2 is deposited into the blood and is carried to cells throughout the body for use in cellular respiration.

Cellular respiration produces CO_2, which leaves cells and is deposited in the blood. It is then carried to the lungs where it is released from the body.

Section 33.2 What Are Some Evolutionary Adaptations for Gas Exchange?

Gas exchange occurs by diffusion across moist, thin respiratory surfaces that have a large surface area.

Some animals that live in moist environments lack a specialized respiratory system, and instead use their small or thin flattened bodies as respiratory surfaces (**Figure 33-1**). Animals with low metabolic demands and/or those without a well-developed circulatory system may also lack specialized respiratory structures.

Larger, more active animals have developed respiratory systems that facilitate gas exchange by alternating **bulk flow** and diffusion (**Figure 33-2**).

Many aquatic animals use **gills**, which are thin projections from the body surface into the surrounding water (**Figures 33-3** and **33-5**). Gills are often elaborately branched or folded to maximize respiratory surface area. Some animals use **countercurrent exchange** mechanisms to maximize the amount of extracted O_2 from the water (**Figure E33-2**).

Terrestrial animals typically use one of two forms of internal respiratory structures: **tracheae** or **lungs**.

Insects use tracheae, which are a series of internal branching tubes that transport air from the outside environment (through **spiracle** openings) directly to the cells for gas exchange (**Figure 33-4**).

Most terrestrial vertebrates use lungs, which are protected internal chambers containing moist respiratory surfaces. Amphibians supplement lung breathing with the use of larval gills and cutaneous respiration (**Figures 33-5a** and **b**). Reptiles, which possess water-impermeable scales, rely solely on their well-developed lungs for respiration (**Figure 33-5c**). Birds use lungs that are composed of hollow, thin-walled tubes (parabronchi) that allow continuous movement of air through them (**Figure 33-6**). This allows for the uninterrupted exchange of gases and exhalation, unlike in the lungs of other vertebrates.

Section 33.3 How Does the Human Respiratory System Work?

The human respiratory system is composed of two parts: a **conducting portion** and a **gas-exchange portion**.

The conducting portion of the human respiratory system consists of passageways that carry air in and out of the lungs (the gas-exchange portion). It is composed of the nose and mouth, **pharynx**, **larynx**, **trachea**, **bronchi**, and **bronchioles** (**Figure 33-7a**).

The gas-exchange portion of the human respiratory system exchanges gases between the air in microscopic lung **alveoli** and the blood (**Figures 33-7b** and **33-9**). Specifically, O_2 diffuses into the blood from alveolar air and CO_2 diffuses out of the blood into alveolar air.

Most of the O_2 in the blood is transported bound to hemoglobin within red blood cells, but some is also dissolved in plasma (**Figure 33-10**).

Breathing involves drawing air into the lungs (**inhalation**) because of chest cavity expansion resulting from the contraction of the diaphragm and rib muscles. Relaxation of these muscles causes them to recoil, which expels air (**exhalation**) because of the partial collapse of the chest cavity (**Figure 33-11**).

Breathing rate is controlled by the respiratory center of the brain, which is influenced by the concentration of respiratory gases (primarily CO_2) within the blood.

FLASH CARDS

To use the flash cards, tear the page from the book and cut along the dashed lines. The key term appears on one side of the flash card, and its definition appears on the opposite side.

alveolus (plural, alveoli)	epiglottis
bronchiole	exhalation
bronchus (plural, bronchi)	gas-exchange portion
bulk flow	gill
conducting portion	Heimlich maneuver
countercurrent exchange	hemoglobin
diaphragm	inhalation

A flap of cartilage in the lower pharynx that covers the opening to the larynx during swallowing, and directs food into the esophagus.

A tiny air sac within the lungs, surrounded by capillaries, where gas exchange with the blood occurs.

The act of releasing air from the lungs, which results from a relaxation of the respiratory muscles.

A narrow tube, formed by repeated branching of the bronchi, that conducts air into the alveoli.

The portion of the respiratory system in lung-breathing vertebrates where gas is exchanged in the alveoli of the lungs.

A tube that conducts air from the trachea to each lung.

In aquatic animals, a branched tissue richly supplied with capillaries around which water is circulated for gas exchange.

The movement of many molecules of a gas or liquid in unison (in bulk, hence the name) from an area of higher pressure to an area of lower pressure.

A method of dislodging food or other obstructions that have entered the airway.

The portion of the respiratory system in lung-breathing vertebrates that carries air to the alveoli.

The iron-containing protein that gives red blood cells their color; binds to oxygen in the lungs and releases it in the tissues.

A mechanism for the transfer of some property, such as heat or a dissolved substance, from one fluid to another, generally without the two fluids actually mixing; in countercurrent exchange, the two fluids flow past one another in opposite directions, and they transfer heat or solute from the fluid with the higher temperature or higher solute concentration to the fluid with the lower temperature or lower solute concentration.

The act of drawing air into the lungs by enlarging the chest cavity.

In the respiratory system, a dome-shaped muscle forming the floor of the chest cavity; when the diaphragm contracts, it flattens, enlarging the chest cavity and causing air to be drawn into the lungs.

larynx

lung

pharynx

respiration

respiratory center

respiratory membrane

spiracle

trachea

tracheae

vocal cords

Within the lungs, the fusion of the epithelial cells of the alveoli and the endothelial cells that form the walls of surrounding capillaries.

The portion of the air passage between the pharynx and the trachea; contains the vocal cords.

An opening in the body wall of insects through which air enters the tracheae.

In terrestrial vertebrates, one of the pair of respiratory organs in which gas exchange occurs; consists of inflatable chambers within the chest cavity.

In terrestrial vertebrates, a flexible tube, supported by rings of cartilage, that conducts air between the larynx and the bronchi.

In vertebrates, a chamber that is located at the back of the mouth, shared by the digestive and respiratory systems; in some invertebrates, the portion of the digestive tube just posterior to the mouth.

The respiratory organ of insects, consisting of a set of air-filled tubes leading from openings in the body called *spiracles* and branching extensively throughout the body.

In terrestrial vertebrates, the act of moving air into the lungs (inhalation) and out of the lungs (exhalation); during respiration, oxygen diffuses from the air in the lungs into the circulatory system and carbon dioxide diffuses from the circulatory system into the air in the lungs.

A pair of bands of elastic tissue that extend across the opening of the larynx and produce sound when air is forced between them. Muscles alter the tension on the vocal cords and control the size and shape of the opening, which in turn determines whether sound is produced and what its pitch will be.

A cluster of neurons, located in the medulla of the brain, that sends rhythmic bursts of nerve impulses to the respiratory muscles, resulting in breathing.

SELF TEST

1. Spiracles are associated with
   a. gills.
   b. alveoli.
   c. tracheae.
   d. surfactant.

2. A major event in animal evolution associated with the respiratory system might be
   a. the development of a method to breathe through the body surface.
   b. the development of an efficient mechanism for cellular respiration.
   c. the development of a circulatory system within the respiratory system.
   d. the development of a method for internal gas exchange closely associated with the circulatory system.

3. After blood passes through the body's tissues, it will become
   a. enriched in oxygen and carbon dioxide.
   b. enriched with carbon dioxide and its oxygen content will decrease.
   c. enriched in oxygen and depleted of carbon dioxide.
   d. depleted of both oxygen and carbon dioxide.

4. Bird lungs differ from mammalian lungs in that
   a. bird lungs are structured for circular airflow.
   b. bird lungs have very efficient sacs for gas exchange called alveoli.
   c. parabronchi in bird lungs keep air flowing through the lungs and in contact with air spaces for gas exchange.
   d. bird lungs do not contain any epithelial tissue since gas exchange takes place in air spaces next to the capillaries of the circulatory system.

5. The vocal cords are found within the
   a. larynx.
   b. trachea.
   c. bronchi.
   d. parabronchi.

6. An advantage of gas exchange in aquatic habitats, as compared to terrestrial, is that
   a. water contains more dissolved oxygen than air.
   b. gills are more protected from environmental damage.
   c. keeping respiratory membranes moist is easy.
   d. body heat is more easily maintained.

7. If the cilia that line the respiratory tract were paralyzed
   a. air could not move into the lungs and the individual would suffocate.
   b. the parabronchi could not remain open, leading to inefficient gas exchange.
   c. the particles that are trapped in the mucus could not be expelled.
   d. the alveoli could not remain open and gas exchange could not occur.

8. A branch of the _____ brings deoxygenated blood to an alveolus.
   a. pulmonary artery
   b. pulmonary vein
   c. aorta
   d. inferior vena cava

9. Which of the following represents the role of the epiglottis in the human respiratory system?
   a. It blocks the esophagus during breathing to prevent air from going into the stomach.
   b. It contracts to force air out of the trachea.
   c. It blocks the larynx during swallowing to prevent food from entering the lungs.
   d. It acts as a gas-exchange surface for the exchange of carbon dioxide and oxygen.

10. The common passageway for food and air is the
    a. esophagus.
    b. pharynx.
    c. larynx.
    d. trachea.

11. Air moves into and out of the lungs as a result of
    a. osmosis.
    b. the pressure gradient of gases.
    c. diffusion.
    d. turbulence.

12. Which of the following structures represents the smallest tubes within the conducting portion of the respiratory system?
    a. bronchioles
    b. bronchi
    c. trachea
    d. larynx

13. The major criterion that determines rate of breathing is the
    a. blood oxygen level.
    b. blood carbon dioxide level.
    c. nitrogen in the atmosphere.
    d. blood carbon monoxide level.

14. When you perform strenuous exercise,
    a. the diaphragm and rib muscles will contract to increase chest volume.
    b. excess carbon monoxide will stimulate the respiratory center of the brain so the rate of breathing increases.
    c. oxygen will diffuse into the capillaries around the alveoli in addition to diffusing through the walls of the bronchi.
    d. additional oxygen and carbon dioxide are exchanged through the skin to support the increase in metabolism.

15. Carbon monoxide poisoning kills by
    a. preventing the production of neurotransmitters in brain cells.
    b. binding to oxygen, thereby preventing it from getting to hemoglobin.
    c. depriving cells of oxygen by competing with it for hemoglobin.
    d. enzymatically destroying the cells of the body.

16. The process depicted in this figure describes blood flow and gas exchange in which organism?
    a. humans
    b. insects
    c. snails
    d. fish

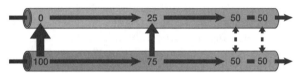

(a) Concurrent exchange

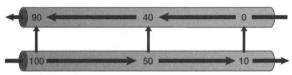

(b) Countercurrent exchange

17. When you break or bruise a rib it is difficult to breathe because
    a. muscle contraction around the ribs is painful, so it is difficult to increase and decrease the size of the chest cavity.
    b. the diaphragm pulls on the ribs making breathing painful.
    c. the broken rib keeps the lung from fully expanding.
    d. blood flow to the injury compromises the exchange of gases in the alveoli.

18. Oxygen diffuses from the blood into the tissues because
    a. the tissues have depleted their oxygen, and the oxygen in the blood is at a higher concentration than in the tissues.
    b. tissues produce carbon dioxide, which attracts oxygen.
    c. tissues produce oxygen and the oxygen in the blood wants to mix with the oxygen in the tissues.
    d. oxygen is charged and is attracted to oppositely charged molecules in the tissues.

19. Which of the following is the correct pathway of carbon dioxide from the blood to the atmosphere?
    a. mouth—pharynx—trachea—bronchi—alveoli
    b. mouth—larynx—bronchi—trachea—alveoli
    c. alveoli—bronchi—trachea—larynx—mouth
    d. alveoli—bronchi—trachea—pharynx—larynx—mouth

20. Which of the following animals use tracheae for respiration?
    a. mollusks
    b. snails
    c. insects
    d. fish

21. The "gas-exchange portion" of the respiratory system is the part of the anatomy in which
    a. the blood mixes with air.
    b. gases in the air mix.
    c. the only function is to carry air.
    d. respiratory gases are exchanged between the inhaled air within the alveoli and the blood flowing through the pulmonary capillaries.

22. Take a huge breath while sitting up straight in your chair, and then do this again while slumped over. Compare the air volume entering the lungs in each situation. Why is there a difference?
    a. The chest and rib cage cannot be increased in size effectively when you are slumped over in a chair.
    b. You cannot contract your chest muscles when slumped in a chair.
    c. Less air is moved into the lungs when you are slumped because the tissues need less oxygen for cellular respiration.
    d. The rib cage stays locked in an upward and outward orientation.

23. Compare the lungs to other respiratory structures such as gills and skin. What is the major advantage of the lungs?
    a. Oxygen diffuses faster across lung respiratory membranes than gill respiratory membranes.
    b. The lungs have more surface area, with an accompanying greater respiratory membrane for gas exchange.
    c. Lungs are easier to inflate than gills.
    d. The alveoli of lungs are more protected than other respiratory membranes.

24. Sleep apnea is a cessation of breathing that can result in lack of sleep, irritability, lack of concentration, tiredness, and, in severe cases, heart attacks. The apnea is prompted either by a failure of the brain to communicate with the respiratory muscles or an obstructed airway associated with the tongue or muscles of the pharynx or larynx. The more efficient treatment, although one that many people prefer not to use, is CPAP—continuous positive airway pressure. Fitting over the nose, a mask is attached to a tube that is connected to a portable low-pressure generator. How is this breathing different from normal breathing?
    a. There is no difference between the breathing mechanics. The generator produces negative pressure breathing, causing the air to be sucked into the lung in a similar way to normal breathing.
    b. The generator increases the air pressure so that the air is actually forced into the lungs. This is the opposite of what happens in a normal breath.
    c. The generator sucks the air out of the lungs actively, whereas in a normal expiration it is passively accomplished.
    d. The generator provides enough air pressure to force the pharynx in a back and open position so that air can flow freely into and out of the lungs.

25. Which mechanism is used to transport both carbon dioxide and oxygen in the blood?
    a. hemoglobin
    b. blood plasma
    c. sodium bicarbonate ions
    d. capillaries

26. If humans did not have lungs, but instead used their thin and moist skin for gas exchange, why would they struggle to meet the oxygen demand of the tissues?
    a. The connection to the circulatory system would be lost.
    b. Temperature control would become very difficult.
    c. The external air does not contain a sufficient amount of oxygen for sufficient gas exchange.
    d. The skin would not provide a great enough surface area for gas exchange.

27. What would you expect if the layer in this figure labeled surfactant fluid were destroyed or absent?
    a. Oxygen would not be able to diffuse across the membrane.
    b. The epithelial tissue of the alveoli would die due to lack of nutrients.
    c. The respiratory membranes would fuse and make diffusion of gases difficult.
    d. Red blood cells would be able to escape across the membrane into the alveoli of the lungs.

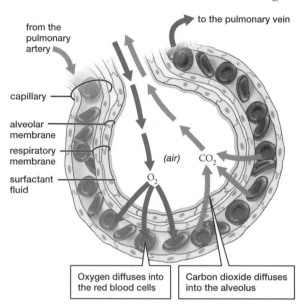

from the pulmonary artery

to the pulmonary vein

capillary

alveolar membrane

respiratory membrane

surfactant fluid

(air) CO_2

O_2

Oxygen diffuses into the red blood cells

Carbon dioxide diffuses into the alveolus

28. Why do many emphysema patients voluntarily contract muscles that help them increase their chest volume when they inhale?
    a. Emphysema patients have less respiratory surface area because of the damaged alveoli. To compensate, they must bring in larger volumes of air when they inhale.
    b. Emphysema patients have a higher metabolism. In order to meet the higher oxygen demand of the tissues, they must bring in larger volumes of air.
    c. The diaphragms of emphysema patients become paralyzed. The voluntary actions used to increase chest volume help compensate for this problem.
    d. The tumors associated with this disease obstruct the airways. The expansion of the chest cavity helps draw air past these obstructions.

29. A patient is rushed to the hospital unconscious. The doctor notes that the lips, extremities, and nailbeds of this patient have turned blue. The patient's friend tells the doctor that he thinks it is carbon monoxide poisoning. Why should the doctor be skeptical of this diagnosis?
    a. Carbon monoxide would cause harm to humans only if it were at extremely high concentrations in the air we breathe.
    b. Carbon monoxide breaks apart into carbon and oxygen. The oxygen can then bind to the hemoglobin in the red blood cells.
    c. The blue color in the nailbeds and lips means that the person has either lost a large amount of blood or blood is just not reaching these areas of the body.
    d. When carbon monoxide binds to hemoglobin, it produces the bright red color seen when oxygen binds to this molecule.

30. When oxygen and carbon dioxide are carried in the blood, very little is carried in the plasma in a dissolved form. How does this help facilitate gas exchange at the lungs and tissues?
    a. This helps pull oxygen into the blood from the air and carbon dioxide into the blood from the tissues.
    b. The gases travel faster when they are bound to hemoglobin.
    c. By competing for hemoglobin, the carbon dioxide can force the oxygen into the tissues. At the lungs, the oxygen forces the carbon dioxide off of this molecule so that it can be released into the air.
    d. By removing the oxygen and carbon dioxide from solution in the plasma, a gradient for these gases can be maintained to favor diffusion.

ANSWER KEY

1. c	16. d
2. d	17. a
3. b	18. a
4. c	19. c
5. a	20. c
6. c	21. d
7. c	22. a
8. a	23. b
9. c	24. b
10. b	25. a
11. b	26. d
12. a	27. c
13. b	28. a
14. a	29. d
15. c	30. d

CHAPTER 34 NUTRITION AND DIGESTION

OUTLINE

Section 34.1 What Nutrients Do Animals Need?

Nutrients provide the body with energy (measured as **calories**) and the raw materials needed to synthesize the molecules necessary for life. The six major forms of nutrients are (1) lipids, (2) carbohydrates, (3) proteins, (4) **minerals**, (5) **vitamins**, and (6) water.

Lipids are used as an energy source, help form cell membranes, and are used in the synthesis of hormones and bile. Animals can store energy in the form of fat. **Essential fatty acids** cannot be synthesized by humans for the production of specific lipids and must instead be ingested.

Carbohydrates are used as a source of quick energy but may be stored as starch or glycogen. The polysaccharide cellulose is the primary structural component in plants.

Proteins play many structural and functional roles in the body and are composed of amino acids. Essential amino acids cannot be synthesized by humans and must instead be ingested.

Minerals are elements that play an essential role in animal nutrition (**Table 34-2**).

Vitamins are organic compounds that are required for normal cell function and control many metabolic reactions throughout the body (**Table 34-3**). Vitamins can be grouped as water-soluble vitamins and fat-soluble vitamins.

Water is an important medium for internal transport and chemical reactions. Two-thirds of the human body is water.

Section 34.2 How Does Digestion Occur?

Digestion acts to break down food into absorbable forms. Animal **digestive systems** are organ systems that help optimize digestive processes. This is accomplished through **ingestion**, **mechanical digestion**, **chemical digestion**, **absorption**, and **elimination**.

Sponges are the only animals that digest their food by **intracellular digestion**, through the use of collar cells, **food vacuoles**, and **lysosomes** (**Figure 34-6**). Larger organisms digest their food by **extracellular digestion**.

The simplest digestive systems (e.g., those found in cnidarians) consist of a sac with one opening (**gastrovascular cavity**) in which all extracellular digestive processes take place (**Figure 34-7**).

Active animals that must eat frequently digest their food using a tubular digestive tract, composed of a one-way tube with openings at either end. This arrangement allows for the consumption of multiple meals and digestive specialization. Examples of digestive specializations include variation in tooth forms (**Figure 34-9**), bird gizzards (**Figure 34-10**), **ruminant** cellulose digestion (**Figure 34-11**), and variations in **small intestine** length.

Section 34.3 How Do Humans Digest Food?

Human digestive systems are well adapted for processing a wide variety of foods (**Figure 34-12**). This is accomplished, in part, by the actions of powerful digestive secretions (**Table 34-4**).

Mechanical and chemical breakdown of food begins in the **mouth**, through the chewing action of teeth and the secretion of **amylase** (which begins breaking down starch into sugar) in saliva.

Swallowing forces food from the mouth into the **esophagus** (**Figure 34-13**), which then transports it to the **stomach** by **peristalsis**. The stomach is an expansible muscular sac that stores and churns food, continuing its mechanical breakdown (**Figure 34-14**). The stomach also continues the chemical breakdown of food through the secretion of **proteases** and acidic stomach juices that start the protein digestion process.

Most chemical digestion occurs in the small intestine, through the action of **bile** (which emulsifies fat droplets), **pancreatic juices** (which neutralize acidic chyme and continue carbohydrate, lipid, and protein breakdown), and cells of the intestinal wall (which contain peptidases and disaccharidases). These processes are summarized in **Figure 34-15**. Most absorption takes place in the small intestine through the use of **villi** and **microvilli** in the intestinal lining (**Figure 34-16**). Most nutrients are absorbed into villus capillaries, but glycerol and fatty acids are absorbed by lymphatic capillaries (**lacteals**).

The **large intestine** is responsible for water absorption and the formation of semisolid **feces**, which are composed of indigestible wastes, bacteria, and leftover nutrients.

Stimuli from food (taste, smell, and visual appeal) trigger nervous system responses that prepare the digestive system for food digestion. The hormone **gastrin** is involved in the regulation of stomach acidity (**Figure 34-17**), the hormones **secretin** and **cholecystokinin** stimulate the release of bile and pancreatic juices into the small intestine, and the hormone gastric inhibitory peptide stimulates insulin release by the pancreas and inhibits stomach acid production and peristalsis.

FLASH CARDS

To use the flash cards, tear the page from the book and cut along the dashed lines. The key term appears on one side of the flash card, and its definition appears on the opposite side.

absorption	chemical digestion
amylase	cholecystokinin
bile	chylomicron
body mass index	chyme
calorie	colon
Calorie	digestion
carnivore	digestive system

The process by which particles of food within the digestive tract are exposed to enzymes and other digestive fluids that break down large molecules into smaller subunits.

The process by which nutrients enter the body through the cells lining the digestive tract.

A digestive hormone produced by the small intestine that stimulates the release of pancreatic enzymes.

An enzyme, found in saliva and pancreatic secretions, that catalyzes the breakdown of starch.

A particle produced by cells of the small intestine, consisting of proteins, triglycerides, and cholesterol; transports the products of lipid digestion into the lymphatic system and ultimately into the circulatory system.

A liquid secretion, produced by the liver, that is stored in the gallbladder and released into the small intestine during digestion; consists of a complex mixture of bile salts, water, other salts, and cholesterol.

An acidic, souplike mixture of partially digested food, water, and digestive secretions that is released from the stomach into the small intestine.

A number derived from an individual's weight and height that is used to estimate body fat. The formula is weight (in kg) / height^2 (in meters^2).

The longest part of the large intestine; does not include the rectum.

The amount of energy required to raise the temperature of 1 gram of water by 1 degree Celsius.

The process by which food is physically and chemically broken down into molecules that can be absorbed by cells.

A unit used to measure the energy content of foods; it is the amount of energy required to raise the temperature of 1 liter of water 1 degree Celsius; also called a *kilocalorie*, equal to 1,000 calories.

A group of organs responsible for ingesting food, digesting food into simple molecules that can be absorbed into the circulatory system, and expelling undigested wastes from the body.

Literally, "meat eater"; a predatory organism that feeds on herbivores or on other carnivores; a secondary (or higher) consumer.

duodenum	feces
elimination	food vacuole
epiglottis	gallbladder
esophagus	gastric gland
essential amino acid	gastrin
essential fatty acid	gastrovascular cavity
essential nutrient	herbivore
extracellular digestion	ingestion

Semisolid waste material that remains in the intestine after absorption is complete and is voided through the anus. Feces consist principally of indigestible wastes and bacteria.

The first section of the small intestine, in which most food digestion occurs; receives chyme from the stomach, buffers and digestive enzymes from the pancreas, and bile from the liver and gallbladder.

A membranous sac within a cell in which food is enclosed. Digestive enzymes are released into the vacuole, where intracellular digestion occurs.

The expulsion of indigestible materials from the digestive tract, through the anus, and outside the body.

A small sac located next to the liver that stores and concentrates the bile secreted by the liver. Bile is released from the gallbladder to the small intestine through the bile duct.

A flap of cartilage in the lower pharynx that covers the opening to the larynx during swallowing; directs food into the esophagus.

One of numerous small glands in the stomach lining; contains cells that secrete mucus, hydrochloric acid, or pepsinogen (the inactive form of the protease pepsin).

A muscular, tubular portion of the mammalian digestive tract located between the pharynx and the stomach; no digestion occurs in the esophagus.

A hormone produced by the stomach that stimulates acid secretion in response to the presence of food.

An amino acid that is a required nutrient; the body is unable to manufacture essential amino acids, so they must be supplied in the diet.

A saclike chamber with digestive functions, found in some invertebrates such as cnidarians (sea jellies, anemones, and related animals); a single opening serves as both mouth and anus.

A fatty acid that is a required nutrient; the body is unable to manufacture essential fatty acids, so they must be supplied in the diet.

Literally, "plant-eater"; an organism that feeds directly and exclusively on producers; a primary consumer.

Any nutrient that cannot be synthesized by the body, including certain fatty acids and amino acids, vitamins, minerals, and water.

The movement of food into the digestive tract, usually through the mouth.

The physical and chemical breakdown of food that occurs outside a cell, normally in a digestive cavity.

intracellular digestion	mechanical digestion
lacteal	metabolic rate
lactose intolerance	microvillus (plural, microvilli)
large intestine	mineral
lipase	mouth
liver	nutrient
lysosome	omnivore

The process by which food in the digestive tract is physically broken down into smaller pieces.

The chemical breakdown of food within single cells.

The speed at which cellular reactions that release energy occur.

A lymph capillary; found in each villus of the small intestine.

A microscopic projection of the plasma membrane of a cell that increases the cell's surface area; in the digestive tract, microvilli increase the surface area of the villi.

Symptoms such as bloating, gas pains, and diarrhea, caused by the inability to digest lactose (the principal sugar in milk); occurs because lactase, the enzyme that digests lactose, is not produced in sufficient amounts.

An inorganic substance, especially one in rocks or soil, or dissolved in water. In nutrition, minerals such as sodium, calcium, and potassium are essential nutrients that must be obtained in the diet.

The final section of the digestive tract; consists of the colon and the rectum, where feces are formed and stored.

The opening through which food enters a tubular digestive system.

An enzyme that catalyzes the breakdown of lipids such as fats.

A substance acquired from the environment and needed for the survival, growth, and development of an organism.

An organ with varied functions, including bile production, glycogen storage, and the detoxification of poisons.

An organism that consumes both plants and other animals.

A membrane-bound organelle containing intracellular digestive enzymes.

pancreas

secretin

pancreatic juice

small intestine

peristalsis

sphincter muscle

pharynx

stomach

protease

villus (plural, villi)

rectum

vitamin

ruminant

A hormone produced by the small intestine that stimulates the production and release of digestive secretions by the pancreas and liver.

A combined exocrine and endocrine gland located in the abdominal cavity next to the stomach. The endocrine portion secretes the hormones insulin and glucagon, which regulate glucose concentrations in the blood. The exocrine portion secretes pancreatic juice (a mixture of water, enzymes, and sodium bicarbonate) into the small intestine; the enzymes digest fat, carbohydrate, and protein; the bicarbonate neutralizes acidic chyme entering the intestine from the stomach.

The portion of the digestive tract, located between the stomach and large intestine, in which most digestion and absorption of nutrients occur.

A mixture of water, sodium bicarbonate, and enzymes released by the pancreas into the small intestine.

A circular ring of muscle surrounding a tubular structure, such as the esophagus, stomach, or intestine; contraction and relaxation of a sphincter muscle controls the movement of materials through the tube.

Rhythmic coordinated contractions of the smooth muscles of the digestive tract that move substances through the digestive tract.

The muscular sac between the esophagus and small intestine where food is stored and mechanically broken down and in which protein digestion begins.

In vertebrates, a chamber that is located at the back of the mouth and is shared by the digestive and respiratory systems; in some invertebrates, the portion of the digestive tube just posterior to the mouth.

A finger-like projection of the wall of the small intestine that increases its absorptive surface area.

An enzyme that digests proteins.

One of a group of diverse chemicals that must be present in trace amounts in the diet to maintain health; used by the body in conjunction with enzymes in a variety of metabolic reactions.

The terminal portion of the vertebrate digestive tube where feces are stored until they are eliminated.

An herbivorous animal with a digestive tract that includes multiple stomach chambers, one of which contains cellulose-digesting bacteria, and that regurgitates the contents ("cud") of the first chamber for additional chewing ("ruminating").

SELF TEST

1. The best way to lose weight is through
   a. exercise, which uses more calories than the calories consumed.
   b. the elimination of fats from the diet.
   c. the consumption of supplements containing high levels of chromium.
   d. the elimination of carbohydrates from the diet.

2. Why can vitamins A, D, E, and K be toxic, but C and B cannot be toxic?
   a. Fat-soluble vitamins such as C and B are stored in the body.
   b. Vitamins B and C do not affect brain chemistry.
   c. Vitamins A, D, E, and K are used faster than C and B.
   d. Humans get more vitamins A, D, E, and K than C and B in their diets.

3. Which of the following BMI values would be considered to be healthy for an adult female?
   a. 12.5
   b. 20
   c. 26
   d. 30

4. Which of the following explains why the average herbivore has a longer intestine than the average carnivore?
   a. Carnivores do not have to digest their food mechanically.
   b. Carnivores absorb many of their nutrients from the stomach.
   c. The cell walls in plants are difficult to digest. For herbivores, the longer intestine provides more opportunity to extract nutrients from plant material.
   d. Carnivores produce the enzyme cellulase. This enzyme helps in the digestion of cellulose found in the cell walls of plant cells. This efficient digestion of plant material means that carnivores can have shorter intestines.

5. Digestion in cnidarians, as in hydra, occurs in a
   a. crop.
   b. coelom.
   c. digestive tract.
   d. gastrovascular cavity.
   e. pharynx.

6. Collar cells are found in
   a. sponges.
   b. fish.
   c. insects.
   d. tapeworms.
   e. hydra.

7. The role of the liver in the digestion of food is to produce
   a. lipase for the digestion of fat.
   b. bile for the emulsification of fat.
   c. pepsin for the digestion of protein.
   d. bile for the digestion of carbohydrates.

8. A totally fat-free diet is unrealistic and unhealthy. A fat-free diet would be dangerous because
   a. there are essential fatty acids acquired only through diet that you would not get in a fat-free diet.
   b. the body needs insulation to maintain a constant internal body temperature.
   c. the bile produced by the liver to break down fats would build up in the gallbladder and become toxic.
   d. only plant material could be eaten, which would lead to too much fiber for the colon to digest.

9. A sudden increase in the amount of secretin circulating in the blood is an indication that food has recently been introduced into the
   a. mouth.
   b. pharynx.
   c. stomach.
   d. small intestine.
   e. large intestine.

10. Which organs produce secretions involved in the chemical breakdown of carbohydrates?
    a. salivary glands
    b. stomach
    c. large intestine
    d. pancreas
    e. both a and d
    f. both b and c

11. If you found a skull in the woods while you were hiking, and the skull had no canines, you could say that the organism
    a. could eat any type of food.
    b. was an herbivore.
    c. utilized a strategy of intracellular digestion.
    d. was a ruminant.

12. This child has
    a. a vitamin B$_{12}$ deficiency that leads to diges-tive disorders related to nutrient absorption.
    b. a blockage in the intestine that causes waste material to build up in the abdominal cavity.
    c. a lack of protein, which causes fluid to build up in the abdominal cavity.
    d. starvation, which causes fluids to remain trapped in the intestine due to an inability to absorb fluids through osmosis.

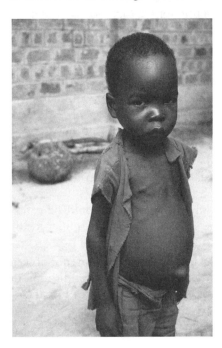

13. Peristalsis is important in
    a. the movement of chyme.
    b. protein absorption.
    c. movement of food in the esophagus.
    d. the regulation of the release of digestive enzymes.

14. Intracellular digestion occurs in
    a. humans.
    b. earthworms.
    c. sea anemones.
    d. sponges.

15. The diet of a raccoon would best be described as
    a. herbivorous.
    b. omnivorous.
    c. carnivorous.
    d. detritivorous.

16. A(n) _____ raises the temperature of a gram of water by 1 degree Celsius.
    a. Calorie   c. fat gram
    b. calorie   d. ATP

17. "Water-soluble compound that works primarily as an enzyme helper" would be a good definition of
    a. vitamin B.   c. vitamin E.
    b. vitamin A.   d. vitamin D.

18. Food particles are broken down into smaller pieces to make enzyme activity more efficient. This would best be described as
    a. chemical digestion.
    b. absorption.
    c. elimination.
    d. mechanical digestion.

19. Which is the correct sequence of events that all digestive systems must follow in the processing of foodstuffs?
    a. ingestion—chemical digestion—mechanical digestion—absorption—elimination
    b. ingestion—mechanical digestion—chemical digestion—absorption—elimination
    c. ingestion—absorption—mechanical digestion—chemical digestion—elimination
    d. ingestion—absorption—chemical digestion—mechanical digestion—elimination

20. An earthworm's digestive anatomy is designed so that this animal can feed more frequently than an animal such as a sea anemone. Which of the following explains this relationship?
    a. A sea anemone does not produce any diges-tive enzymes.
    b. A sea anemone cannot ingest its food.
    c. An earthworm has a highly efficient rumi-nant stomach.
    d. An earthworm has a tubular digestive tract, whereas a sea anemone has a digestive sys-tem with only one opening.

21. Chyme is
    a. a protein-digesting enzyme.
    b. the thickened food material after digestion in the stomach.
    c. enzymatic juice secreted by the small intestine.
    d. the fatty material absorbed into lacteals, also called chylomicrons.
    e. an enzyme that breaks down complex car-bohydrates into disaccharides.

22. The cause of ulcers is
    a. *Helicobacter pylori.* d. allergies.
    b. stress.   e. smoking.
    c. alcoholism.

23. The function of bile is the
    a. facilitation of sugar and amino acid absorp-tion.
    b. mechanical digestion of fats for better break-down with lipase, a lipid-digestive enzyme.
    c. production of bile pigments.
    d. production of glucose.
    e. maintenance of blood glucose level.

24. The nasal cavities and oral cavity merge into the pharynx. What keeps food and drink from going into the nasal cavity rather than down the esophagus?
    a. The tongue blocks the upper pharynx leading into the nasal cavities.
    b. The soft palate moves up against the opening to the nasal cavities, blocking the entryway.
    c. Breathing is temporarily stopped so that the nasal cavities can close when swallowing.
    d. The esophagus constricts at the upper end of the pharynx so the entry to the nasal cavities is blocked.

25. Weight loss, anemia, and fatigue are symptoms of
    a. gallstones. The hardened bile in the gallbladder affects lipid metabolism, causing weight loss and fatigue. The anemia is due to a lack of bile pigments available to produce red blood cells.
    b. peptic ulcer. The bleeding ulcer causes loss of blood, resulting in anemia and fatigue. The weight loss is caused by the inability of the stomach to absorb nutrients.
    c. intestinal cancer. There is less surface area for nutrient absorption, hence the weight loss. The lack of nutrients also accounts for the inability to make the hemoglobin and red cells, leading to anemia, and as a result of the lack of oxygen-carrying capacity, fatigue.
    d. colitis. The inflammation of the colon makes it impossible for nutrients to be properly absorbed, hence the weight loss and fatigue. The inflammation also disrupts the absorption of various vitamins required to make hemoglobin and red blood cells and reduces oxygen-carrying capacity.

26. Food entering the _____ triggers the swallowing reflex.
    a. mouth   c. pharynx
    b. larynx   d. trachea

27. Which of the following is NOT a function of the stomach?
    a. Protein digestion begins due to the actions of pepsin.
    b. Rhythmic contractions known as peristaltic movements move food into the stomach and aid in the mixing of the partially digested contents and the gastric secretions (chyme) of the stomach.
    c. It produces a thick mucous secretion along the stomach lining that protects the

stomach from being digested by its own secretions (e.g., pepsin).
    d. It is one of the primary sites for the absorption of digested protein fragments.

28. Why does the small intestine absorb more nutrients than any other region of the digestive tract?
    a. It is the site of segmentation movements.
    b. The exposed surface area approaches that of a tennis court in size.
    c. It contains a concentrated supply of capillaries that lie in intimate contact with this region of the digestive tract.
    d. All of the above are correct.

29. Ruminant physiology works because
    a. the digestive enzymes produced are better able to digest material than in organisms with traditional physiology.
    b. organisms with that digestive physiology have symbiotic organisms in their digestive system that break down cellulose.
    c. organisms with that digestive physiology produce special enzymes for the breakdown of cellulose.
    d. it is more complex, and organisms that utilize that strategy have evolved to feed on a variety of foods to avoid competition.

30. This child is suffering from a deficiency of
    a. calcium.   c. sodium.
    b. vitamin B$_{12}$.   d. vitamin D.

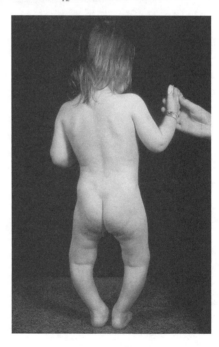

ANSWER KEY

1. a
2. a
3. b
4. c
5. d
6. a
7. b
8. a
9. d
10. e
11. b
12. c
13. c
14. d
15. b

16. b
17. a
18. d
19. b
20. d
21. b
22. a
23. b
24. b
25. c
26. c
27. d
28. d
29. b
30. d

CHAPTER 35 THE URINARY SYSTEM

OUTLINE

Section 35.1 What Are the Basic Functions of Urinary Systems?

All **urinary (excretory) systems** perform three functions: (1) filter blood or **extracellular fluid**, thus removing water and small dissolved molecules; (2) selectively reabsorb nutrients from filtrate; and (3) excrete remaining water and dissolved wastes from body.

Section 35.2 What Are Some Examples of Invertebrate Excretory Systems?

Flatworm excretory systems consist of branching **protonephridia** that carry fluids and wastes to excretory pores that lead out of the body (**Figure 35-1a**). Flatworms also have a large body surface area through which most cellular wastes diffuse out.

Insects use **Malphigian tubules** that filter hemocoel blood and release wastes, salts, and water into the filtrate that is deposited into the intestine. The intestine and rectum reabsorb salts and water, producing concentrated urine that is released with feces (**Figure 35-1b**).

Earthworms, mollusks, and several other invertebrates use **nephridia** that filter the interstitial fluid that fills the body cavity. Salts and nutrients are reabsorbed, leaving water and wastes that are excreted through the nephridiopore opening in the body wall (**Figure 35-1c**).

Section 35.3 What Are the Functions of the Human Urinary System?

Vertebrate urinary systems regulate blood ion, water, and nutrient levels; maintain blood pH; secrete hormones; and eliminate cellular wastes.

The **kidney** is the organ in humans that produces **urea** from the nitrogen waste produced as a by-product of the breakdown of proteins (**Figure 35-2**).

Section 35.4 What Are the Structures of the Human Urinary System?

The human urinary system is composed of kidneys, **ureters**, a **bladder**, and a **urethra** (**Figure 35-3**). Each kidney is composed of more than 1 million urine-forming nephrons that are located in the **renal cortex** and **renal medulla** (**Figure 35-4**). Each nephron is composed of a **glomerulus, Bowman's capsule**, and **tubule** (**Figure 35-5**). Fluid within the nephron tubule is received by a **collecting duct**, which deposits urine in the **renal pelvis** of the kidney. The renal pelvis then funnels urine into the ureter and then into the bladder where it is collected. Distention of the muscular bladder triggers urination, which causes urine to flow through the urethra on its way out of the body.

Section 35.5 How Is Urine Formed and Concentrated?

During **filtration**, blood is filtered under pressure through capillaries in the glomerulus, which deposits water and dissolved substances as **filtrate** in Bowman's capsule (**Figure 35-6**). Next, during **tubular reabsorption**, filtrate in Bowman's capsule flows into the nephron tubule, where the active transport of nutrients and osmotic movement of water into the capillary blood take place. During the final stage, **tubular secretion**, some wastes are actively transported into the distal tubule filtrate from capillary blood.

The nephron **loop of Henle** generates a salt gradient in the kidney interstitial fluid surrounding the nephron. This causes the osmotic movement of water from the filtrate into the interstitial fluid of the kidney, thus concentrating urine (**Figure E35-2**).

Section 35.6 How Do Vertebrate Kidneys Help Maintain Homeostasis?

Mammalian kidneys regulate the water content of the blood by the release of **antidiuretic hormone (ADH)**, which increases the permeability of the collecting ducts to water. ADH release is stimulated by water loss and dehydration (**Figure 35-7**).

The kidneys regulate blood pressure by releasing the hormone **renin** into the bloodstream. When blood pressure is low, renin is released by the kidney and catalyzes the formation of the hormone **angiotensin**. Angiotensin causes arterioles to constrict, elevating blood pressure.

The kidneys regulate blood oxygen levels by releasing the hormone **erythropoietin** when blood oxygen levels are low, which stimulates bone marrow to produce red blood cells.

The kidneys regulate the concentration of dissolved substances in the blood by controlling tubular secretion and reabsorption.

Loops of Henle determine the urine-concentrating capacity of mammals. Mammals that live in water-abundant habitats (e.g., beavers) have short loops of Henle, resulting in the production of dilute urine. Mammals that live in desert environments have long loops of Henle, resulting in the production of concentrated urine (**Figure 35-8**). Fish lack loops of Henle due to excretion regulated by **osmoregulation** (**Figure 35-9**).

FLASH CARDS

To use the flash cards, tear the page from the book and cut along the dashed lines. The key term appears on one side of the flash card, and its definition appears on the opposite side.

ammonia	distal tubule
angiotensin	erythropoietin
antidiuretic hormone	excretion
aquaporin	extracellular fluid
bladder	filtrate
Bowman's capsule	filtration
collecting duct	glomerulus

The last section of a mammalian nephron, following the loop of Henle and emptying urine into a collecting duct; most tubular secretion and a small amount of tubular reabsorption occur in the distal tubule.

A highly toxic nitrogen-containing waste product of amino acid breakdown. In the mammalian liver, it is converted to urea.

A hormone produced by the kidneys in response to oxygen deficiency that stimulates the production of red blood cells by the bone marrow.

A hormone that functions in water regulation in mammals by stimulating physiological changes that increase blood volume and blood pressure.

The elimination of waste substances from the body; can occur from the digestive system, skin glands, urinary system, or lungs.

A hormone produced by the hypothalamus and released into the bloodstream by the posterior pituitary when blood volume is low; increases the permeability of the distal tubule and the collecting duct to water, allowing more water to be reabsorbed into the bloodstream.

Fluid that bathes the cells of the body; in mammals, extracellular fluid leaks from capillaries and is similar in composition to blood plasma, but lacking the large proteins found in plasma.

A channel protein in the plasma membrane of a cell that is selectively permeable to water.

The fluid produced by filtration; in the kidneys, the fluid produced by the filtration of blood through the glomerular capillaries.

A hollow muscular organ that stores urine.

Within Bowman's capsule in each nephron of a kidney, the process by which blood is pumped under pressure through permeable capillaries of the glomerulus, forcing out water and small solutes, including wastes and nutrients.

The cup-shaped portion of the nephron in which blood filtrate is collected from the glomerulus.

A dense network of thin-walled capillaries, located within the Bowman's capsule of each nephron of the kidney, where blood pressure forces water and small solutes, including wastes and nutrients, through the capillary walls into the nephron.

A tube within the kidney that collects urine from many nephrons and conducts it through the renal medulla into the renal pelvis. Urine may become concentrated in the collecting ducts if antidiuretic hormone (ADH) is present.

hemodialysis	osmolarity
homeostasis	osmoregulation
kidney	protonephridium (plural, protonephridia)
loop of Henle	proximal tubule
Malpighian tubule	renal artery
nephridium (plural, nephridia)	renal cortex
nephron	renal medulla

A measure of the total number of dissolved solute particles in a solution.

A procedure that simulates kidney function in individuals with damaged or ineffective kidneys; blood is diverted from the body, artificially filtered, and returned to the body.

Homeostatic maintenance of the water and salt content of the body within a limited range.

Maintenance of the relatively constant environment that is required for the optimal functioning of cells; maintained by the coordinated activity of numerous regulatory mechanisms, including the respiratory, endocrine, circulatory, and excretory systems.

The functional unit of the excretory system of some invertebrates, such as flatworms; consists of a tubule that has an external opening to the outside of the body, but lacks an internal opening within the body. Fluid is filtered from the body cavity into the tubule by a hollow cell at the end of the tubule, such as a flame cell in flatworms, and released outside the body.

One of a pair of organs of the excretory system that is located on either side of the spinal column and filters blood, removing wastes and regulating the composition and water content of the blood.

The initial part of a mammalian nephron, between Bowman's capsule and the loop of Henle; most tubular reabsorption and a small amount of tubular secretion occur in the proximal tubule.

A specialized portion of the tubule of some nephrons in birds and mammals that creates an osmotic concentration gradient in the fluid immediately surrounding it. This gradient in turn makes possible the production of urine that is more osmotically concentrated than blood plasma.

The artery carrying blood to each kidney.

The functional unit of the excretory system of insects, consisting of a small tube protruding from the intestine; wastes and nutrients move from the surrounding blood into the tubule, which drains into the intestine, where nutrients are reabsorbed back into the blood, while the wastes are excreted along with feces.

The outer layer of the kidney, in which the largest portion of each nephron is located, including Bowman's capsule and the distal and proximal tubules.

An excretory organ found in earthworms, mollusks, and certain other invertebrates; somewhat resembles a single vertebrate nephron.

The layer of the kidney just inside the renal cortex, in which loops of Henle produce a highly concentrated extracellular fluid, allowing the production of concentrated urine.

The functional unit of the kidney; where blood is filtered and urine is formed.

renal pelvis

renal vein

renin

tubular reabsorption

tubular secretion

tubule

urea

ureter

urethra

urinary system

urine

A water-soluble, nitrogen-containing waste product of amino acid breakdown; one of the principal components of mammalian urine.

The inner chamber of the kidney, in which urine from the collecting ducts accumulates before it enters the ureter.

A tube that conducts urine from a kidney to the urinary bladder.

The vein carrying blood away from each kidney.

The tube leading from the urinary bladder to the outside of the body; in males, the urethra also receives sperm from the vas deferens and conducts both sperm and urine (at different times) to the tip of the penis.

In mammals, an enzyme that is released by the kidneys when blood pressure falls. Renin catalyzes the formation of angiotensin, which causes arterioles to constrict, thereby elevating blood pressure.

The organ system that produces, stores, and eliminates urine, which contains cellular wastes, excess water and nutrients, and toxic or foreign substances. The urinary system is critical for maintaining homeostatic conditions within the bloodstream. In mammals, it includes the kidneys, ureters, bladder, and urethra.

The process by which cells of the tubule of the nephron remove water and nutrients from the filtrate within the tubule and return those substances to the blood.

The fluid produced and excreted by the urinary system, containing water and dissolved wastes, such as urea.

The process by which cells of the tubule of the nephron remove wastes from the blood, actively secreting those wastes into the tubule.

The tubular portion of the nephron; includes a proximal portion, the loop of Henle, and a distal portion. Urine is formed from the blood filtrate as it passes through the tubule.

SELF TEST

1. Freshwater trout are urinating all of the time and producing many times their own blood volume over the course of a 24-hour day, whereas the desert-dwelling kangaroo rat might excrete only a few milliliters over the same time period. The excretory mechanisms in both animals are NOT attempting to
   a. regulate the water volume within the body.
   b. maintain a constant level of acidity within the body.
   c. govern the levels of the same ions (e.g., Na^+ and K^+).
   d. maintain blood pressure through osmosis.

2. Flame cells are the specialized excretory structures found in
   a. flatworms.
   b. earthworms.
   c. protozoa.
   d. insects.

3. For affected species, what physiological advantage does the ability to excrete uric acid have over the ability to excrete urea?
   a. Fewer electrolytes (ions) are lost.
   b. Excretion of uric acid enables a greater volume of urine to be produced.
   c. More electrolytes are lost.
   d. Very little water loss occurs as a result of excreting uric acid.
   e. Both a and d are correct.

4. Blood in the urine would
   a. indicate a problem with filtration.
   b. indicate a problem with tubular secretion.
   c. be normal in species like fish that depend on osmoregulation.
   d. indicate a problem with erythropoietin production.

5. Which of the following must be reclaimed from the filtrate during the process of reabsorption?
   a. hydrogen ions
   b. glucose
   c. drugs
   d. toxins

6. Alcohol inhibits production of ADH. This would mean that
   a. without ADH there would be no way to regulate the amount of salts in the medulla to create a gradient for the movement of water.
   b. water would not be reabsorbed and urine volume would increase.

   c. the regulation of aldosterone release for regulation of salt balance would also be inhibited.
   d. the water that would be excreted in urine would be excreted in sweat instead.

7. Which portion of the loop of Henle actively transports salt into the kidney medulla?
   a. descending portion
   b. thick portion of the ascending loop
   c. thin portion of the ascending loop

8. Diuretics are substances that cause the production of dilute urine. Which of the following descriptions represents a possible mechanism of action for a diuretic?
   a. It increases the release of ADH from the pituitary of the brain.
   b. It promotes salt reabsorption.
   c. It has a direct effect on the collecting duct by increasing the permeability of this part of the nephron.
   d. It blocks the effects of ADH.

9. What effect would constriction of the incoming arteriole have on the filtration pressure in the glomerulus?
   a. The glomerular pressure and the amount of filtrate produced would increase as a result of the constricted afferent arteriole.
   b. The constricted afferent arteriole would not change the already smaller size of the outgoing arteriole, thus the glomerular pressure and amount of filtrate would not change.
   c. The glomerular pressure would decrease and the amount of filtrate produced would decrease as a result of the constricted afferent arteriole.
   d. The glomerular pressure would decrease and the amount of filtrate produced would increase as a result of the constricted afferent arteriole.

10. After leaving the proximal tubule, the filtrate enters
    a. Bowman's capsule.
    b. the loop of Henle.
    c. the distal tubule.
    d. the collecting duct.

11. Based on the information shown in the figure, if the rate of active transport of salt is increased, then
    a. the rate of water absorption would be greater.
    b. the rate of chloride absorption would be greater.
    c. more drugs can move from the blood to the nephron tubules.
    d. more urine can be produced.

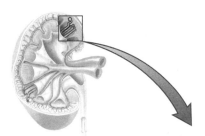

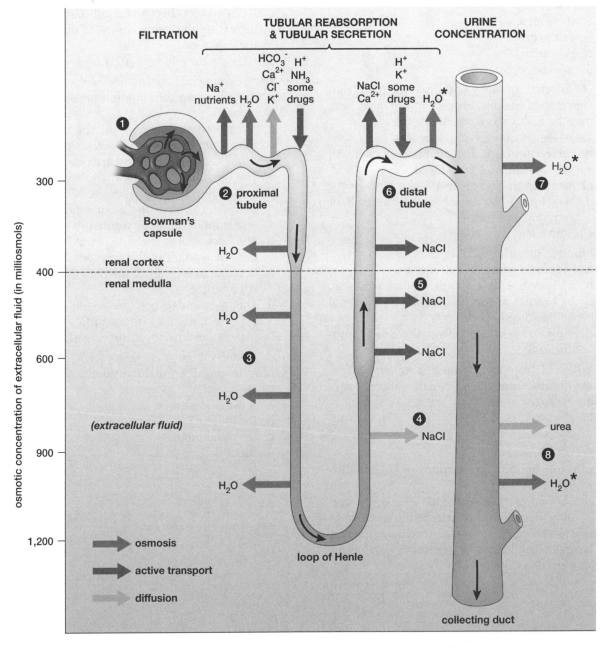

12. In urine formation, the role of ADH (antidiuretic hormone) is to cause
    a. the collecting duct to become highly permeable to water.
    b. glomerular capillaries to dilate, which increases blood flow into the glomerulus.
    c. the collecting duct to become impermeable to water.
    d. the collecting duct to constrict to reduce the flow of filtrate out of a nephron.

13. Based on the figure below, a nice cold beer on a hot day would
    a. decrease blood osmolarity and cause the hypothalamus to shut down.
    b. increase the amount of nitrogen waste excreted in sweat due to the retention of water in the blood.
    c. cause the person to become even more dehydrated since concentrated urine could not be produced.
    d. cool the body temperature and lead to more concentrated urine.

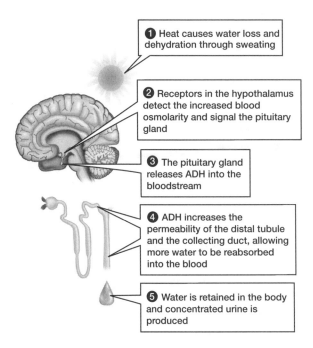

① Heat causes water loss and dehydration through sweating

② Receptors in the hypothalamus detect the increased blood osmolarity and signal the pituitary gland

③ The pituitary gland releases ADH into the bloodstream

④ ADH increases the permeability of the distal tubule and the collecting duct, allowing more water to be reabsorbed into the blood

⑤ Water is retained in the body and concentrated urine is produced

14. Which of the following is NOT a sign of a failing kidney?
    a. a drop in the number of red blood cells
    b. dramatic changes in systemic blood pressure
    c. water retention and swelling in the tissues
    d. increased volume of urine

15. A very common type of drug used to control high blood pressure is an ACE inhibitor. This drug inhibits an enzyme involved in the production of a hormone called *angiotensin*. Which of the following explains why blood pressure would drop if you were taking this drug?
    a. Angiotensin production drops and more water is retained in the blood. The increase in blood volume helps lower blood pressure.
    b. Angiotensin production drops and more water is removed from the blood. The decrease in blood volume helps lower blood pressure.
    c. There is less angiotensin traveling to the kidneys. This leads to a constriction of the blood vessels leading to the kidneys and a lower filtration rate.
    d. There is less angiotensin traveling to the kidneys. This leads to a constriction of the blood vessels leading to the kidneys and a higher pressure in the glomerulus.

16. What similar functions do the protonephridia of flatworms and the nephridia of roundworms have in common?
    a. Almost none; the species are too different.
    b. Both act to filter out wastes for subsequent excretion.
    c. Both act to filter out nutrients to retain them inside the body.
    d. Both b and c are correct.
    e. None of the above is correct.

17. Based on your knowledge of osmoregulation, osmosis, and kidney function, if a freshwater fish were placed in salt water, you might expect that
    a. it would produce more concentrated urine to keep from losing too much water to the environment.
    b. the nephron would not be able to set up a concentration gradient so kidney function would cease.
    c. the fish would begin to produce uric acid rather than urine.
    d. the fish would begin to take in more water to the tissues, and the cells would rupture and die.

18. When a person takes a drug, the drug will eventually be eliminated from the body. One of the primary mechanisms for this removal is tubular secretion. Which of the following would produce the greatest reduction in the ability of our kidneys to remove drugs?
    a. a reduction in the salt gradient within the renal medulla
    b. damage to Bowman's capsule
    c. damage to the wall of the urinary bladder
    d. damage to the distal tubule of the nephron

19. Kangaroo rats are adapted to desert environments and never drink water because they are able to utilize metabolic water. What structure assists them in desert environment adaptation?
    a. the loosened coil of the glomerulus for an increase in surface area for filtration of more salts
    b. the long loop of Henle, which allows more absorption of water
    c. the increased release of aldosterone, which creates a greater salt gradient so more water can be reabsorbed
    d. the tightly convoluted tubules, which create a water and salt gradient right next to the blood vessels for increased water absorption

20. The muscular tube that transports urine formed by a kidney to the bladder is called a(n)
    a. ureter.
    b. urethra.
    c. vena cava.
    d. glomerulus.

21. A loop of Henle would be missing in which of the following animals?
    a. fish
    b. bird
    c. reptile
    d. mammal

22. What happens to a substance that is neither filtered nor reabsorbed but is secreted?
    a. The body retains it.
    b. It is eliminated from the body.
    c. Its concentration in the blood gradually decreases.
    d. Its concentration in the blood gradually increases.
    e. Both b and c are correct.
    f. Both a and d are correct.

23. Blood is filtered between
    a. the proximal tubule and the distal tubule.
    b. the descending limb of the loop of Henle and the ascending limb of the loop of Henle.
    c. the collecting duct and Bowman's capsule.
    d. Bowman's capsule and the capillary bed known as the glomerulus.

24. Which of the following hormone levels might be elevated in patients who have recently had a rapid blood loss due to injury or surgery?
    a. aldosterone
    b. erythropoietin
    c. antidiuretic hormone
    d. angiotensin

25. Tubular reabsorption is the
    a. process by which wastes and excess substances that were not initially filtered out into Bowman's capsule are removed from the blood.
    b. concentration of the filtrate resulting from the concentration of salts and urea.
    c. passage of blood cells and proteins from the glomerulus.
    d. process by which cells of the proximal tubule remove water and nutrients and pass them back into the blood.

26. What initiates glomerular pressure filtration?
    a. the permeability of the glomerular walls
    b. the inability of large proteins to pass through the capillary walls
    c. diameter differences between the incoming and outgoing arterioles
    d. osmotic pressure differences between the glomerulus and Bowman's capsule

27. The difference between the blood plasma that exits the kidney by way of the renal vein and the plasma that enters the kidney by way of the renal artery is that the exiting plasma
    a. contains fewer dissolved nutrients.
    b. contains more dissolved nutrients.
    c. contains fewer dissolved wastes.
    d. contains more dissolved wastes.

28. Which would you not expect to find in filtrate?
    a. albumin
    b. chloride ions
    c. glucose
    d. sodium ions

29. Angiotensin is a hormone that
    a. causes arterioles to constrict.
    b. regulates the amount of water reabsorbed from the urine.
    c. catalyzes the formation of angiotensin.
    d. stimulates the production of red blood cells.

30. In which organism would you expect to find Malphigian tubules?
    a. kangaroo rat
    b. grasshopper
    c. earthworm
    d. sponge

ANSWER KEY

1. d	16. d
2. a	17. a
3. e	18. d
4. a	19. b
5. b	20. a
6. b	21. a
7. b	22. e
8. d	23. d
9. c	24. b
10. b	25. d
11. a	26. c
12. a	27. c
13. c	28. a
14. d	29. a
15. b	30. b

CHAPTER 36 IMMUNITY: DEFENSES AGAINST DISEASE

OUTLINE

Section 36.1 What Are the Mechanisms of Defense Against Disease?

Vertebrates have three major lines of defense against disease: (1) nonspecific external barriers, (2) nonspecific internal defenses, and (3) specific internal defenses (**Figure 36-1**).

Invertebrates have two major lines of defense against disease: (1) nonspecific external barriers and (2) nonspecific internal barriers.

Section 36.2 How Do Nonspecific Defenses Function?

Nonspecific external defenses, which include the skin and mucous membranes, keep **pathogens** from entering the body. The skin is a physical barrier for pathogens, while its secretions prevent bacterial growth. Mucous membranes secrete mucus and antimicrobial substances that trap and destroy pathogens. Respiratory mucous membranes contain cilia that sweep microbes trapped in mucus to the nose and mouth where they can be removed from the body (**Figure 36-2**).

Nonspecific internal defenses attack pathogens that penetrate nonspecific external defenses. **Phagocytes** engulf pathogens, whereas **natural killer cells** secrete proteins that kill virally infected or cancerous cells.

Tissue damage stimulates the **inflammatory response (Figure 36-4)** in which secreted chemicals, including **histamine**, stimulate phagocyte activity and increase blood flow and capillary permeability. Clotting eventually seals off the wound site, preventing additional microbe entry. **Fever** occurs when endogenous pyrogens are released by white blood cells in response to infection. Elevated temperature inhibits bacterial growth, accelerates immune cell activity, and causes the secretion of virus-inhibiting interferons by infected cells.

Section 36.3 What Are the Key Components of the Adaptive Immune System?

The immune response consists of three major components: (1) immune cells, which include the white blood cells, the **B cells**, and the **T cells** (**Table 36-2**); (2) tissues and organs, which include the **spleen**, **lymph nodes**, **thymus**, and **tonsils**; and (3) secreted proteins called **cytokines**.

The immune response has three steps: (1) recognition, (2) attack, and (3) memory.

Section 36.4 How Does the Adaptive Immune System Recognize Invaders?

In the recognition step, an **antibody** (made by B cells or plasma cells) or **T-cell receptors** recognize foreign **antigens** and trigger the immune response. Antibodies are Y-shaped proteins that bind to and help destroy antigens (**Figures 36-6 and 36-7**). Each B cell produces only one type of antibody. Each antibody binds to only one or a few types of antigens.

There are genes that code for parts of antibodies: (1) constant regions, (2) variable regions, (3) joining regions, and (4) diversity regions. Antibodies are composed from various combinations of the coded regions (**Figure 36-8**).

The **major histocompatibility complex** on the surface of body cells acts as foreign antigens in the body of other individuals, which is why transplant patients are required to take immune system–suppressing medication.

Section 36.5 How Does the Adaptive Immune System Launch an Attack?

In the attack step, antigens from a pathogen bind to B or T cells with complementary membrane-bound antibodies or T-cell receptors. In **humoral immunity**, B cells bound to a specific antigen undergo **clonal selection**, resulting in the production of plasma cells that produce massive amounts of identical antibodies complementary to the original antigen (**Figure 36-10**).

In **cell-mediated immunity**, T cells (with proper receptors) bind to complementary antigens and divide rapidly. Activated **cytotoxic T cells** bind to antigens on microbes, cancer cells (**Figure 36-12**), or infected cells and kill the cells. Activated helper T cells chemically stimulate B-cell and cytotoxic T-cell responses. **Figure 36-13** compares the cell-mediated and humoral responses.

Section 36.6 How Does the Adaptive Immune System Remember Its Past Victories?

In the memory step, some of the B and T cells that are produced form long-lived memory cells that are activated if the same antigen reappears in the bloodstream. This produces a fast and effective immune response (**Figure 36-14**).

Section 36.7 How Does Medical Care Assist the Immune Response?

Antibiotics are drugs that slow the growth and reproduction of bacteria, fungi, and protists. This allows the body more time to generate an immune response that will destroy these invaders.

A **vaccine** is an injection of pathogenic antigens or weakened pathogenic microbes. This stimulates the development of memory cells that rapidly fight off future infections.

Section 36.8 What Happens When the Immune System Malfunctions?

Allergies occur when an immune response forms against a normally harmless substance. B cells respond to these substances (which are treated like antigens) by producing antibodies that bind to **mast cells**. Mast cells release histamine when exposed to these harmless antigens, which causes a local inflammatory response (**Figure 36-15**).

Autoimmune diseases occur when the body forms an immune response against its own molecules. Immune deficiency diseases occur when the body cannot form an immune response strong enough to ward off pathogens.

Acquired immune deficiency syndrome (AIDS) occurs when **human immunodeficiency viruses (HIV)** destroy helper T cells, causing an individual to become extremely susceptible to infections.

Section 36.9 How Does the Immune System Combat Cancer?

Cancer occurs when body cells multiply without control. These cells can be destroyed by natural killer cells or cytotoxic T cells, but if cells multiply too quickly for the immune system to keep up, a tumor develops.

Vaccinations can prevent some forms of cancer, and research is being done on vaccines that could possibly cure cancer.

FLASH CARDS

To use the flash cards, tear the page from the book and cut along the dashed lines. The key term appears on one side of the flash card, and its definition appears on the opposite side.

acquired immune deficiency syndrome (AIDS)	autoimmune disease
adaptive immune response	B cell
adaptive immune system	cancer
allergy	cell-mediated immunity
antibiotic	clonal selection
antibody	complement
antigen	constant region

A disorder in which the immune system attacks the body's own cells or molecules.

An infectious disease caused by the human immunodeficiency virus (HIV); attacks and destroys T cells, thus weakening the immune system.

A type of lymphocyte that matures in the bone marrow, and that participates in humoral immunity; gives rise to plasma cells, which secrete antibodies into the circulatory system, and to memory cells.

A response to invading toxins or microbes in which immune cells are activated by a specific invader, selectively destroy that invader, and then "remember" the invader, allowing a faster response if that type of invader reappears in the future; see also *innate immune response*.

A disease in which some of the body's cells escape from normal regulatory processes and divide without control.

A widely distributed system of organs (including the thymus, bone marrow, and lymph nodes), cells (including macrophages, dendritic cells, B cells, and T cells), and molecules (including cytokines and antibodies) that work together to combat microbial invasion of the body; the adaptive immune system responds to and destroys specific invading toxins or microbes; see also *innate immune response*.

An adaptive immune response in which foreign cells or substances are destroyed by contact with T cells.

An inflammatory response produced by the body in response to invasion by foreign materials, such as pollen, that are themselves harmless.

The mechanism by which the adaptive immune response gains specificity; an invading antigen elicits a response from only a few lymphocytes, which proliferate to form a clone of cells that attack only the specific antigen that stimulated their production.

Chemicals that help to combat infection by destroying or slowing down the multiplication of bacteria, fungi, or protists.

A group of blood-borne proteins that participate in the destruction of foreign cells, especially those to which antibodies have bound.

A protein, produced by cells of the immune system, that combines with a specific antigen and normally facilitates the destruction of the antigen.

The part of an antibody molecule that is similar in all antibodies of a given class.

A complex molecule, normally a protein or polysaccharide, that stimulates the production of a specific antibody.

cytokine

cytotoxic T cell

dendritic cell

emerging infectious disease

fever

helper T cell

histamine

human immunodeficiency virus (HIV)

humoral immunity

immune system

inflammatory response

innate immune response

leukocyte

lymph node

A pathogenic virus that causes acquired immune deficiency syndrome (AIDS) by attacking and destroying the immune system's helper T cells.

Any of several chemical messenger molecules released by cells that facilitate communication with other cells and transfer signals within and between the various systems of the body. Cytokines are important in cellular differentiation and the adaptive immune response.

An immune response in which foreign substances are inactivated or destroyed by antibodies that circulate in the blood.

A type of T cell that, upon contacting foreign cells, directly destroys them.

A system of cells, including macrophages, B cells and T cells, and molecules, such as antibodies and cytokines, that work together to combat microbial invasion of the body.

A type of phagocytic leukocyte that presents antigen to T and B cells, thereby stimulating an adaptive immune response to an invading microbe.

A nonspecific, local response to injury to the body, characterized by the phagocytosis of foreign substances and tissue debris by white blood cells and by the walling off of the injury site by the clotting of fluids that escape from nearby blood vessels.

A previously unknown infectious disease (one caused by a microbe), or a previously known infectious disease whose frequency or severity has significantly increased in the past two decades.

Nonspecific defenses against many different invading microbes, including phagocytic white blood cells, natural killer cells, the inflammatory response, and fever.

An elevation in body temperature caused by chemicals (pyrogens) that are released by white blood cells in response to infection.

Any of the white blood cells circulating in the blood.

A type of T cell that helps other immune cells recognize and act against antigens.

A small structure located on a lymph vessel, containing macrophages and lymphocytes (B and T cells). Macrophages filter the lymph by removing microbes; lymphocytes are the principal components of the adaptive immune response to infection.

A substance released by certain cells in response to tissue damage and invasion of the body by foreign substances; promotes the dilation of arterioles and the leakiness of capillaries and triggers some of the events of the inflammatory response.

lymphocyte	natural killer cell
macrophage	neutrophil
major histocompatibility complex (MHC)	pathogen
mast cell	phagocyte
memory B cell	plasma cell
memory T cell	regulatory T cell
microbe	severe combined immune deficiency (SCID)

A type of white blood cell that destroys some virus-infected cells and cancerous cells on contact; part of the innate immune system's nonspecific internal defense against disease.

A type of white blood cell (natural killer cell, B cell, or T cell) that is important in either the innate or adaptive immune response.

A type of white blood cell that engulfs invading microbes and contributes to the nonspecific defenses of the body against disease.

A type of white blood cell that engulfs microbes and destroys them by phagocytosis; also presents microbial antigens to T cells, helping stimulate the immune response.

An organism (or a toxin) capable of producing disease.

A group of proteins, normally located on the surfaces of body cells, that identify the cell as "self"; also important in stimulating and regulating the immune response.

A type of immune system cell that destroys invading microbes by using phagocytosis to engulf and digest the microbes. Also called a *phagocytic cell*.

A cell of the immune system that releases histamine and other molecules used in the body's response to trauma and that are a factor in allergic reactions.

An antibody-secreting descendant of a B cell.

A type of white blood cell that is produced by clonal selection as a result of the binding of an antibody on a B cell to an antigen on an invading microorganism. Memory B cells persist in the bloodstream and provide future immunity to invaders bearing that antigen.

A type of T cell that suppresses the adaptive immune response, especially by self-reactive lymphocytes, and appears to be important in the prevention of autoimmune disorders.

A type of white blood cell that is produced by clonal selection as a result of the binding of a receptor on a T cell to an antigen on an invading microorganism. Memory T cells persist in the bloodstream and provide future immunity to invaders bearing that antigen.

A disorder in which no immune cells, or very few, are formed; the immune system is incapable of responding properly to invading disease organisms, and the individual is very vulnerable to common infections.

A microorganism.

spleen

T cell

T-cell receptor

thymus

tonsil

vaccine

variable region

A patch of lymphatic tissue, located at the entrance to the pharynx, that contains macrophages and lymphocytes; destroys many microbes entering the body through the mouth and stimulates an adaptive immune response to them.

The largest organ of the lymphatic system, located in the abdominal cavity; contains macrophages that filter the blood by removing microbes and aged red blood cells, and lymphocytes (B and T cells) that reproduce during times of infection.

An injection into the body that contains antigens characteristic of a particular disease organism and that stimulates an immune response appropriate to that disease organism.

A type of lymphocyte that matures in the thymus, and that recognizes and destroys specific foreign cells or substances or that regulates other cells of the immune system.

The part of an antibody molecule that differs among antibodies; the ends of the variable regions of the light and heavy chains form the specific binding site for antigens.

A protein receptor, located on the surface of a T cell, that binds a specific antigen and triggers the immune response of the T cell.

An organ of the lymphatic system, located in the chest just behind the sternum, that is the site of T-cell maturation; produces the hormone thymosin, which stimulates the development of T cells.

SELF TEST

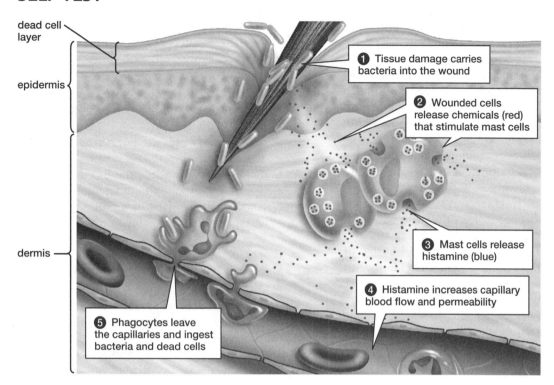

dead cell layer

epidermis

dermis

1 Tissue damage carries bacteria into the wound

2 Wounded cells release chemicals (red) that stimulate mast cells

3 Mast cells release histamine (blue)

4 Histamine increases capillary blood flow and permeability

5 Phagocytes leave the capillaries and ingest bacteria and dead cells

1. Which of the following is explained by the figure above?
   a. cell-mediated immune response
   b. nonspecific defenses
   c. adaptive immune response
   d. allergic reactions

2. Your sister has been diagnosed with lupus and doesn't understand the disorder. She asks you if she should take her daughter to the doctor for a vaccination so her daughter will not get the disease in the future. You tell her
   a. that a vaccination only works for bacterial infections, and lupus doesn't fit this category.
   b. that a vaccination is not needed because children cannot contract lupus.
   c. that lupus is an autoimmune disorder and cannot be prevented with vaccinations.
   d. that a vaccination for lupus has not yet been developed.

3. An emergent disease would be
   a. one that is found only in non-human vertebrates.
   b. Ebola.
   c. a disease that would be more prevalent in the human populations found in regions where habitats are being converted from tropical forests to urban areas.
   d. a disease that was once rare but is now increasing in prevalence in human populations.

4. If you went to the doctor with a fever and she took a sample of blood, what might she be looking for?
   a. She would be looking at the number of white blood cells to determine if you have an infection.
   b. She would be looking for virus particles to determine what infection you have.
   c. She would be trying to determine your histamine level.
   d. She would be attempting to determine the type of antigens you have at that particular time.

5. Your 5-year-old daughter has just come home from day care feeling poorly and not eating. The next day, she has vomiting, a rash, diarrhea, and some neurological symptoms. The diagnosis is bacterial meningitis, a contagious bacterial infection, usually not life threatening. Within a week, she is up and about again, apparently healthy. The little boy who lives next door to you gets this same disease 3 months later, and you become worried that your daughter will again get meningitis. They play together all the time, and he could easily transmit the bacterium to her during play. Is this a potential problem?

   a. Yes, your daughter is at risk for contracting meningitis again. Kids under the age of 10 years have immature immune systems that will not protect them. Maternal antibodies have long since dissipated in her body, leaving her exposed to all sorts of pathogenic microorganisms.

   b. Yes, your daughter is at risk for meningitis again. Bacteria do not elicit as strong an immune response as viruses do, and her immune system will not be able to contain the infection.

   c. No, your daughter is not at risk for contracting meningitis again, particularly since the time span between the first exposure and second exposure to this infectious agent is not very long. Upon "seeing" the bacteria again, her memory cells from the previous exposure will become activated quickly, again making large amounts of protective antibody.

   d. No, your daughter is not at risk for meningitis again. Once you have contacted an infectious agent and memory cells have been produced, that infectious agent will always be identified and eradicated before the infection can take place.

6. Which of the following is a limitation of the humoral response to microbial invasion?

   a. B cells are capable of attacking only microbial invaders that have taken over host cells.

   b. The humoral response is capable of effectively attacking invading microbes only before they enter a host cell.

   c. B cells have a limited ability to stop the immune response once the microbes have been neutralized.

   d. B cells indiscriminately devour invading microbes.

7. Can you have immunity to a microbe that you have never encountered?

   a. Yes, if you have encountered the same or a very similar antigen of another microbe.

   b. Yes. Your immune system has as many memory cells as it has antibodies.

   c. No. You must encounter an entire microbe before you can become immune to it.

   d. No. You cannot establish immunity against any microbes.

8. Why is the initial immune response to an antigen slower than and not as strong as the second exposure to the antigen?

   a. The memory cells are not active during the first exposure.

   b. There is a delay while the B-cell clones are selected, multiply, and differentiate.

   c. The B cells involved in the initial exposure are slower than those involved in the second exposure.

   d. The helper T cells involved in the initial exposure are slower than those involved in the second exposure.

9. Which of the following "selects" which B cells will become active during the immune response to an antigen?

   a. T cells   c. antibodies
   b. other B cells   d. the antigen

10. During differentiation of B cells, which type of cell becomes filled with rough endoplasmic reticulum, which enables it to "crank out" huge numbers of antibodies?

    a. plasma cells   c. helper T cells
    b. macrophages   d. memory cells

11. A high neutrophil count would indicate

    a. a bacterial infection. c. an HIV infection.
    b. a viral infection.   d. an emergent disease.

12. Vaccinations are good at providing immunity against

    a. bacteria.
    b. protists.
    c. viruses.
    d. bacteria and protists.
    e. protists and viruses.
    f. bacteria, protists, and viruses.

13. Because allergies are such a nuisance and allergens are not harmful to us, why do we even make allergy antibodies?

    a. Allergens used to be harmful, but we have evolved resistances to them, so allergy antibodies are just evolutionary baggage.

    b. Allergy antibodies help defend us against parasites that typically enter the body through the mouth, nose, and throat by increasing mucous secretions and coughing or sneezing to expel the parasites.

    c. Allergy antibodies are just a mistake of the immune system and have no real use.

    d. Allergens are harmful to us and allergy antibodies protect us from them.

14. A moderate fever or inflammatory response could be beneficial because
    a. the process is natural rather than man-made.
    b. the processes increase the body's ability to fight the infection.
    c. these methods are the body's way of killing viruses.
    d. it gives the body time to develop memory cells and antibodies.

15. Since lymph vessels have valves they are similar to
    a. arteries.   c. nerves.
    b. veins.   d. capillaries.

16. Fever
    a. decreases interferon production.
    b. decreases the concentration of iron in the blood.
    c. decreases the activity of phagocytes.
    d. increases the reproduction rate of invading bacteria.

17. The role of histamine in the inflammatory response is to cause
    a. the dilation of arterioles, leading to increased blood flow to the injured tissue.
    b. larger numbers of white blood cells to be released from the bone marrow.
    c. capillary permeability to increase as capillaries become "leaky."
    d. both a and c.

18. An inflamed tissue turns red because of
    a. an increased movement of fluid from blood into tissues across "leaky" capillary walls.
    b. an increased number of white blood cells circulating in the blood.
    c. an increased blood flow to the injured tissues resulting from the dilation of arterioles in the injured area.
    d. the recruitment of macrophages to the injured tissue.

19. An inflamed tissue swells because of
    a. an increased movement of fluid from blood into tissues across "leaky" capillary walls.
    b. an increased number of white blood cells circulating in the blood.
    c. an increased blood flow to the injured tissues resulting from the dilation of arterioles in the injured area.
    d. the recruitment of macrophages to the injured tissue.

20. What is an antigen?
    a. a substance produced by B cells to attack microbes
    b. a receptor found on a B cell that is important for B-cell activation

    c. a molecule released by T cells to activate B cells
    d. a molecule found on the surface of invading microbes that is recognized by the immune system

21. Your daughter has the flu. Your husband wants you to take her in to get an antibiotic. You tell him you won't get the antibiotic because
    a. your daughter will have an allergic reaction.
    b. antibiotics will only work with bacterial infections.
    c. the flu vaccine will prevent the illness from getting worse.
    d. the flu mutates so fast that it becomes resistant to antibiotics.

22. Which of the following cells is responsible for producing antibodies?
    a. red blood cell   c. B cell
    b. T cell   d. macrophage

23. Major histocompatibility complex (MHC) antigens are
    a. antigenic determinants on microorganisms.
    b. proteins produced by newly formed cancer cells.
    c. a type of antibody molecule.
    d. the binding sites of the HIV virus.
    e. recognition markers on cells used for identification by one's own immune system.

24. Which of the following statements regarding antibodies is true? An antibody
    a. has one binding site for antigens.
    b. is found in a variety of body fluids and organ systems.
    c. can be given to treat bacterial infections as an antibiotic.
    d. is produced by platelets.
    e. once produced, protects for the rest of one's life.

25. In order for transplanted organs to be successfully accepted by the recipient's body, the donor and recipient must be matched and the recipient must be placed on medication, even for a long period after the transplant. Why are organ transplants such a problem?
    a. We do not yet have the technology to transplant organs successfully.
    b. Unless every donor and recipient are close relatives, their MHC proteins are bound to differ.
    c. Although scientists can move cells into a recipient body, they cannot do this with an entire organ.
    d. The recipient's disease may have progressed to such an extent that the transplant cannot succeed even when matched properly.

26. You and your friend each have a little boy. The kids are always playing in dirt, climbing trees, falling off bikes, and generally getting into everything. When your little Johnny comes home with a runny nose or a small cut, your philosophy is to give him a decongestant or to clean the wound well, and then let it take its own course. Your friend's philosophy is the exact opposite of yours. When her little Tommy comes home with similar problems, she takes him immediately to the doctor for antibiotics. Who is right and who is wrong? Why?
   a. Your friend is right to take her son for antibiotics. One can never know how pathogenic an infectious microorganism can be.
   b. You are. It is probably better to allow his immune system to take care of these minor infections.
   c. Neither of you is doing the best thing. Both of you should vaccinate the boys with every known immunization, thereby protecting them from everything.

27. Which of the following is designed to confer immunity to a particular disease?
   a. antibiotics
   b. neuraminidase inhibitors
   c. vaccines
   d. gene therapy

28. If allergic reactions are caused by antibodies binding to allergens, why are the responses localized (limited to a given tissue, such as a runny nose or an upset stomach)?
   a. Antibodies that respond to allergens are attached to mast cells, which are located in tissues, not circulating through the bloodstream.

b. Allergens interact directly with tissues and cause inflammatory response wherever they fall.
   c. Allergic responses are not tissue specific.

29. The surface of your skin is colonized by bacteria. These bacteria
   a. cause infection if not depleted with antibacterial soaps.
   b. are harmless.
   c. are a form of nonspecific defense.
   d. are the first that your body will produce antibodies for during illness.

30. Based on the figure below, you could say that
   a. antibody levels are highest immediately after the first exposure to a pathogen.
   b. the pattern of the immune response is the same after each exposure.
   c. the immune response is highest after the second exposure.
   d. the production of antibodies takes place after the second exposure to a pathogen.

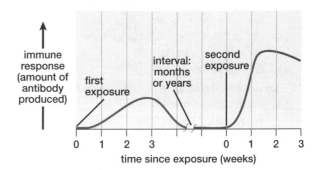

ANSWER KEY

CHAPTER 37 CHEMICAL CONTROL OF THE ANIMAL BODY: THE ENDOCRINE SYSTEM

OUTLINE

Section 37.1 How Do Animal Cells Communicate?

Cells communicate when messenger molecules that bind to target cell **receptors** are released, triggering a change within the **target cell**.

Local hormones, like **prostaglandins** and histamines, have short-range reactions and degrade rather quickly.

Endocrine glands produce **endocrine hormones** that are released directly into the bloodstream.

Section 37.2 How Do Animal Hormones Work?

Hormones are found in many organisms not limited to vertebrates and mammals.

The three classes of hormones are (1) **peptide hormones**, (2) **amino acid–derived hormones**, and (3) **steroid hormones** (**Table 37-2**). The **endocrine system** is composed of glands responsible for the production and secretion of hormones into the bloodstream (**Figure 37-1**).

Hormones bind to specific receptors on target cells (**Figure 37-2**). Peptide hormones and amino acid–derived hormones bind to membrane receptors that result in the production of **second messenger** molecules (e.g., **cyclic adenosine monophosphate** or **cyclic AMP**) within the cell. Second messenger molecules initiate an amplified reaction within a cell, resulting in the synthesis or secretion of a desired product (**Figure 37-2**). Steroid hormones diffuse through the cell membrane to bind to receptors in the nucleus. In the nucleus, the hormone–receptor complex then attaches to a specific gene on DNA, resulting in the production of mRNA that is used to synthesize a protein (**Figure 37-3**).

Most hormones are regulated by **negative feedback**, although a few (e.g., oxytocin) are regulated by **positive feedback**.

Insects have a hormone called **ecdysone** that regulates the molting or shedding of the external skeleton (**Figure 37-4**).

Section 37.3 What Are the Structures and Functions of the Mammalian Endocrine System?

The endocrine glands and the hormones they produce are summarized in **Figure 37-5** and **Table 37-3**.

The **hypothalamus** controls secretions by the **pituitary gland**. The **anterior pituitary gland** is controlled by hypothalamic **releasing hormones** or **inhibiting hormones**, whereas the **posterior pituitary gland** releases secretions from hypothalamic **neurosecretory cells** (**Figure 37-6**).

The anterior pituitary gland releases six types of hormones: (1) **follicle-stimulating hormone (FSH)**, (2) **luteinizing hormone (LH)**, (3) **thyroid-stimulating hormone (TSH)**, (4) **adrenocorticotropic hormone (ACTH)**, (5) **growth hormone (GH)**, and (6) **prolactin (PRL)**.

The posterior pituitary gland releases hormones produced by hypothalamic neurosecretory cells. **Antidiuretic hormone (ADH)** prevents dehydration by increasing the water permeability of the nephron collecting ducts. **Oxytocin** causes muscular contractions that are involved with childbirth and milk "letdown" (**Figure 37-8**).

The **thyroid gland** releases the hormones **thyroxine** and **calcitonin** (**Figure 37-9**). Thyroxine release elevates metabolic rate (**Figure 37-11**), and calcitonin plays a role in calcium uptake from the bloodstream.

The **parathyroid glands** (**Figure 37-9**) release **parathyroid hormone (PTH)**, which elevates blood calcium levels by stimulating calcium release from bone.

The **pancreas** controls blood glucose levels through the release of **insulin** and **glucagon**. When blood glucose is high, insulin is released, which stimulates glucose uptake by cells. When blood glucose is low, glucagons are released, which stimulates the breakdown of fat and glucose storage molecules in the liver, resulting in increased blood glucose (**Figure 37-12**).

The sex organs secrete steroid hormones (sex hormones) that play a key role in puberty and gamete production. **Testes** in males produce **androgens** (e.g., **testosterone**), while **ovaries** in females produce **estrogen** and **progesterone**.

The **adrenal glands** are composed of two layers: the **adrenal medulla** and the **adrenal cortex** (**Figure 37-13**). The adrenal medulla releases the hormones **epinephrine** and **norepinephrine** in response to stress and cause changes that prepare the body for emergency action. The adrenal cortex releases **glucocorticoids** that help the body cope with short-term stress and that stimulate glucose production. The adrenal cortex also releases **aldosterone**, which regulates sodium content in the blood.

The **pineal gland** produces **melatonin**, which helps regulate daily rhythms in animals. The **thymus** produces **thymosin**, which stimulates the development of T cells. The kidneys produce **erythropoietin**, which regulates red blood cell production, and **renin**, which helps the body respond to low blood pressure. The heart produces **atrial natriuretic peptide (ANP)**, which reduces blood volume. The stomach and small intestine produce gastrin, secretin, and cholecystokinin, which all help regulate digestion. Fat cells produce **leptin**, which helps control food consumption.

FLASH CARDS

To use the flash cards, tear the page from the book and cut along the dashed lines. The key term appears on one side of the flash card, and its definition appears on the opposite side.

adrenal cortex	angiotensin
adrenal gland	anterior pituitary
adrenal medulla	antidiuretic hormone (ADH)
adrenocorticotropic hormone (ACTH)	atrial natriuretic peptide (ANP)
aldosterone	calcitonin
amino acid-derived hormone	cortisol
androgen	cyclic adenosine monophosphate (cyclic AMP)

A hormone that functions in water regulation in mammals by stimulating physiological changes that increase blood volume and blood pressure.

The outer part of the adrenal gland, which secretes steroid hormones that regulate metabolism and salt balance.

A lobe of the pituitary gland that produces prolactin and growth hormone as well as hormones that regulate hormone production in other glands.

A mammalian endocrine gland, adjacent to the kidney; secretes hormones that function in water regulation and in the stress response.

A hormone produced by the hypothalamus and released into the bloodstream by the posterior pituitary when blood volume is low; in the kidney, ADH increases the permeability of the distal tubule and the collecting duct to water, allowing more water to be reabsorbed into the bloodstream.

The inner part of the adrenal gland, which secretes epinephrine (adrenaline) and norepinephrine (noradrenaline) in the stress response.

A hormone, secreted by cells in the mammalian heart, that reduces blood volume by inhibiting the release of ADH and aldosterone.

A hormone, secreted by the anterior pituitary, that stimulates the release of hormones by the adrenal cortex, especially in response to stress.

A hormone, secreted by the thyroid gland, that inhibits the release of calcium from bone.

A hormone, secreted by the adrenal cortex, that helps regulate ion concentration in the blood by stimulating the reabsorption of sodium by the kidneys and sweat glands.

A steroid hormone released into the bloodstream by the adrenal cortex in response to stress. Cortisol helps the body cope with short-term stressors by raising blood glucose levels; it also inhibits the immune response.

A class of hormone that is synthesized by the body from single amino acids. Examples include epinephrine and thyroxine.

A cyclic nucleotide, formed within many target cells as a result of the reception of amino acid derivatives or peptide hormones, that causes metabolic changes in the cell.

A male sex hormone.

diabetes mellitus	erythropoietin
ecdysone	estrogen
endocrine disrupter	follicle-stimulating hormone (FSH)
endocrine gland	glucagon
endocrine hormone	glucocorticoid
endocrine system	goiter
epinephrine	growth hormone

A hormone, produced by the kidneys in response to oxygen deficiency, that stimulates the production of red blood cells by the bone marrow.

A disease characterized by defects in the production, release, or reception of insulin; characterized by high blood glucose levels that fluctuate with sugar intake.

A female sex hormone, produced by follicle cells of the ovary, that stimulates follicle development, oogenesis, the development of secondary sex characteristics, and growth of the uterine lining.

A steroid hormone that triggers molting in insects and other arthropods.

A hormone, produced by the anterior pituitary, that stimulates spermatogenesis in males and the development of the follicle in females.

An environmental pollutant that interferes with endocrine function, often by disrupting the action of sex hormones.

A hormone, secreted by the pancreas, that increases blood sugar by stimulating the breakdown of glycogen (to glucose) in the liver.

A ductless, hormone-producing gland consisting of cells that release their secretions into the extracellular fluid from which the secretions diffuse into nearby capillaries.

A class of hormones, released by the adrenal cortex in response to the presence of ACTH, that makes additional energy available to the body by stimulating the synthesis of glucose.

A molecule produced by cells of endocrine glands and released into the circulatory system. An endocrine hormone causes changes in target cells that bear specific receptors for the hormone.

A swelling of the thyroid gland caused by iodine deficiency, which affects the functioning of the thyroid gland and its hormones.

An animal's organ system for cell-to-cell communication; composed of hormones and the cells that secrete them.

A hormone, released by the anterior pituitary, that stimulates growth, especially of the skeleton.

A hormone, secreted by the adrenal medulla, that is released in response to stress and that stimulates a variety of responses, including the release of glucose from the liver and an increase in heart rate.

hypothalamus	melatonin
inhibiting hormone	mineralocorticoids
insulin	negative feedback
islet cell	neurosecretory cell
leptin	norepinephrine
local hormone	ovary
luteinizing hormone (LH)	oxytocin

A hormone, secreted by the pineal gland, that is involved in the regulation of circadian cycles.

A region of the brain that controls the secretory activity of the pituitary gland; synthesizes, stores, and releases certain peptide hormones; and directs autonomic nervous system responses.

Steroid hormones produced by the adrenal cortex that regulate salt retention in the kidney, thereby regulating the salt concentration in the blood and other extracellular fluids.

A hormone, secreted by the neurosecretory cells of the hypothalamus, that inhibits the release of specific hormones from the anterior pituitary.

A situation in which a change initiates a series of events that tend to counteract the change and restore the original state. Negative feedback in physiological systems maintains homeostasis.

A hormone, secreted by the pancreas, that lowers blood sugar by stimulating many cells to take up glucose and by stimulating the liver to convert glucose to glycogen.

A specialized nerve cell that synthesizes and releases hormones.

A cell in the endocrine portion of the pancreas that produces either insulin or glucagon.

A neurotransmitter, released by neurons of the parasympathetic nervous system, that prepares the body to respond to stressful situations; also called *noradrenaline*.

A peptide hormone released by fat cells that helps the body monitor its fat stores and regulate weight.

The gonad of female animals.

A general term for messenger molecules produced by most cells and released into the cells' immediate vicinity. Local hormones, which include prostaglandins and cytokines, influence nearby cells bearing appropriate receptors.

A hormone, released by the posterior pituitary, that stimulates the contraction of uterine and mammary gland muscles.

A hormone, produced by the anterior pituitary, that stimulates testosterone production in males and the development of the follicle, ovulation, and the production of the corpus luteum in females.

pancreas	posterior pituitary
parathyroid gland	progesterone
parathyroid hormone	prolactin
peptide hormone	prostaglandin
pineal gland	receptor
pituitary gland	releasing hormone
positive feedback	renin

A lobe of the pituitary gland that is an outgrowth of the hypothalamus and that releases antidiuretic hormone and oxytocin.

A combined exocrine and endocrine gland located in the abdominal cavity next to the stomach. The endocrine portion secretes the hormones insulin and glucagon, which regulate glucose concentrations in the blood. The exocrine portion secretes enzymes for fat, carbohydrate, and protein digestion into the small intestine and neutralizes the acidic chyme.

A hormone, produced by the corpus luteum in the ovary, that promotes the development of the uterine lining in females.

One of four small endocrine glands, embedded in the surface of the thyroid gland, that produces parathyroid hormone, which (with calcitonin from the thyroid gland) regulates calcium ion concentration in the blood.

A hormone, released by the anterior pituitary, that stimulates milk production in human females.

A hormone released by the parathyroid gland that stimulates the release of calcium from bone.

A family of modified fatty acid hormones manufactured by many cells of the body.

A hormone consisting of a chain of amino acids; includes small proteins that function as hormones.

A protein molecule in a plasma membrane that binds to another molecule (hormone, neurotransmitter), triggering metabolic or electrical changes in a cell.

A small gland within the brain that secretes melatonin; controls the seasonal reproductive cycles of some mammals.

A hormone, secreted by the hypothalamus, that causes the release of specific hormones by the anterior pituitary.

An endocrine gland, located at the base of the brain, that produces several hormones, many of which influence the activity of other glands.

An enzyme that is released (in mammals) when blood pressure and/or sodium concentration in the blood drops below a set point; initiates a cascade of events that restores blood pressure and sodium concentration.

A situation in which a change initiates events that tend to amplify the original change.

second messenger	thymosin
steroid hormone	thymus
target cell	thyroid gland
testis (plural, testes)	thyroid-stimulating hormone (TSH)
testosterone	thyroxine

A hormone, secreted by the thymus, that stimulates the maturation of T lymphocytes of the immune system.

An intracellular chemical, such as cyclic AMP, that is synthesized or released within a cell in response to the binding of a hormone or neurotransmitter (the first messenger) to receptors on the cell surface; brings about specific changes in the metabolism of the cell.

An organ of the lymphatic system that is located in the upper chest in front of the heart and that secretes thymosin, which stimulates maturation of T lymphocytes of the immune system.

A class of hormone whose chemical structure (four fused carbon rings with various functional groups) resembles cholesterol; steroids, which are lipids, are secreted by the ovaries and placenta, the testes, and the adrenal cortex.

An endocrine gland, located in front of the larynx in the neck, that secretes the hormones thyroxine (affecting metabolic rate) and calcitonin (regulating calcium ion concentration in the blood).

A cell on which a particular hormone exerts its effect.

A hormone, released by the anterior pituitary, that stimulates the thyroid gland to release hormones.

The gonad of male animals.

A hormone, secreted by the thyroid gland, that stimulates and regulates metabolism.

In vertebrates, a hormone produced by the interstitial cells of the testis; stimulates spermatogenesis and the development of male secondary sex characteristics.

SELF TEST

1. Which hormone would you think might be active when your body is trying to maintain your body temperature on a very hot day?
   a. glucocorticoids
   b. insulin
   c. aldosterone
   d. antidiuretic hormone

2. Which hormone is responsible for the process seen in this figure?
   a. leptin
   b. ecdysone
   c. growth hormone
   d. renin

Emerging cicada

Old cuticle

3. Which of the following hormones would be regulated by a positive feedback system?
   a. thyroid hormone
   b. insulin
   c. oxytocin
   d. antidiuretic hormone

4. Which of the following represents the "first messenger" when the mechanism of action for a hormone involves the production of a "second messenger"?
   a. mRNA
   b. cyclic AMP
   c. an enzyme activated by an increased level of cyclic AMP in the cytoplasm
   d. the hormone

5. _____ from the hypothalamus stimulate the actions of the anterior pituitary.
   a. Nerve impulses
   b. Exocrine hormones
   c. ADH hormones
   d. Releasing hormones

6. Undersecretion of thyroid hormones can produce _____, characterized by retarded mental and physical development.
   a. cretinism
   b. precocious development
   c. goiter
   d. Graves' disease

7. The hypothalamus responds to decreasing levels of thyroxine by
   a. stimulating the thyroid gland with a nerve impulse.
   b. releasing more thyroxine.
   c. secreting hormones that stimulate the release of TSH from the anterior pituitary.
   d. increasing body temperature.

8. The hormone that stimulates the development of specialized white blood cells is
   a. parathormone.
   b. estrogen.
   c. thymosin.
   d. aldosterone.

9. Which hormone secreted in a daily rhythm is thought to influence sleep–wake cycles?
   a. insulin
   b. melatonin
   c. thymosin
   d. glucagon

10. Gigantism is a result of
    a. overproduction of growth hormone.
    b. underproduction of thyroxine.
    c. a genetic abnormality that causes excess calcitonin to be produced, resulting in the increased growth of bone.
    d. an imbalance in the production of glucocorticoids.

11. Androgen insensitivity is a rare condition where a genetic (XY) male develops as a female except that the individual has testes that remain in the body cavity and no uterus or ovaries. Blood samples show that the levels of testosterone and other androgens are normal for a male body. Where is the defect in hormonal signaling likely to be in this condition?
    a. LH levels from the anterior pituitary must be below average.
    b. FSH levels from the anterior pituitary must be below average.
    c. Androgen receptors must not be functioning properly and are unable to respond to the circulating androgens.
    d. The cyclic AMP intracellular second messenger pathway must not be functioning properly.

12. Which of the following statements about the role of the hypothalamus is true?
    a. The hypothalamus controls the release and synthesis of hormones by the anterior pituitary through a combination of several hormones, in part because of a unique linking of capillary beds—one in the hypothalamus and one in the anterior pituitary.
    b. The hypothalamic hormones affect the anterior pituitary by stimulating or inhibiting other hormones synthesized by the anterior pituitary.
    c. Modified neurons (neurosecretory cells) with their cell bodies in the hypothalamus, and their endings in the anterior pituitary, release hormones that enter a capillary bed and carry blood to the rest of the body.
    d. Both a and b are correct.

13. The concentration of thyroxine in the bloodstream is regulated by
    a. negative feedback mechanisms.
    b. the presence of thyroid-stimulating hormone (TSH).
    c. the presence of thyroid-releasing hormone (TRH).
    d. the parathyroid gland.

14. Which of the following statements about the adrenal gland is true?
    a. The adrenal glands are similar to the pancreas and pituitary in being composed of two different tissues and, therefore, the two parts of the adrenal gland have different functions.
    b. The adrenal cortex secretes adrenaline.
    c. The hormones produced by the adrenal cortex are important in the "fight-or-flight" response.
    d. The adrenal medulla produces three very different functional classes of hormones.

15. The men in the figure at right are suffering from an abnormality in the production of hormones from which gland?
    a. testes
    b. adrenal gland
    c. thyroid gland
    d. pituitary gland

16. Which of the following hormones is based on the structure of cholesterol?
    a. testosterone
    b. glucocorticoid
    c. oxytocin
    d. both a and b
    e. both a and c

17. Ibuprofen is often prescribed for relief of menstrual cramps because it
    a. blocks electrical signals from pain nerves to the brain.
    b. acts directly on the muscle cells in the uterus to cause relaxation.
    c. blocks production of prostaglandins.
    d. blocks the oxytocin production responsible for menstrual cramps.

18. Why are positive feedback loops much rarer than negative feedback loops?
    a. Positive feedback loops amplify the initial response that triggered them, so without a self-limiting end result, they would be "explosive."
    b. Negative feedback loops control a single variable, whereas positive feedback loops can control more than one variable. Thus, negative feedback loops are simpler and use less energy.
    c. Negative feedback loops act to negate the response that triggers them, which helps to keep important variables such as body temperature or blood sugar levels within acceptable ranges.
    d. both a and b
    e. both a and c

19. Which of the following is NOT a hormone associated with the anterior pituitary?
    a. oxytocin
    b. growth hormone
    c. thyroid-stimulating hormone
    d. follicle-stimulating hormone

20. Insulin is a small (51-amino-acid) protein that is produced by the pancreas, which signals the body's cells to take up glucose from the blood. Where is the insulin receptor located in the cell?
    a. in the cell membrane
    b. inside the cell (intracellular)
    c. in the nucleus
    d. none of the above

21. Why does ordinary table salt prevent the development of simple goiters?
    a. Salt is required to stimulate the feedback mechanism that controls the amount of thyroxine produced.
    b. Salts and other minerals inhibit the feedback mechanism, thereby creating the conditions that allow goiters to develop.
    c. Table salt is iodized.
    d. Salt contains the nutrients required for the production of thyroxine.

22. Identify the series of events that is initiated when thyroxine levels are low.
    a. The anterior pituitary causes the hypothalamus to secrete TSH, which inhibits the continued production of thyroxine.
    b. The hypothalamus secretes releasing hormones that cause the anterior pituitary to secrete TSH, which stimulates the thyroid to secrete thyroxine.
    c. The thyroid gland produces increased thyroxine when it stimulates the posterior pituitary to secrete its releasing hormones.
    d. The posterior pituitary stimulates the anterior pituitary, which responds by stimulating the thyroid to secrete TSH.

23. Which of the following statements is a correct description of why a particular gland is very large when we are young, but much smaller when we reach adulthood?
    a. The thyroid is most active when we are young to prevent the development of conditions such as cretinism.
    b. The thymus gland is large during youth and decreases with age as our immune system matures.
    c. The pituitary decreases in size after we reach adult stature, and growth hormone is no longer needed to stimulate growth.

    d. The pancreas is most active when we are young and decreases in size because less insulin is required when we mature.

24. Gigantism occurs when there is an oversecretion of the hormone that causes exceptionally rapid growth. In some cases, tumor growth can cause oversecretion of hormones. Where would the tumor have to be to give rise to gigantism?
    a. thyroid gland
    b. anterior pituitary
    c. pancreas
    d. adrenal medulla

25. Insulin and glucagon from the pancreas work in concert to keep blood sugar levels under tight control. In addition, hormones such as thyroxine and the glucocorticoids from other glands also play a role in keeping glucose levels in the appropriate range. Why are so many different hormones used to regulate blood glucose levels?
    a. Glucose is used exclusively for energy by all the tissues of the body.
    b. Glucose is the only source of energy the brain can use.
    c. Glucose plays a role in keeping ions inside and outside of cells in balance.
    d. Glucose is used to help carry oxygen in the blood.

26. Prolactin would not work in a kidney because
    a. the target cells for prolactin are not found in the kidney.
    b. the cells in the kidney require releasing hormones in order to work properly and prolactin is not a releasing hormone.
    c. prolactin is not a steroid hormone.
    d. the receptors for prolactin in the kidney have been blocked so the protein cannot bind and cause action in kidney cells.

27. The posterior pituitary is not a true endocrine gland because it
    a. is only part of the nervous system and has nothing to do with hormones.
    b. is only a hormone storage area and receives the hormones that it releases from the hypothalamus.
    c. has only an exocrine function.
    d. is not involved in homeostasis.

28. If your dog began to lose weight rapidly and kept shedding her coat much longer than she normally would you might suspect she was having trouble with her
    a. ovaries.
    b. adrenal gland.
    c. thyroid gland.
    d. pituitary gland.

29. Which hormone might be given to chemotherapy patients who are anemic?
    a. renin
    b. melatonin
    c. cortisol
    d. erythropoietin

30. Atrial natriuretic peptide
    a. regulates heart rate.
    b. regulates water retention.
    c. lowers blood volume.
    d. regulates blood pressure.

ANSWER KEY

1. d
2. b
3. c
4. d
5. d
6. a
7. c
8. c
9. b
10. a
11. c
12. d
13. a
14. a
15. d

16. d
17. c
18. e
19. a
20. a
21. c
22. b
23. b
24. b
25. b
26. a
27. b
28. c
29. d
30. c

Chapter 38 The Nervous System

OUTLINE

Section 38.1 What Are the Structures and Functions of Nerve Cells?

Neuron **dendrites** receive electrical signals from other **neurons**. The neuron **cell body** integrates these signals and generates an electrical signal (**action potential**) that is transmitted down an **axon**. The axon electrical stimulus communicates with other cells at **synapses** that form between **synaptic terminals** and the dendrites or cell body of another cell (**Figure 38-1**).

Section 38.2 How Do Neurons Produce and Transmit Information?

If a resting neuron plasma membrane is stimulated, its membrane potential becomes less negative, and if **threshold** is reached, an action potential will form (**Figure 38-2**). The action potential is conducted down the axon plasma membrane, which occurs more rapidly in myelinated axons (**Figure 38-3**).

When an action potential reaches the presynaptic membrane of a synaptic terminal, neurotransmitters are released into the **synaptic cleft** of a synapse, which then bind to receptors on the postsynaptic membrane (**Figure 38-4**). Neurotransmitter binding causes the formation of **inhibitory postsynaptic potentials (IPSPs)** or **excitatory postsynaptic potentials (EPSPs)** on the postsynaptic membrane, which undergo summation that influences the formation of a new action potential (**Figure E38-2**).

Section 38.3 How Do Nervous Systems Process Information?

All nervous systems must determine the type of stimulus, signal the **intensity** of a stimulus (**Figure 38-5**), integrate information from many sources, and initiate and direct appropriate responses.

Section 38.4 How Are Nervous Systems Organized?

Most behaviors are controlled by neuron-to-muscle pathways composed of **sensory neurons**, **interneurons**, **motor neurons**, and **effectors**. The simplest type of behavior is the reflex, which is an involuntary movement of a body part in response to a stimulus (**Figure 38-10**).

Nervous systems are composed of either a diffuse network of neurons (e.g., a **nerve net**; **Figure 38-6a**) or are centralized (**Figures 38-6b, c**).

Section 38.5 What Are the Structures and Functions of the Human Nervous System?

The vertebrate nervous system consists of the **central nervous system (CNS)** and the **peripheral nervous system (PNS; Figure 38-7)**.

The CNS includes the **brain** and **spinal cord**; the PNS consists of neurons that lie outside the CNS and the axons that connect these neurons with the CNS.

The PNS includes peripheral nerves that form the **somatic nervous system** (controls voluntary movements) and the **autonomic nervous system** (controls involuntary responses).

The autonomic nervous system consists of the **sympathetic division** and **parasympathetic division**, both of which innervate most organs but cause opposite effects (**Figure 38-8**). The sympathetic division helps prepare the body for stressful or energetic activity, whereas the parasympathetic division directs maintenance activities during periods of rest.

The central nervous system is protected by the skull and vertebral column, meninges, and the **blood–brain barrier**.

The spinal cord has nerve axons that extend from the dorsal and ventral portions of the spinal cord (containing sensory and motor neurons, respectively; **Figure 38-9**). The center of the spinal cord is made of **gray matter** (composed of motor neuron cell bodies and interneurons) and **white matter** (composed of myelinated axons of neurons that extend up or down the spinal cord).

The vertebrate brain is divided into the **hindbrain, midbrain**, and **forebrain** (**Figure 38-12**).

The hindbrain includes the **medulla** and **pons**, which control many involuntary functions, and the **cerebellum**, which coordinates complex motor activities (**Figure 38-12**).

The human midbrain contains the majority of the **reticular formation**, a filter and relay for sensory stimuli (**Figure 38-12**).

The forebrain includes the **thalamus**, **limbic system**, and the **cerebral cortex** (**Figure 38-12**). The thalamus is a sensory relay station that directs information to and from the conscious centers of the forebrain. The limbic system (**Figure 38-13**) is a group of structures that work together to produce the most basic emotions, drives, and behaviors. The cerebral cortex is the outer layer of the forebrain and is the center for information processing, memory, and initiation of voluntary actions. It is composed of four pairs of lobes: frontal, parietal, occipital, and temporal (**Figure 38-14**).

The brain has two cerebral hemispheres that are specialized (**Figure 38-16**). The left hemisphere dominates logic, speech, reading, writing, and language comprehension. The right hemisphere specializes in spatial perception, music, emotions, and facial recognition.

Memory takes two forms: **working memory** and **long-term memory**. Working memory is both electrical and biochemical in nature and lasts for only several seconds. Long-term memory involves structural changes in the effectiveness or number of synapses and allows for near-permanent memory retention. The **hippocampus** is important in converting working memory to long-term memory, whereas the temporal lobes are important in the memory recognition of objects and faces and in understanding language.

FLASH CARDS

To use the flash cards, tear the page from the book and cut along the dashed lines. The key term appears on one side of the flash card, and its definition appears on the opposite side.

action potential	cell body
amygdala	central nervous system (CNS)
autonomic nervous system	cerebellum
axon	cerebral cortex
basal ganglia	cerebral hemisphere
blood-brain barrier	cerebrum
brain	convolution

The part of a nerve cell in which most of the common cellular organelles are located; typically a site of integration of inputs to the nerve cell.

A rapid change from a negative to a positive electrical potential in a nerve cell. An action potential travels along an axon without a change in amplitude.

In vertebrates, the brain and spinal cord.

Part of the forebrain of vertebrates that is involved in the production of appropriate behavioral responses to environmental stimuli.

The part of the hindbrain of vertebrates that is concerned with coordinating movements of the body.

The part of the peripheral nervous system of vertebrates that synapses on glands, internal organs, and smooth muscle and produces largely involuntary responses.

A thin layer of neurons on the surface of the vertebrate cerebrum, in which most neural processing and coordination of activity occurs.

A long extension of a nerve cell, extending from the cell body to synaptic endings on other nerve cells or on muscles.

One of two nearly symmetrical halves of the cerebrum, connected by a broad band of axons, the corpus callosum.

Several clusters of neurons in the interior of the cerebrum, plus the substantial nigra in the midbrain, that function in the control of movement. Damage to or degeneration of one or more basal ganglia causes disorders such as Parkinson's disease and Huntington's disease.

The part of the forebrain of vertebrates that is concerned with sensory processing, the direction of motor output, and the coordination of most of the body's activities; consists of two nearly symmetrical halves (the hemispheres) connected by a broad band of axons, the corpus callosum.

Relatively impermeable capillaries of the brain that protect the cells of the brain from potentially damaging chemicals that reach the bloodstream.

A folding of the cerebral cortex of the vertebrate brain.

The part of the central nervous system of vertebrates that is enclosed within the skull.

corpus callosum

glia

dendrite

gray matter

dorsal root ganglion

hindbrain

effector

hippocampus

excitatory postsynaptic
potential (EPSP)

hypothalamus

forebrain

inhibitory postsynaptic
potential (IPSP)

ganglion

integration

Cells of the nervous system that provide nutrients for neurons, regulate the composition of the extracellular fluid in the brain and spinal cord, modulate communication between neurons, and insulate axons, thereby speeding up the conduction of action potentials. Also called *glial cells*.

The band of axons that connects the two cerebral hemispheres of vertebrates.

The outer portion of the brain and inner region of the spinal cord; composed largely of neuron cell bodies, which give this area a gray color.

A branched tendril that extends outward from the cell body of a neuron; specialized to respond to signals from the external environment or from other neurons.

The posterior portion of the brain, containing the medulla, pons, and cerebellum.

A ganglion, located on the dorsal (sensory) branch of each spinal nerve, that contains the cell bodies of sensory neurons.

The part of the forebrain of vertebrates that is important in emotion and especially learning.

A part of the body (normally a muscle or gland) that carries out responses as directed by the nervous system.

A region of the brain that controls the secretory activity of the pituitary gland; synthesizes, stores, and releases certain peptide hormones; directs autonomic nervous system responses.

An electrical signal produced in a postsynaptic cell that makes the resting potential of the postsynaptic neuron less negative and, hence, makes the neuron more likely to produce an action potential.

An electrical signal produced in a postsynaptic cell that makes the resting potential more negative and, hence, makes the neuron less likely to fire an action potential.

During development, the anterior portion of the brain. In mammals, the forebrain differentiates into the thalamus, the limbic system, and the cerebrum. In humans, the cerebrum contains about half of all the neurons in the brain.

The process of adding up all of the electrical signals in a neuron, including sensory inputs and postsynaptic potentials, to determine the output of the neuron (action potentials and/or synaptic transmission).

A cluster of neurons.

intensity	myelin
interneuron	Na^+-K^+ pump
limbic system	nerve
long-term memory	nerve net
medulla	neuron
midbrain	neurotransmitter
motor neuron	parasympathetic division

A wrapping of insulating membranes of specialized nonneural cells around the axon of a vertebrate nerve cell; increases the speed of conduction of action potentials.

The strength of stimulation or response.

An active transport protein that uses the energy of ATP to transport Na^+ out of a cell and K^+ into a cell; produces and maintains the concentration gradients of these ions across the plasma membrane, such that the concentration of Na^+ is higher outside a cell than inside, and the concentration of K^+ is higher inside a cell than outside.

In a neural network, a nerve cell that is postsynaptic to a sensory neuron and presynaptic to a motor neuron. In actual circuits, there may be many interneurons between individual sensory and motor neurons.

A bundle of axons of nerve cells, bound together in a sheath.

A diverse group of brain structures, mostly in the lower forebrain, that includes the thalamus, hypothalamus, amygdala, hippocampus, and parts of the cerebrum and is involved in basic emotions, drives, behaviors, and learning.

A simple form of nervous system, consisting of a network of neurons that extends throughout the tissues of an organism such as a cnidarian.

The second phase of learning; a more-or-less permanent memory formed by a structural change in the brain, brought on by repetition.

A single nerve cell.

The part of the hindbrain of vertebrates that controls automatic activities such as breathing, swallowing, heart rate, and blood pressure.

A chemical that is released by a nerve cell close to a second nerve cell, a muscle, or a gland cell and that influences the activity of the second cell.

During development, the central portion of the brain; contains an important relay center, the reticular formation.

The division of the autonomic nervous system that produces largely involuntary responses related to the maintenance of normal body functions, such as digestion; often called the *parasympathetic nervous system.*

A neuron that receives instructions from sensory neurons or interneurons and activates effector organs, such as muscles or glands.

peripheral nervous system (PNS)

reticular formation

pons

sensory neuron

postsynaptic neuron

somatic nervous system

postsynaptic potential (PSP)

spinal cord

presynaptic neuron

sympathetic division

reflex

synapse

resting potential

synaptic cleft

A diffuse network of neurons extending from the hindbrain, through the midbrain, and into the lower reaches of the forebrain; involved in filtering sensory input and regulating what information is relayed to conscious brain centers for further attention.

In vertebrates, the part of the nervous system that connects the central nervous system to the rest of the body.

A nerve cell that responds to a stimulus from the internal or external environment.

.

A portion of the hindbrain, just above the medulla, that contains neurons that influence sleep and the rate and pattern of breathing.

That portion of the peripheral nervous system that controls voluntary movement by activating skeletal muscles.

At a synapse, the nerve cell that changes its electrical potential in response to a chemical (the neurotransmitter) released by another (presynaptic) cell.

The part of the central nervous system of vertebrates that extends from the base of the brain to the hips and is protected by the bones of the vertebral column; contains the cell bodies of motor neurons that form synapses with skeletal muscles, the circuitry for some simple reflex behaviors, and axons that communicate with the brain.

An electrical signal produced in a postsynaptic cell by transmission across the synapse; it may be excitatory (EPSP), making the cell more likely to produce an action potential, or inhibitory (IPSP), tending to inhibit an action potential.

The division of the autonomic nervous system that produces largely involuntary responses that prepare the body for stressful or highly energetic situations; often called the *sympathetic nervous system*.

A nerve cell that releases a chemical (the neurotransmitter) at a synapse, causing changes in the electrical activity of another (postsynaptic) cell.

The site of communication between nerve cells. At a synapse, one cell (presynaptic) normally releases a chemical (the neurotransmitter) that changes the electrical potential of the second (postsynaptic) cell.

A simple, stereotyped movement of part of the body that occurs automatically in response to a stimulus.

In a synapse, a small gap between the presynaptic and postsynaptic neurons.

An electrical potential, or voltage, in unstimulated nerve cells; the inside of the cell is always negative with respect to the outside.

synaptic terminal

white matter

thalamus

working memory

threshold

The portion of the brain and spinal cord that consists largely of myelin-covered axons and that gives these areas a white appearance.

A swelling at the branched ending of an axon; where the axon forms a synapse.

The first phase of learning; short-term memory that is electrical or biochemical in nature.

The part of the forebrain that relays sensory information to many parts of the brain.

The electrical potential at which an action potential is triggered; the threshold is usually about 10 to 20 mV less negative than the resting potential.

SELF TEST

1. Which portion of a neuron conducts impulses toward the cell body?
   a. axon
   b. Schwann cells
   c. dendrites
   d. synaptic terminals

2. The role of the axon is to
   a. integrate signals from the dendrites.
   b. release neurotransmitters.
   c. conduct the action potential to the synaptic terminal.
   d. synthesize cellular components.
   e. stimulate a muscle, gland, or another neuron.

3. When a neuron is maintaining a resting membrane potential,
   a. the extracellular (outside the cell) level of sodium ions is higher than that inside the cell.
   b. the inside of the cell is positive relative to the outside.
   c. the cells are more "leaky" to sodium ions than they are to potassium ions.
   d. there is an equal distribution of charge across the membrane.

4. During the first part of the action potential, the inside of the cell becomes less negative because of an
   a. influx of sodium ions.
   b. influx of chloride ions.
   c. influx of potassium ions.
   d. efflux of potassium ions.

5. Action potentials do NOT
   a. represent a brief reversal in the resting potential, so that the inside of the cell becomes positive relative to the outside of the cell.
   b. carry information down the axon to the synapse.
   c. generate when the membrane potential of a cell reaches its threshold potential.
   d. carry information across a synapse.

6. The point at which the action potential is triggered is called the
   a. resting potential.
   b. threshold.
   c. the repolarization point.
   d. hyperpolarization.

7. Depending on the type of synapse, the effect of the neurotransmitter can be _____ to the postsynaptic membrane, making the membrane _____.
   a. excitatory; more negative
   b. inhibitory; less negative
   c. excitatory; less negative
   d. inhibitory; more negative

8. Which component of the reflex arc responds to stimuli from the environment and carries it to the central nervous system?
   a. sensory neurons
   b. motor neurons
   c. association neurons
   d. effectors

9. The synapse that transforms incoming information about the environment into outgoing commands to direct a behavior is found in the
   a. peripheral nervous system.
   b. dorsal root ganglion.
   c. brain.
   d. spinal cord.

10. Which of these divisions of the peripheral nervous system will innervate skeletal muscle?
    a. autonomic nervous system
    b. sympathetic nervous system
    c. somatic nervous system
    d. parasympathetic system

11. When looking at a cross section of the human spinal cord, the gray, butterfly-shaped area you see is made up mostly of
    a. myelinated axons.
    b. cell bodies.
    c. meninges.
    d. white matter.

12. If the myelin of the axon were to be destroyed you would expect
    a. an increase in the speed of the action potential.
    b. a disruption in the action potential leading to poor innervations of muscle cells.
    c. no change in nervous system function.
    d. the myelinated axons to take over stimulation of target cells.

13. The receptors for neurotransmitters would be found
    a. on the presynaptic membrane.
    b. on the postsynaptic membrane.
    c. in the synapse.
    d. along the surface of the axon.

14. Many recreational drugs utilize the same receptors in the brain as those for which neurotransmitter?
    a. dopamine
    b. serotonin
    c. glutamate
    d. GABA

15. You might expect to find receptors for norepinephrine
    a. in the heart.
    b. in skeletal muscle.
    c. in the pineal gland.
    d. in the spinal cord.

16. Which part of a neuron is responsible for receiving signals from the environment or from other neurons?
    a. cell body
    b. axon
    c. dendrite
    d. synapse

17. _____ are integration centers in neurons.
    a. Dendrites
    b. Axons
    c. Cell bodies
    d. Ion channels
    e. Synapses

18. During the second part of the action potential, the membrane potential returns to its resting negative state because of an
    a. influx of sodium ions.
    b. influx of chloride ions.
    c. influx of potassium ions.
    d. efflux of potassium ions.

19. To prevent a cell from reaching its threshold potential and generating an action potential, you can
    a. block the potassium leak channels.
    b. open some sodium ion channels.
    c. open some chloride ion channels.
    d. block the sodium channels.

20. How does a drug that causes the neuron's membrane to become a little more permeable to all ions affect the resting and threshold potentials?
    a. The resting and threshold potentials will be unchanged.
    b. The resting potential will move closer to the threshold potential.
    c. The resting potential will move further from the threshold potential.
    d. The resting potential will move as though it were an EPSP.

21. If a neurotransmitter were allowed to remain in the synaptic cleft bound to receptors, the size of the synaptic signal would
    a. remain the same.
    b. be smaller.
    c. be larger.
    d. be changed over stimulation time.

22. You are recording from a neuron with an electrode and determine that the resting membrane potential is around −70 millivolts. If you added three times more sodium to the extracellular solution, the membrane potential will
    a. stay the same.
    b. become more positive.
    c. become more negative.
    d. would be unresponsive to the electrode since the change in the solution would cause a reaction with the electrode.

23. You have just discovered a new terrestrial species. It is tube shaped, like an earthworm, but it moves at a right angle to the long axis of its body by rolling (like a rolling pin rolling across a table). It has chemoreceptors all over its body, but no obvious eyes or other sensory structures. Which type of nervous system does this new species have?
    a. centralized
    b. nerve net
    c. none
    d. peripheral

24. Based on the figure below, which sense do you think has a close association with the limbic system?
    a. taste
    b. hearing
    c. smell
    d. touch

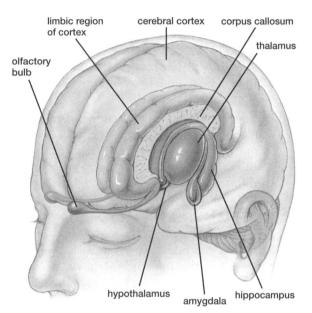

25. A patient has suffered a small stroke and reports that he can no longer feel the touch of his hand on a surface. He knows he is touching the surface, but only because he sees his hand there. In which region of the brain did this stroke occur?
    a. frontal lobe
    b. occipital lobe
    c. parietal lobe
    d. hippocampus

26. You would expect to find a nerve net in which organism?
    a. jellyfish
    b. crab
    c. horse
    d. human

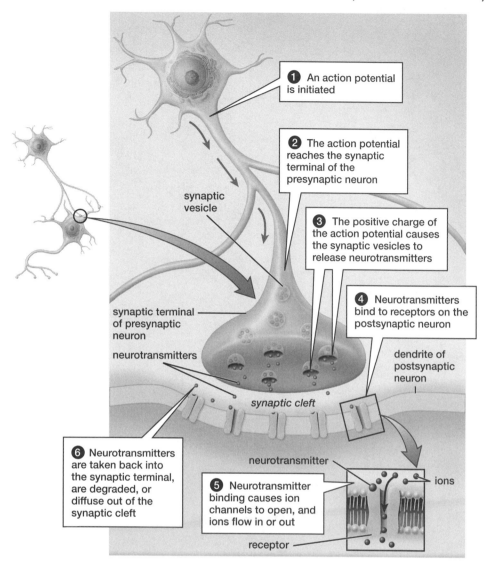

1 An action potential is initiated

2 The action potential reaches the synaptic terminal of the presynaptic neuron

3 The positive charge of the action potential causes the synaptic vesicles to release neurotransmitters

4 Neurotransmitters bind to receptors on the postsynaptic neuron

synaptic vesicle

synaptic terminal of presynaptic neuron

neurotransmitters

dendrite of postsynaptic neuron

synaptic cleft

6 Neurotransmitters are taken back into the synaptic terminal, are degraded, or diffuse out of the synaptic cleft

neurotransmitter

ions

5 Neurotransmitter binding causes ion channels to open, and ions flow in or out

receptor

27. If the postsynaptic cell in this figure were a muscle cell instead of a neuron, you would expect to see which neurotransmitter in the synapse?
    a. dopamine
    b. norepinephrine
    c. endorphins
    d. acetylcholine

28. In which structure would you expect to find interneurons?
    a. heart
    b. spinal cord
    c. muscle
    d. kidney

29. The fight-or-flight response is a function of the
    a. central nervous system.
    b. sympathetic nervous system.
    c. somatic nervous system.
    d. parasympathetic nervous system.

30. The blood–brain barrier would be permeable to
    a. glucose.
    b. viruses.
    c. neurotransmitters.
    d. plasma.

ANSWER KEY

1. c	16. c
2. c	17. c
3. a	18. d
4. a	19. c
5. d	20. b
6. b	21. c
7. c	22. a
8. a	23. b
9. d	24. c
10. c	25. c
11. b	26. a
12. b	27. d
13. b	28. b
14. a	29. b
15. a	30. a

Chapter 39 The Senses

OUTLINE

Section 39.1 How Does the Nervous System Sense the Environment?

Sensory receptors pick up environmental stimuli and convert the information into nervous system language (**Table 39-1**).

When a receptor is stimulated (**Figure 39-1**), it generates a **receptor potential** proportional in strength to that of the stimulus. If the receptor potential is strong enough, an action potential will form in a sensory neuron.

Section 39.2 How Are Mechanical Stimuli Detected?

Mechanical stimuli are detected by mechanoreceptors, which respond to touch, vibration, pressure, stretch, or sound (**Figure 39-3**). Most mechanoreceptors generate receptor potentials when their plasma membrane is deformed or stretched.

Section 39.3 How Does the Ear Detect Sound?

In the vertebrate ear, sound waves vibrate the **tympanic membrane**, which transmits these vibrations to the **middle ear** bones and then to the **oval window** of the fluid-filled **cochlea** that is part of the **inner ear** (**Figure 39-4a**).

Within the cochlea, vibrations cause the **basilar membrane** to move, bending hairs on **hair cell** mechanoreceptors against the **tectorial membrane** (**Figure 39-4b, c**). The bending of the hairs on hair cells causes action potential formation in the axons of the **auditory nerve**, which leads to the brain.

Loudness is perceived when large vibrations bend hair cell hairs to a greater degree, causing increased neurotransmitter release, which results in a higher frequency of action potential formation (**Figure 39-5**). Pitch is perceived based on which hair cells along the cochlear tubes are activated. High-pitch sounds activate hair cells near the oval window, whereas low-pitch sounds activate hair cells at the tip of the cochlea.

The vestibular apparatus is responsible for equilibrium or the detection of gravity (**Figure 39-6**).

Section 39.4 How Does the Eye Detect Light?

Arthropods use **compound eyes**, consisting of many light-sensitive **ommatidia** that produce a mosaic image (**Figure 39-7**).

In the vertebrate eye, light enters the **cornea** and passes through the **pupil** to the **lens**, which focuses an image on the **fovea** of the **retina** (**Figure 39-9a**). **Rod** and **cone photoreceptors** (which are sensitive to light intensity and color wavelengths of light, respectively) produce receptor potentials when exposed to light. These receptor potentials are processed by retinal neuron layers (**Figure 39-9b**) that cause an action potential in the **optic nerve** that leads to the brain.

Section 39.5 How Are Chemicals Sensed?

Terrestrial vertebrates detect chemicals in the external environment by smell (olfaction) or taste. Airborne chemicals are detected by olfactory receptors located in the tissues lining the nasal cavity (**Figure 39-13**). Chemicals that contact the tongue are detected by taste receptors located in clusters called **taste buds** (**Figure 39-14**). In both smell and taste, chemical receptors are sensitive to specific types of molecules, allowing for discrimination among odors and tastes.

Section 39.6 How Is Pain Perceived?

Pain is a specialized type of chemical sense that involves the detection of chemicals released from damaged tissues near **pain receptors**.

FLASH CARDS

To use the flash cards, tear the page from the book and cut along the dashed lines. The key term appears on one side of the flash card, and its definition appears on the opposite side.

anvil	blind spot
aqueous humor	choroid
auditory canal	cochlea
auditory nerve	compound eye
auditory tube	cone
basilar membrane	cornea
binocular vision	farsighted

The area of the retina at which the axons of the ganglion cells merge to form the optic nerve; because there are no photoreceptors in the blind spot, objects focused at the blind spot cannot be seen.

The second of the small bones of the middle ear, linking the tympanic membrane (eardrum) to the oval window of the cochlea; also called the *incus*.

A darkly pigmented layer of tissue behind the retina; contains blood vessels and also pigment that absorbs stray light.

The clear, watery fluid between the cornea and lens of the eye; nourishes the cornea and lens.

A coiled, bony, fluid-filled tube found in the mammalian inner ear; contains mechanoreceptors (hair cells) that respond to the vibration of sound.

A relatively large-diameter tube within the outer ear that conducts sound from the pinna to the tympanic membrane.

A type of eye, found in many arthropods, that is composed of numerous independent subunits called *ommatidia*. Each ommatidium contributes a piece of a mosaic-like image perceived by the animal.

The nerve leading from the mammalian cochlea to the brain; it carries information about sound.

A cone-shaped photoreceptor cell in the vertebrate retina; not as sensitive to light as are the rods. The three types of cones are most sensitive to different colors of light and provide color vision; see also *rod*.

A thin tube connecting the middle ear with the pharynx, which allows pressure to equilibrate between the middle ear and the outside air; also called the *Eustachian tube*.

The clear outer covering of the eye, located in front of the pupil and iris; begins the focusing of light on the retina.

A membrane in the cochlea that bears hair cells that respond to the vibrations produced by sound.

The inability to focus on nearby objects, caused by the eyeball being slightly too short or the cornea too flat.

The ability to see objects simultaneously through both eyes, providing greater depth perception and more accurate judgment of the size and distance of an object than can be achieved by vision with one eye alone.

fovea	lens
ganglion cell	middle ear
hair cell	nearsighted
hammer	ommatidium (plural, ommatidia)
inner ear	optic nerve
intensity	outer ear
iris	oval window

A clear object that bends light rays; in eyes, a flexible or movable structure used to focus light on the photoreceptor cells of the retina.

In the vertebrate retina, the central region on which images are focused; contains closely packed cones.

The part of the mammalian ear composed of the tympanic membrane, the Eustachian tube, and three bones (hammer, anvil, and stirrup) that transmit vibrations from the auditory canal to the oval window.

In the vertebrate retina, the nerve cells whose axons form the optic nerve.

The inability to focus on distant objects caused by an eyeball that is slightly too long or a cornea that is too curved.

A type of mechanoreceptor cell found in the inner ear that produces an electrical signal when stiff hairlike cilia projecting from the surface of the cell are bent. Hair cells in the cochlea respond to sound vibrations; those in the vestibular system respond to motion and gravity.

An individual light-sensitive subunit of a compound eye; consists of a lens and several receptor cells.

The second of the small bones of the middle ear, linking the tympanic membrane (eardrum) with the oval window of the cochlea; also called the *malleus*.

The nerve leading from the eye to the brain; it carries visual information.

The innermost part of the mammalian ear; composed of the bony, fluid-filled tubes of the cochlea and the vestibular apparatus.

The outermost part of the mammalian ear, including the external ear and auditory canal leading to the tympanic membrane.

The strength of stimulation or response.

The membrane-covered entrance to the cochlea.

The pigmented muscular tissue of the vertebrate eye that surrounds and controls the size of the pupil, through which light enters the eye.

pain receptor	retina
photopigment	rod
photoreceptor	round window
pinna	saccule
pupil	sclera
receptor	semicircular canal
receptor potential	sensory receptor

A multilayered sheet of nerve tissue at the rear of camera-type eyes, composed of photoreceptor cells plus associated nerve cells that refine the photoreceptor information and transmit it to the optic nerve.

A receptor cell that stimulates activity in the brain that is perceived as the sensation of pain; responds to very high or very low temperatures, mechanical damage (such as extreme stretching of tissue), and/or certain chemicals, such as potassium ions or bradykinin, that are produced as a result of tissue damage; also called *nociceptor*.

A rod-shaped photoreceptor cell in the vertebrate retina, sensitive to dim light but not involved in color vision; see also *cone*.

A chemical substance in a photoreceptor cell that, when struck by light, changes shape and produces a response in the cell.

A flexible membrane at the end of the cochlea opposite the oval window that allows the fluid in the cochlea to move in response to movements of the oval window.

A receptor cell that responds to light; in vertebrates, rods and cones.

A patch of hair cells in the vestibule of the inner ear; bending of the hairs of the hair cells permits detection of the direction of gravity and the degree of tilt of the head.

A flap of skin-covered cartilage on the surface of the head that collects sound waves and funnels them to the auditory canal.

A tough, white connective tissue layer that covers the outside of the eyeball and forms the white of the eye.

The adjustable opening in the center of the iris, through which light enters the eye.

In the inner ear, one of three fluid-filled tubes, each with a bulge at one end containing a patch of hair cells; movement of the head moves fluid in the canal and consequently bends the hairs of the hair cells.

A cell that responds to an environmental stimulus (chemicals, sound, light, pH, and so on) by changing its electrical potential; also, a protein molecule in a plasma membrane that binds to another molecule (hormone, neurotransmitter), triggering metabolic or electrical changes in a cell.

A cell (typically, a neuron) specialized to respond to particular internal or external environmental stimuli by producing an electrical potential.

An electrical potential change in a receptor cell, produced in response to the reception of an environmental stimulus (chemicals, sound, light, heat, and so on). The size of the receptor potential is proportional to the intensity of the stimulus.

stirrup

taste bud

tectorial membrane

tympanic membrane

utricle

vestibular apparatus

vitreous humor

A patch of hair cells in the vestibule of the inner ear; bending of the hairs of the hair cells permits detection of the direction of gravity and the degree of tilt of the head.

Part of the inner ear, consisting of the vestibule and the semicircular canals, involved in the detection of gravity, tilt of the head, and movement of the head.

A clear, jelly-like substance that fills the large chamber of the eye between the lens and the retina; helps to maintain the shape of the eyeball.

The third of the small bones of the middle ear, linking the tympanic membrane with the oval window; the stirrup is directly connected to the oval window; also called the *stapes*.

A cluster of taste receptor cells and supporting cells that is located in a small pit beneath the surface of the tongue and that communicates with the mouth through a small pore.

One of the membranes of the cochlea in which the hairs of the hair cells are embedded. In sound reception, movement of the basilar membrane relative to the tectorial membrane bends the hairs.

The eardrum; a membrane that stretches across the opening of the middle ear and transmits vibrations to the bones of the middle ear.

SELF TEST

1. Which receptor type would you expect to find in your bladder?
   a. mechanoreceptors
   b. thermoreceptors
   c. chemoreceptors
   d. nociceptors

2. Sensory receptors that don't have axons work similarly to
   a. action potentials.
   b. IPSPs.
   c. EPSPs.
   d. threshold potentials.

3. The blind spot corresponds to the
   a. location where the optic nerve leaves the eye.
   b. location where there are no cones so color vision cannot be detected.
   c. area just to the right of the fovea.
   d. region of the retina where there are no ganglion cells.

4. Your son came down with an ear infection shortly after having a bad cold. This can be explained by
   a. the movement of bacteria through the body.
   b. the connection of the nasopharynx to the Eustachian tubes.
   c. the buildup of pressure around the tympanic membrane while the sinuses are clogged.
   d. the ineffective antibiotic treatment from the cold, which caused the bacteria to build up in the ear.

5. The receptor type found here would be
   a. mechanoreceptors.
   b. thermoreceptors.
   c. chemoreceptors.
   d. nociceptors.

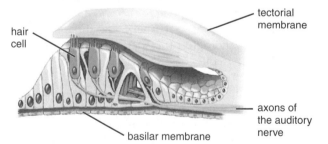

hair cell
tectorial membrane
axons of the auditory nerve
basilar membrane

6. The eyeshine (eyes that glow in the dark when you shine a flashlight on them) of nocturnal mammals is due to the
   a. rods and cones working in dark light.
   b. the reflection of the light on the retina.
   c. shine of the light through the humors of the eye.
   d. the reflective choroid layer behind the retina.

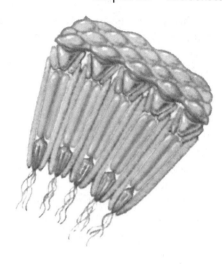

7. You would expect to find the structures shown here in
   a. an owl.
   b. a human.
   c. a grasshopper.
   d. a bird.

8. Predators tend to have binocular vision, which is beneficial because
   a. it provides depth perception to better detect prey.
   b. it helps the predators see the prey at a longer distance.
   c. it provides the predator with the ability to see the prey from any direction.
   d. it provides better color vision.

9. Stem cells are associated with olfactory membranes because
   a. the tissue is derived from mesoderm.
   b. the receptors are damaged frequently and need to be replaced.
   c. the stem cells are the receptor cells.
   d. they are also associated with the taste receptors and need to be replaced often because the taste and smell receptors to work together.

10. Monosodium glutamate (MSG) would be perceived as which taste?
    a. sweet
    b. sour
    c. salty
    d. umami

11. You are not as sensitive to touch on your back as you are on your arm because
    a. the mechanoreceptors are too far apart on your back to detect touch.
    b. the skin on you back is thicker, so it is more difficult to detect touch.
    c. there are fewer receptors in your back than in your arm.
    d. there are no Pacinian corpuscles on your back.

12. In which of its parts does the ear convert the mechanical energy called sound into electrical signals that are sent to the brain?
    a. outer ear
    b. middle ear
    c. inner ear
    d. Eustachian tube

13. The fovea is the
    a. blind spot.
    b. clear area in front of the pupil and iris.
    c. tough outer covering of the eyeball.
    d. substance that gives the eyeball its shape.
    e. central focal region of the vertebrate retina.

14. Which of the following shows the path that light entering the eye and striking the choroid takes?
    a. lens, vitreous humor, cornea, aqueous humor, retina
    b. retina, aqueous humor, lens, vitreous humor, cornea
    c. cornea, aqueous humor, retina, vitreous humor, lens
    d. cornea, aqueous humor, lens, vitreous humor, retina
    e. lens, aqueous humor, cornea, vitreous humor, retina

15. Chemoreceptors for taste are activated when
    a. chemicals dissolved in the saliva bind to receptors on the tongue.
    b. chemicals in air dissolve in mucus and bind to receptors in the olfactory epithelium.
    c. cell contents from damaged tissue stimulate receptor neurons.
    d. mechanoreceptors detect pressure on the taste buds.

16. Fish are able to swim in large groups, called schools, without bumping into each other because they have lateral lines that contain
    a. mechanoreceptors.
    b. hair cells.
    c. ommatidia.
    d. multiple fovea.

17. As we age, we tend to need glasses for reading because the muscles have a harder time changing the shape of the lens. Because of this we tend to become
    a. nearsighted.
    b. farsighted.
    c. colorblind.
    d. hampered by an astigmatism.

18. A opossum has
    a. rods with few or no cones.
    b. cones with few or no rods.
    c. equal numbers of rods and cones.
    d. no photoreceptors since they are nocturnal.

19. When an odor is detected by the olfactory receptors, the next structure to receive information is the
    a. olfactory nerve.
    b. temporal lobe of the brain.
    c. olfactory bulb.
    d. olfactory dendrites.

20. Which sense is most closely associated with the limbic system (the emotional center) of the brain?
    a. smell
    b. vision
    c. taste
    d. touch

21. Photons of light that hit the photoreceptors must travel through what structure before reaching the optic nerve?
    a. fovea
    b. vitreous humor
    c. retina
    d. ganglion cells

22. The Eustachian tubes function to
    a. move fluid out of the ear into the nasopharynx.
    b. equalize pressure in the ear.
    c. move sound waves from the throat to the ear so you can hear yourself talk.
    d. amplify the resonating chamber of the middle ear.

23. You have just discovered a new terrestrial species that has a nerve net. You would expect that this organism would have or utilize which of structures listed below?
    a. malleus
    b. chemoreceptors
    c. axons
    d. retinas

24. How can you distinguish between a light touch and a moderate poke to the arm?
    a. The action potentials are different sizes depending on the strength of the stimulus.
    b. The frequency of action potentials changes with the strength of the stimulus.
    c. Different neurotransmitters are released depending upon the strength of the stimulus.
    d. The brain makes the distinction based on the mechanical, visual, and auditory stimuli.

25. Getting dizzy on a merry-go-round means that mechanoreceptors are activated in the
    a. cochlea.
    b. basilar membrane.
    c. semicircular canals.
    d. vestibule.

26. Which structure contains mechanoreceptors?
    a. ommatidia
    b. Pacinian corpuscle
    c. taste bud
    d. olfactory epithelium

27. Sound reception is carried out by hair cells, which are a special type of
    a. chemoreceptor.
    b. photoreceptor.
    c. mechanoreceptor.
    d. magnetoreceptor.
    e. thermoreceptor.

28. What happens when an image (or a portion of an image) falls on the spot where the ganglion cell axons exit the eye?
    a. Transduction of light into an electrical signal by rods and cones occurs.
    b. Transduction of light energy to electrical energy does not occur here.
    c. Only black-and-white vision exists at this point.
    d. There is no perception of objects in the field of view.

29. During sudden bursts of intense light, the pupils of the mammal eye constrict because the
    a. aqueous humor will break down when exposed to intense light.
    b. iris constricts to reduce the amount of light projected on the retina.
    c. lens changes shape in an effort to focus the light properly.
    d. ommatidia cannot function properly when exposed to high light intensities.

30. Does having only one tongue present the same disadvantages as if you had only one eye or one ear?
    a. Yes, because having two ears allows us to localize sound.
    b. Yes, because having two eyes gives us a bigger visual range and binocular vision, which again gives us the ability to accurately localize objects in our environment.
    c. Yes, because having two tongues would allow us to better locate food sources in our environment.
    d. No, having two tongues would not allow us to better locate food sources in our environment.

ANSWER KEY

1. a
2. c
3. a
4. b
5. a
6. d
7. c
8. a
9. b
10. d
11. c
12. c
13. e
14. d
15. a
16. a
17. b
18. a
19. c
20. a
21. d
22. b
23. b
24. b
25. c
26. b
27. c
28. b
29. b
30. d

FLASH CARDS

To use the flash cards, tear the page from the book and cut along the dashed lines. The key term appears on one side of the flash card, and its definition appears on the opposite side.

actin	cartilage
antagonistic muscles	chondrocyte
appendicular skeleton	compact bone
axial skeleton	endoskeleton
ball-and-socket joint	exoskeleton
bone	extensor
cardiac muscle	flexor

A form of connective tissue that forms portions of the skeleton; consists primarily of an extracellular matrix of collagen and the cells (chondrocytes) that secrete it.

A major muscle protein whose interactions with myosin produce contraction; found in the thin filaments of the muscle fiber; see also *myosin*.

A living cell of cartilage. Together with their extracellular secretions of collagen, chondrocytes form cartilage.

A pair of muscles, one of which contracts and in so doing extends the other, relaxed muscle; this arrangement makes possible movement of the skeleton at joints.

The hard and strong outer bone; composed of osteons. Compare with *spongy bone*.

The portion of the skeleton consisting of the bones of the extremities and their attachments to the axial skeleton; the appendicular skeleton therefore consists of the pectoral and pelvic girdles and the arms, legs, hands, and feet.

A rigid internal skeleton with flexible joints that allow for movement.

The skeleton forming the body axis, including the skull, vertebral column, and rib cage.

A rigid external skeleton that supports the body, protects the internal organs, and has flexible joints that allow for movement.

A joint in which the rounded end of one bone fits into a hollow depression in another, as in the hip; allows movement in several directions.

A muscle that straightens (increases the angle of) a joint.

A hard, mineralized connective tissue that is a major component of the vertebrate endoskeleton; provides support and sites for muscle attachment.

A muscle that flexes (decreases the angle of) a joint.

The specialized muscle of the heart, able to initiate its own contraction, independent of the nervous system.

hinge joint	myofibril
hydrostatic skeleton	myosin
insertion	myosin head
joint	neuromuscular junction
ligament	origin
motor unit	osteoblast
muscle fiber	osteoclast

A cylindrical subunit of a muscle cell, consisting of a series of sarcomeres, surrounded by sarcoplasmic reticulum.

A joint at which the bones fit together in a way that allows movement in only two dimensions, as at the elbow or knee.

One of the major proteins of muscle, the interaction of which with the protein actin produces muscle contraction; found in the thick filaments of the muscle fiber; see also *actin*.

In invertebrate animals, a body structure in which fluid-filled compartments provide support for the body and change shape when acted on by muscles, which alters the animal's body shape and position, or causes the animal to move.

The part of a myosin protein that binds to the actin subunits of a thick filament; flexion of the myosin head moves the thin filament toward the center of the sarcomere, causing muscle fiber contraction.

The site of attachment of a muscle to the relatively movable bone on one side of a joint.

The synapse formed between a motor neuron and a muscle fiber.

A flexible region between two rigid units of an exoskeleton or endoskeleton, allowing for movement between the units.

The site of attachment of a muscle to the relatively stationary bone on one side of a joint.

A tough connective tissue band connecting two bones.

A cell type that produces bone.

A single motor neuron and all the muscle fibers on which it forms synapses.

A cell type that dissolves bone.

An individual muscle cell.

osteocyte	spongy bone
osteoporosis	T tubule
sarcomere	tendon
sarcoplasmic reticulum (SR)	thick filament
skeletal muscle	thin filament
skeleton	Z line
smooth muscle	

Porous, lightweight bone tissue in the interior of bones; the location of bone marrow. Compare with *compact bone.*

A mature bone cell.

A deep infolding of the muscle plasma membrane; conducts the action potential inside a cell.

A condition in which bones become porous, weak, and easily fractured; most common in elderly women.

A tough connective tissue band connecting a muscle to a bone.

The unit of contraction of a muscle fiber; a subunit of the myofibril, consisting of thick and thin filaments, bounded by Z lines.

In the sarcomere, a bundle of myosin that interacts with thin filaments, producing muscle contraction.

The specialized endoplasmic reticulum in muscle cells; forms interconnected hollow tubes. The sarcoplasmic reticulum stores calcium ions and releases them into the interior of the muscle cell, initiating contraction.

In the sarcomere, a protein strand that interacts with thick filaments, producing muscle contraction; composed primarily of actin, plus the accessory proteins troponin and tropomyosin.

The type of muscle that is attached to and moves the skeleton and is under the direct, normally voluntary, control of the nervous system; also called *striated muscle.*

A fibrous protein structure to which the thin filaments of skeletal muscle are attached; forms the boundary of a sarcomere.

A supporting structure for the body, on which muscles act to change the body configuration; may be external or internal.

The type of muscle that surrounds hollow organs, such as the digestive tract, bladder, and blood vessels; not striped in appearance (hence the name "smooth") and normally not under voluntary control.

SELF TEST

1. If the sliding filament model of contraction is dependent on the availability of ATP molecules, then why would a dead body undergo rigor mortis?
   a. Cross-bridges of myosin cannot pull actin toward the midline of the sarcomere without ATP.
   b. A dead body's cells cannot make ATP.
   c. Cross-bridges cannot turn loose of the binding sites on actin and, thus, the muscle cells cannot return to a relaxed state.

2. Sarcomeres are composed of
   a. repeating units of thick and thin filaments.
   b. many myofibrils.
   c. repeating muscle fibers.
   d. many motor neurons.

3. Which type of tissue is involved in the attachment of a skeletal muscle to bone?
   a. ligament
   b. tendon
   c. adipose
   d. cartilage

4. Which muscle type does NOT contain striations (striping) in its cells?
   a. skeletal muscle
   b. cardiac muscle
   c. smooth muscle
   d. all muscle contains striations

5. Which cells would be active in step 3 of the process depicted in the figure below?
   a. osteoblasts
   b. osteoclasts
   c. osteons
   d. osteocytes

6. Muscles contract uncontrollably when exposed to a strong electric shock because the electric shock
   a. directly causes myosin cross-bridges to bind with actin.
   b. is carried throughout the muscle fibers and stimulates the sarcoplasmic reticula to release their calcium stores.
   c. speeds up the rate at which ATP binds to myosin cross-bridges.

7. Functionally speaking, smooth muscle could NOT be used to generate skeletal movements because its contractions
   a. could be faster than skeletal muscle contractions, leading to added stress on the bones.
   b. would not allow an animal to generate voluntary movements.
   c. would not be simultaneous for maximum skeletal movements.
   d. would be stimulated by acetylcholine.

8. The condition shown in the figure above is likely a result of
   a. osteoblast malfunction.
   b. inadequate osteoclasts.
   c. osteoclasts outfunctioning osteoblasts.
   d. the aging of bones in the human body.

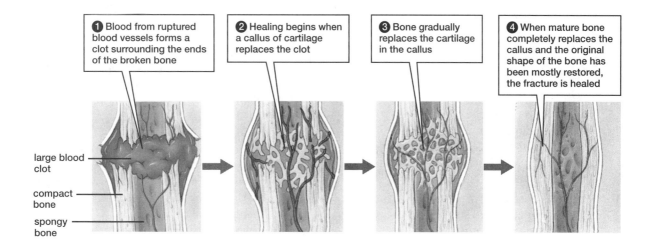

❶ Blood from ruptured blood vessels forms a clot surrounding the ends of the broken bone

❷ Healing begins when a callus of cartilage replaces the clot

❸ Bone gradually replaces the cartilage in the callus

❹ When mature bone completely replaces the callus and the original shape of the bone has been mostly restored, the fracture is healed

large blood clot

compact bone

spongy bone

9. An advantage of an exoskeleton would be that it
   a. is a site for muscle attachment.
   b. would grow with the organism.
   c. allows organisms to move through tight spaces.
   d. is made of protein instead of minerals.

10. Which muscle pair would be considered antagonistic?
    a. rectus abdominus and oblique abdominus
    b. deltoid and latissimus dorsi
    c. biceps and triceps
    d. deltoid and hamstrings

11. Which skeleton type would be most desirable for a small animal that needed to change its body shape to squeeze into/through small cracks and crevices in the ground?
    a. endoskeleton
    b. exoskeleton
    c. hydrostatic
    d. no skeleton

12. Osteoporosis is a disease in which calcium is removed from bone with a resultant loss of bone mass and strength. Why are women eight times more likely than men to suffer from this disease?
    a. Women's bones are less massive than men's bones and density decreases after menopause.
    b. Women consume more calcium than men, so more is lost.
    c. Women suffer physiological stress during childbirth that causes more bone loss than men.
    d. Women tend to do less weight-bearing exercise, which would lead to less bone mass.

13. If you saw muscle tissue under a microscope, what kind would you be looking at if it had one nucleus, striations, and branching fibers?
    a. skeletal muscle
    b. smooth muscle
    c. cardiac muscle
    d. This is not a description of muscle tissue.

14. Which bone would not be considered part of the axial skeleton?
    a. skull
    b. sternum
    c. vertebrae
    d. clavicle

15. A torn biceps muscle would keep you from
    a. flexing your forearm.
    b. lifting your arm at the shoulder.
    c. extending your forearm.
    d. rotating your arm at the shoulder.

16. A striated muscle
    a. has a striped appearance.
    b. is an involuntary muscle.

17. c. takes part in muscle contractions of the heart.
    d. is a very large muscle.

17. Cartilage takes a long time to heal because
    a. it lacks a blood supply.
    b. there are no living cells for replacement and repair.
    c. it is too pliable to damage.
    d. it is hard to damage due to pliability so it would therefore be hard to heal.

18. The nervous system controls the strength and degree of muscle contraction because of the
    a. number of fibers stimulated.
    b. frequency of action potentials.
    c. size of action potentials.
    d. both a and b.

19. Which of the following joints is a ball-in-socket joint?
    a. elbow
    b. knee
    c. shoulder
    d. the thumb joint

20. Which protein makes up the thick filament of the sarcomere?
    a. collagen
    b. troponin
    c. actin
    d. myosin

21. Which of the following is the BEST description of a cross-bridge?
    a. the connection of the head of the myosin molecules of the thick filament to the actin of the thin filament
    b. deep indentations along the plasma membrane of the muscle fiber (the sarcolemma)
    c. thin lines composed of fibrous protein bands that attach to the thin filaments of the sarcomere and represent the boundary of an individual sarcomere
    d. flattened, membrane-enclosed compartments that contain high concentrations of ions that are released during the contraction of the muscle.

22. Which type of ion is released from the sarcoplasmic reticulum during a muscle contraction?
    a. hydrogen
    b. potassium
    c. calcium
    d. chloride

23. Why does a muscle produce a stronger contraction after extensive weight training, even though the number of muscle cells does not change?
    a. The number of thick and thin filaments increases in each muscle cell, allowing a sarcomere to form more cross-bridges and thus a stronger contraction.

b. A greater number of action potentials is sent to a trained muscle, thus generating stronger contractions.

c. A trained muscle has more motor units than an untrained one, resulting in the potential for stronger contractions.

d. The repetitive nature of weight training makes the sarcomeres more efficient, which leads to a stronger contraction.

24. What happens to tropomyosin after calcium is released from the sarcoplasmic reticulum?

a. It forms a cross-bridge with the myosin of the thick filament.

b. It rotates out of the way of the myosin binding sites found on the actin molecules of the thin filament.

c. It rotates around the thin filament and blocks the myosin binding sites found on the actin molecules of the thin filament.

d. ATP binds to the tropomyosin to provide energy for the contraction.

25. When a molecule of ATP binds to the myosin head while it is attached to the actin of the thin filament, the

a. myosin forms a cross-bridge.

b. myosin head detaches from the actin.

c. myosin head changes its configuration and pulls the actin of the thin filament during the power stroke.

d. tropomyosin rotates out of the way of the myosin binding sites on the actin of the thin filament.

26. If you were to look at connective tissue under a microscope, you would know you were looking at bone if you saw

a. large circular cells with large dark nuclei.

b. collagen fibers.

c. ring-like structures with large central openings.

d. striations with mineral deposits.

27. John has fallen and broken his femur. Which of the following will deposit new bone to mend the fracture?

a. osteocytes

b. osteoblasts

c. osteoclasts

d. chondrocytes

28. If you tear the ACL (anterior cruciate ligament) in your knee, you are tearing the structure that

a. holds two bones together.

b. attaches a muscle to a bone.

c. holds the two hamstring muscles together in your thigh.

d. connects the muscles of your thigh to the muscles of your calf.

29. Osteons contain all of the following EXCEPT

a. blood vessels.

b. intervertebral discs.

c. osteocytes.

d. calcium phosphate crystals.

e. concentric layers of bone.

30. What do you call the muscle attachment on the hip bone (pelvis) if the other end of the muscle is attached to the femur near the knee? Assume the knee is elevated when the muscle contracts.

a. flexor

b. extensor

c. insertion

d. origin

ANSWER KEY

1. c
2. a
3. b
4. c
5. a
6. b
7. b
8. c
9. a
10. c
11. c
12. a
13. c
14. d
15. a

16. a
17. a
18. c
19. c
20. d
21. a
22. c
23. a
24. b
25. b
26. c
27. b
28. a
29. b
30. d

CHAPTER 41 ANIMAL REPRODUCTION

OUTLINE

Section 41.1 How Do Animals Reproduce?

Asexual reproduction involves a single individual that produces genetically identical offspring through repeated mitotic divisions of cells from some part of its body. Asexual reproduction can occur through **budding** (**Figure 41-1**), **fission**, or **parthenogenesis** (**Figure 41-2**).

Sexual reproduction involves the meiotic production of haploid gametes (**sperm** and **eggs**), which can then fuse by the process of **fertilization** to form a diploid cell called a **zygote**. This cell will divide mitotically, producing a diploid individual. Most species have separate sexes, but some are **hermaphrodites** (**Figure 41-4**).

External fertilization occurs in water outside the bodies of the parents. It may be synchronized by environmental cues (**Figure 41-5**), courtship rituals (**Figure 41-6**), or pheromones.

Internal fertilization occurs within the female body, either by **copulation** or the transfer of **spermatophores** (**Figure 41-8**).

Section 41.2 What Are the Structures and Functions of Human Reproductive Systems?

Humans copulate and fertilize eggs internally. Sperm and egg cells are produced by paired **gonads**.

The ability to reproduce occurs at **puberty**, when hormones produced by the hypothalamus (**gonadotropin-releasing hormone, GnRH**) stimulate the anterior pituitary to produce **luteinizing hormone (LH)** and **follicle-stimulating hormone (FSH)**. LH and FSH stimulate the gonads to produce either **estrogen** (by the **ovaries**) or **testosterone** (by the **testes**), which results in development of gametes and secondary sexual characteristics.

The male reproductive tract is composed of the testes (which produce sperm and testosterone) and accessory structures that produce **semen** and facilitate sperm transfer to the female reproductive tract (**Figure 41-9**). These structures are summarized in **Table 41-1**.

Sperm production (**spermatogenesis**) occurs by meiotic cell divisions within the **seminiferous tubules** of the testes (**Figure 41-11**). Spermatogenesis and testosterone production are controlled by hormones produced by the hypothalamus (GnRH) and anterior pituitary (LH and FSH; **Figure 41-13**).

The female reproductive tract is composed of the ovaries (which produce eggs, estrogen, and progesterone) and accessory structures that accept sperm and nourish the developing embryo (**Figure 41-14**). These structures are summarized in **Table 41-2**.

Egg production (**oogenesis; Figure 41-15**), hormone production, and development of the uterine lining occur in a monthly cycle controlled by hormones produced by the hypothalamus (GnRH), anterior pituitary (FSH and LH), and ovaries (estrogen and progesterone). Follicle development and the **menstrual cycle** are described in **Figures 41-16** and **41-17**, respectively.

During copulation, the erect **penis** (**Figure 41-18**) of the male ejaculates semen into the **vagina** of the female. The sperm swim through the vagina, uterus, and uterine tubes where fertilization usually takes place.

During fertilization, enzymes released by sperm **acrosomes** digest the outer **corona radiata** and **zona pellucida** layers of the unfertilized egg (**Figure 41-19**). This allows a single sperm cell to penetrate the egg and fertilize it.

Section 41.3 How Can People Limit Fertility?

Permanent **contraception** can be achieved by **sterilization**, either by vasectomy or tubal ligation (**Figure 41-20**).

Most temporary contraception methods prevent ovulation or prevent sperm and egg from meeting. These methods are summarized in **Table 41-3**.

Abortion removes the embryo from the uterus but is not considered a contraceptive device because it terminates, rather than prevents, pregnancy.

FLASH CARDS

To use the flash cards, tear the page from the book and cut along the dashed lines. The key term appears on one side of the flash card, and its definition appears on the opposite side.

acquired immune deficiency syndrome (AIDS)	chorionic gonadotropin (CG)
acrosome	clitoris
asexual reproduction	cloning
budding	contraception
bulbourethral gland	copulation
cervix	corona radiata
chlamydia	corpus luteum

A hormone secreted by the chorion (one of the fetal membranes) that maintains the integrity of the corpus luteum during early pregnancy.

An infectious disease caused by the human immunodeficiency virus (HIV); attacks and destroys T cells, thus weakening the immune system.

An external structure of the female reproductive system that is composed of erectile tissue; a sensitive point of stimulation during sexual response.

A vesicle, located at the tip of the head of an animal sperm, that contains enzymes needed to dissolve protective layers around the egg.

The production of many genetically identical copies of an organism.

Reproduction that does not involve the fusion of haploid gametes.

The prevention of pregnancy.

Asexual reproduction by the growth of a miniature copy, or bud, of the adult animal on the body of the parent. The bud breaks off to begin independent existence.

Reproductive behavior in which the penis of the male is inserted into the body of the female, where it releases sperm.

In male mammals, a gland that secretes a basic, mucous-containing fluid that forms part of the semen.

The layer of cells surrounding an egg after ovulation.

A ring of connective tissue at the outer end of the uterus that leads into the vagina.

In the mammalian ovary, a structure that is derived from the follicle after ovulation and that secretes the hormones estrogen and progesterone.

A sexually transmitted disease, caused by a bacterium, that causes inflammation of the urethra in males and of the urethra and cervix in females.

crab lice

egg

embryo

endometrium

epididymis

estrogen

external fertilization

fertilization

fission

follicle

follicle-stimulating hormone (FSH)

genital herpes

gonad

gonadotropin-releasing hormone (GnRH)

The fusion of male and female haploid gametes, forming a zygote.

An arthropod parasite that can infest humans; can be transmitted by sexual contact.

Asexual reproduction by dividing the body into two smaller, complete organisms.

The haploid female gamete, normally large and nonmotile; contains food reserves for the developing embryo.

In the ovary of female mammals, the oocyte and its surrounding accessory cells.

In animals, the stages of development that begin with the fertilization of the egg cell and end with hatching or birth; in mammals, the early stages in which the developing animal does not yet resemble the adult of the species.

A hormone produced by the anterior pituitary that stimulates spermatogenesis in males and the development of the follicle in females.

The nutritive inner lining of the uterus.

A sexually transmitted disease, caused by a virus, that can cause painful blisters on the genitals and surrounding skin.

A series of tubes that connect with and receive sperm from the seminiferous tubules of the testis, and empty into the vas deferens.

An organ where reproductive cells are formed; in males, the testes, and in females, the ovaries.

In vertebrates, a female sex hormone, produced by follicle cells of the ovary, that stimulates follicle development, oogenesis, the development of secondary sex characteristics, and growth of the uterine lining.

A hormone produced by the neurosecretory cells of the hypothalamus, which stimulates cells in the anterior pituitary to release follicle-stimulating hormone (FSH) and luteinizing hormone (LH). GnRH is involved in the menstrual cycle and in spermatogenesis.

The union of sperm and egg outside the body of either parent.

gonorrhea

menstrual cycle

hermaphrodite

menstruation

human papillomavirus (HPV)

myometrium

internal fertilization

oogenesis

interstitial cell

oogonium (plural, oogonia)

labium (plural, labia)

ovary

luteinizing hormone (LH)

ovulation

In human females, a roughly 28-day cycle during which hormonal interactions among the hypothalamus, pituitary gland, and ovary coordinate ovulation and the preparation of the uterus to receive and nourish a fertilized egg. If pregnancy does not occur, the uterine lining is shed during menstruation.

A sexually transmitted bacterial infection of the reproductive organs; if untreated, can result in sterility.

In human females, the monthly discharge of uterine tissue and blood from the uterus.

An organism that possesses both male and female sexual organs.

The muscular outer layer of the uterus.

A virus that infects the reproductive organs, often causing genital warts; causes most, if not all, cases of cervical cancer.

The process by which egg cells are formed.

The union of sperm and egg inside the body of the female.

In female animals, a diploid cell that gives rise to a primary oocyte.

In the vertebrate testis, a testosterone-producing cell located between the seminiferous tubules.

In animals, the gonad of females.

One of a pair of folds of skin of the external structures of the mammalian female reproductive system.

The release of a secondary oocyte, ready to be fertilized, from the ovary.

A hormone produced by the anterior pituitary that stimulates testosterone production in males and the development of the follicle, ovulation, and the production of the corpus luteum in females.

parthenogenesis	prostate gland
penis	puberty
placenta	regeneration
polar body	scrotum
primary oocyte	secondary oocyte
primary spermatocyte	secondary spermatocyte
progesterone	semen

A gland that produces part of the fluid component of semen; prostatic fluid is basic and contains a chemical that activates sperm movement.

An asexual specialization of sexual reproduction, in which a haploid egg undergoes development without fertilization.

A stage of development characterized by sexual maturation, rapid growth, and the appearance of secondary sexual characteristics.

An external structure of the male reproductive and urinary systems; serves to deposit sperm into the female reproductive system and deliver urine to the outside of the body.

The regrowth of a body part after loss or damage; also, asexual reproduction by means of the regrowth of an entire body from a fragment.

In mammals, a structure formed by a complex interweaving of the uterine lining and the embryonic membranes, especially the chorion; functions in gas, nutrient, and waste exchange between embryonic and maternal circulatory systems, and also secretes the hormones estrogen and progesterone, which are essential to maintaining pregnancy.

In male mammals, the pouch of skin containing the testes.

In oogenesis, a small cell, containing a nucleus but virtually no cytoplasm, produced by both the first meiotic division (of the primary oocyte) and the second meiotic division (of the secondary oocyte).

A large haploid cell derived from the diploid primary oocyte by meiosis I.

A diploid cell, derived from the oogonium by growth and differentiation, that undergoes meiotic cell division, producing the egg.

A large haploid cell derived from the diploid primary spermatocyte by meiosis I.

A diploid cell, derived from the spermatogonium by growth and differentiation, that undergoes meiotic cell division, producing four sperm.

The sperm-containing fluid produced by the male reproductive tract.

A hormone produced by the corpus luteum that promotes the development of the uterine lining in females.

seminal vesicle	spermatid
seminiferous tubule	spermatogenesis
Sertoli cell	spermatogonium (plural, spermatogonia)
sexual reproduction	spermatophore
sexually transmitted disease (STD)	sterilization
spawning	syphilis
sperm	testis (plural, testes)

A haploid cell derived from the secondary spermatocyte by meiosis II; differentiates into the mature sperm.

In male mammals, a gland that produces a basic, fructose-containing fluid that forms part of the semen.

The process by which sperm cells form.

In the vertebrate testis, a series of tubes in which sperm are produced.

A diploid cell, lining the walls of the seminiferous tubules, that gives rise to a primary spermatocyte.

In the seminiferous tubule, a large cell that regulates spermatogenesis and nourishes the developing sperm.

A package of sperm formed by the males of some invertebrate animals; the spermatophore can be inserted into the female reproductive tract, where it releases its sperm.

A form of reproduction in which genetic material from two parent organisms is combined in the offspring; usually, two haploid gametes fuse to form a diploid zygote.

A generally permanent method of contraception in which the pathways through which the sperm (vas deferens) or egg (oviducts) must travel are interrupted; the most effective form of contraception.

A disease that is passed from person to person by sexual contact; also known as *sexually transmitted infection (STI)*.

A sexually transmitted bacterial infection of the reproductive organs; if untreated, can damage the nervous and circulatory systems.

A method of external fertilization in which male and female parents shed gametes into water, and sperm must swim through the water to reach the eggs.

The gonad of male animals.

The haploid male gamete, normally small, motile, and containing little cytoplasm.

testosterone

uterus

trichomoniasis

vagina

urethra

vas deferens

uterine tube

zona pellucida

zygote

In female mammals, the part of the reproductive tract that houses the embryo during pregnancy.

In vertebrates, a hormone produced by the interstitial cells of the testis; stimulates spermatogenesis and the development of male secondary sex characteristics.

The passageway leading from the outside of a female mammal's body to the cervix of the uterus; serves as the receptacle for semen and as the birth canal.

A sexually transmitted disease, caused by the protist *Trichomonas*, that causes inflammation of the mucous membranes that line the urinary tract and genitals.

The tube connecting the epididymis of the testis with the urethra.

The tube leading from the urinary bladder to the outside of the body; in males, the urethra also receives sperm from the vas deferens and conducts both sperm and urine (at different times) to the tip of the penis.

A clear, noncellular layer between the corona radiata and the egg.

The tube leading from the ovary to the uterus, into which the secondary oocyte (egg cell) is released; also called the *oviduct*, or, in humans, the *Fallopian tube*.

In sexual reproduction, a diploid cell (the fertilized egg) formed by the fusion of two haploid gametes.

SELF TEST

1. Which type of asexual reproduction involves haploid eggs maturing into haploid adults?
   a. budding
   b. regeneration
   c. fission
   d. parthenogenesis

2. Which of the following events must occur to maximize the likelihood of a successful mating that takes place outside the bodies of the parents?
   a. A special spermatophore must be produced by the male.
   b. Sperm and eggs must be released at the same time.
   c. A single individual must release both eggs and sperm.
   d. Sperm and eggs must be released within the same limited area.
   e. Both b and d are correct.

3. Enzymes in the _____ of the sperm digest protective layers that surround the egg.
   a. nucleus
   b. acrosome
   c. midpiece
   d. tail

4. Fertilization usually occurs when the egg is in
   a. one of the ovaries.
   b. the cervix.
   c. one of the uterine tubes.
   d. the uterus.

5. A birth control measure for men cuts and ties off the _____, a tube that carries sperm from the testes to the penis.
   a. vas deferens
   b. urethra
   c. epididymis
   d. seminiferous tubule

6. Which of the following sequences of structures represents the pathway of an egg released during ovulation?
   a. ovary, oviduct, fimbriae, uterus
   b. ovary, fimbriae, uterus, oviduct
   c. ovary, uterus, fimbriae, oviduct
   d. ovary, fimbriae, oviduct, uterus

7. Which of the following male structures produces the secretion that makes up the largest volume of semen?
   a. bulbourethral glands
   b. seminal vesicles
   c. epididymis
   d. prostate gland

8. A sea anemone would most likely reproduce through
   a. fission.
   b. parthenogenesis.
   c. budding.
   d. regeneration.

9. Which of the following descriptions represents the process of ovulation?
   a. The lining of the uterus is shed.
   b. The egg buries itself in the wall of the uterus.
   c. A sperm fuses with an egg.
   d. A follicle in the ovary ruptures and releases an egg.

10. The role of FSH in the first few days of the menstrual cycle is to
    a. stimulate several follicles within the ovaries to develop and mature.
    b. stimulate ovulation.
    c. maintain the corpus luteum at the start of the cycle.
    d. promote the shedding of the lining of the uterus.

11. Which of the following organisms would you expect to display external fertilization?
    a. lizards
    b. humans
    c. fish
    d. grasshoppers

12. Which of the following events is timed to the massive "surge" in the amount of LH released from the pituitary?
    a. shedding of the endometrium
    b. disintegration of the corpus luteum
    c. ovulation
    d. fertilization

13. What might happen if the tip of the sperm did not contain an acrosome?
    a. The sperm would not be able to penetrate the egg.
    b. The motility of sperm through the female reproductive tract would be inhibited.
    c. The sperm motility in the seminal fluid would be reduced.
    d. There would no consequences since the acrosome does not play a role in sperm function.

14. The combination birth control pill works by keeping the levels of estrogen and progesterone in the circulation fairly high during much of the menstrual cycle. Which of the following descriptions explains how this would prevent a pregnancy?
    a. High levels of estrogen and progesterone promote the shedding of the uterine lining, which prevents an egg from implanting.
    b. High levels of estrogen and progesterone inhibit the activity of LH and FSH by interfering with their ability to bind to cells in the ovary.

c. High levels of estrogen and progesterone promote the release of FSH and LH from the pituitary, which prevents the development of new follicles and inhibits ovulation.

d. High levels of estrogen and progesterone inhibit the release of FSH and LH from the pituitary, which prevents the development of new follicles and inhibits ovulation.

15. Which of the following would NOT be considered an advantage of asexual reproduction over sexual reproduction?
   a. There is no need to find a mate.
   b. Competition for mates is unnecessary.
   c. There is no waste of sperm or eggs.
   d. Genetic variation would be much greater.

16. Based on the male reproductive anatomy outlined in the figure below, why might an enlarged prostate gland lead to more trips to the restroom to urinate?
   a. The seminal vesicle would be blocked.
   b. An enlarged prostate would lead to a decrease in the size of the bladder.
   c. It would pinch down on the bulbourethral gland.
   d. It plays a role in urinary urge and a larger prostate would lead to an increased urge.

17. Given the obvious advantages of asexual reproduction, why do so many animals have the capacity for sexual reproduction?
   a. Sexual reproduction is easier than asexual reproduction.
   b. Offspring produced through sexual reproduction are larger and thus are better able to survive in the environment.
   c. Offspring produced through sexual reproduction have the advantage of possessing genetic variability, thus enhancing natural selection processes for the species.
   d. Sexual reproduction is more efficient if mates are easy to find.

18. The concentration of _____ represents the hormonal difference between a child and a young person who is going through puberty.
   a. gonadotropin-releasing hormone (GnRH)
   b. FSH
   c. testosterone or estrogen
   d. insulin
   e. a, b, and c

19. If defective sperm were produced during spermatogenesis because of cell division errors in mitosis, which cells would MOST likely be to blame?
   a. interstitial cells
   b. spermatogonia
   c. primary spermatocytes
   d. secondary spermatocytes

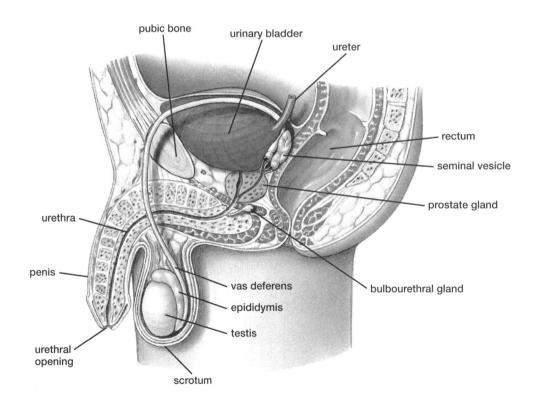

20. The inhibition of the production of _____ would be an effective means of preventing sperm production without affecting testosterone production.
   a. LH
   b. FSH
   c. estrogen
   d. GnRH

21. Based on this figure, home fertility kits would be testing the levels of which hormone?
   a. GnRH
   b. FSH
   c. LH
   d. progesterone

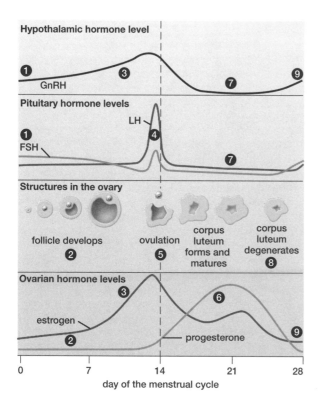

22. During oogenesis, what is the functional advantage of the production of only one "large" viable egg, as opposed to four "small" viable eggs?
   a. A large egg is easier for sperm to penetrate.
   b. It takes less energy to produce one large viable egg.
   c. A large egg has more cytoplasm and nutrients, increasing the chance that a fertilized egg will complete development.
   d. Ovulation is more efficient if the egg is large.

23. The interstitial cells produce
   a. sperm.
   b. ova.
   c. testosterone.
   d. estrogen.

24. Home pregnancy kits test for the presence of which hormone?
   a. chorionic gonadotropin
   b. estrogen
   c. progesterone
   d. FSH

25. Which of the STDs listed below is treatable with antibiotics?
   a. herpes
   b. HPV
   c. HIV
   d. chlamydia

26. Which male reproductive structure maintains the optimum temperature required for sperm production?
   a. vas deferens
   b. scrotum
   c. epididymis
   d. seminiferous tubules

27. Which of the following birth control methods would be most effective when used properly?
   a. diaphragm
   b. sponge
   c. condom
   d. rhythm

28. During a 28-day female menstrual cycle, pituitary hormones, ovarian hormones, and hormones of the hypothalamus interact to drive the cyclic changes in the uterine lining. What might be the outcome if the ovaries of a young woman were removed during a surgical procedure?
   a. GnRH from the hypothalamus and FSH and LH from the pituitary would increase.
   b. GnRH from the hypothalamus would decrease, while the FSH and LH from the pituitary would increase.
   c. GnRH from the hypothalamus would increase, whereas the FSH and LH from the pituitary would decrease.
   d. GnRH from the hypothalamus and FSH and LH from the pituitary would decrease.

29. A reproductive technique for management of endangered species might be
   a. cloning.
   b. tubal ligation.
   c. intracytoplasmic sperm injection.
   d. LH surge monitoring.

30. If there were no zona pellucida, you would expect
   a. that sperm would not penetrate the egg since that region is required for sperm to enter.
   b. multiple sperm to penetrate and fertilize a single egg.
   c. the egg to die and regenerate since that region is required for successful ovulation.
   d. the egg to continue through meiosis instead of stopping after the first division.

ANSWER KEY

1. d	16. b
2. e	17. c
3. b	18. e
4. c	19. b
5. a	20. b
6. d	21. c
7. b	22. c
8. c	23. c
9. d	24. a
10. a	25. d
11. c	26. b
12. c	27. c
13. a	28. a
14. d	29. a
15. d	30. b

CHAPTER 42 ANIMAL DEVELOPMENT

OUTLINE

Section 42.1 What Are the Principles of Animal Development?

Development of organisms leads to growth of the organism and the cells **differentiate** to perform specific functions.

Section 42.2 How Do Indirect and Direct Development Differ?

Indirect development occurs when a small, sexually immature **larva** hatches from an egg and then undergoes **metamorphosis**, becoming a sexually mature adult with a very different body form (**Figure 42-1**).

Direct development occurs when a newborn animal is sexually immature but resembles a miniature version of the sexually mature adult (**Figure 42-2**). These animals typically produce large eggs filled with **yolk** to nourish the embryo within the mother's body.

Section 42.3 How Does Animal Development Proceed?

Embryo formation begins with **cleavage** of the fertilized egg, eventually forming a **morula** and then a hollow **blastula** (**Figure 42-3**).

The blastula forms a **blastopore** through which blastula cells migrate inward. These cells form three embryonic tissue layers (**endoderm**, **mesoderm**, and **ectoderm**) through a process called **gastrulation** (**Figure 42-3**). The resulting embryo is called a **gastrula**.

The embryonic tissue layers are the developmental basis for adult organ structures, a process called **organogenesis** (**Table 42-1**). The embryo will continue to grow until it is born, when it will grow even larger, achieve sexual maturity, and eventually die.

In birds, reptiles, and mammals, extraembryonic membranes encase the embryo in a fluid-filled space and regulate nutrient and waste exchange between it and the environment. The structure and function of these embryonic membranes are listed in **Table 42-2**.

Section 42.4 How Is Development Controlled?

All cells of an animal's body contain a full set of genetic information. However, cells are able to differentiate by stimulating and repressing the transcription of specific genes (**Figure 42-4**).

The developmental fate of most cells is sealed during gastrulation through a process called induction, in which chemical messages produced by other cells direct differentiation.

The position of cells within an embryo may change, a process guided by contact between cell surface proteins and chemical pathways laid out by other cells.

Short sequences of DNA found within larger genes (**homeobox genes**) are important developmental regulators that cause the formation of specific body parts (**Figure 42-5**). This occurs when homeobox genes initiate the production of amino acid sequences that allow transcription factors to bind to specific genes, causing them to be expressed (**Figure 42-6**).

Section 42.5 How Do Humans Develop?

A fertilized egg (**zygote**) develops into a **blastocyst** and implants in the endometrium (**Figure 42-9**). During **implantation**, the outer blastocyst layer forms the **chorion**, whereas the inner cell mass forms the **amnion**, **yolk sac**, and **embryonic disk** (**Figure 42-10**). Gastrulation begins the second week after fertilization.

During the third week of development, **chorionic villi** extend into the endometrium of the uterus.

During the fourth week of development, the endoderm forms a tube that will eventually become the digestive tract. The umbilical cord also forms between the embryo and the **placenta** of the mother (**Figures 42-10** and **42-13**).

During the sixth week of development, the embryo displays prominent chordate features (notochord, tail, gill grooves), rudimentary eyes, and a developing brain.

By the eighth week of development, most of the major organs have developed. The gonads form and develop, producing testosterone or estrogen that will affect future embryonic development. The developing embryo is now called a **fetus** (**Figure 42-12**).

Over the next seven months, the fetus continues to grow and develop, with its organs becoming more functional. The nine months of human embryonic development are summarized in **Figure 42-7**.

Nine months of fetal development culminate in **labor** and delivery (**Figure 42-14**), which occurs through the interplay of estrogen, progesterone, steroid hormones produced by the fetus, oxytocin, and uterine stretching.

During pregnancy, the **mammary glands** within the mother's breasts grow under the influence of estrogen and progesterone (**Figure 42-15**). After birth, infant suckling stimulates the release of prolactin and oxytocin, which triggers milk secretion and release from the mammary glands in a process called **lactation**.

Section 42.6 Is Aging the Final Stage of Human Development?

Aging is a process in which random damage to essential biological molecules (such as DNA) accumulates over time. This impairs the body's ability to repair cell and tissue damage, resulting in the loss of organ function and eventually, death.

FLASH CARDS

To use the flash cards, tear the page from the book and cut along the dashed lines. The key term appears on one side of the flash card, and its definition appears on the opposite side.

adult stem cell (ASC)	blastula
aging	chorion
allantois	chorionic villus (plural, chorionic villi)
amnion	cleavage
amniotic egg	colostrum
blastocyst	development
blastopore	differentiate

In animals, the embryonic stage attained at the end of cleavage, in which the embryo usually consists of a hollow ball with a wall that is one or several cell layers thick.

Any stem cell not found in an early embryo; can divide and differentiate into any of several cell types, but usually not all of the cell types of the body.

The outermost embryonic membrane in reptiles (including birds) and mammals. In reptiles, the chorion functions mostly in gas exchange; in mammals, it forms most of the embryonic part of the placenta.

The gradual accumulation of damage to essential biological molecules, particularly DNA in both the nucleus and mitochondria, resulting in defects in cell functioning, declining health, and ultimately death.

In mammalian embryos, a finger-like projection of the chorion that penetrates the uterine lining and forms the embryonic portion of the placenta.

One of the embryonic membranes of reptiles (including birds) and mammals; in reptiles, serves as a waste-storage organ; in mammals, forms most of the umbilical cord.

The early cell divisions of embryos, in which little or no growth occurs between divisions; reduces the cell size and distributes gene-regulating substances to the newly formed cell.

One of the embryonic membranes of reptiles (including birds) and mammals; encloses a fluid-filled cavity that envelops the embryo.

A yellowish fluid, high in protein and containing antibodies, that is produced by the mammary glands before milk secretion begins.

The egg of reptiles, including birds; contains a membrane, the amnion, that surrounds the embryo, enclosing it in a watery environment and allowing the egg to be laid on dry land.

The process by which an organism proceeds from fertilized egg through adulthood to eventual death.

An early stage of human embryonic development, consisting of a hollow ball of cells, enclosing a mass of cells attached to its inner surface, which becomes the embryo.

In cells, to develop specializations in structure and function.

The site at which a blastula indents to form a gastrula.

direct development

gastrula

ectoderm

gastrulation

embryonic disk

homeobox gene

embryonic stem cell (ESC)

implantation

endoderm

indirect development

extraembryonic membrane

induced pluripotent stem cell (iPSC)

fetal alcohol syndrome (FAS)

induction

fetus

inner cell mass

In animal development, a three-layered embryo with ectoderm, mesoderm, and endoderm cell layers. The endoderm layer usually encloses the primitive gut.

A developmental pathway in which the offspring is born as a miniature version of the adult and does not radically change in body form as it grows and matures.

The process whereby a blastula develops into a gastrula, including the formation of endoderm, ectoderm, and mesoderm.

The outermost embryonic tissue layer, which gives rise to structures such as hair, the epidermis of the skin, and the nervous system.

A sequence of DNA coding for a transcription factor protein that activates or inactivates many other genes that control the development of specific, major parts of the body.

In human embryonic development, the flat, two-layered group of cells that separates the amniotic cavity from the yolk sac; the cells of the embryonic disk produce most of the developing embryo.

The process whereby the early embryo embeds itself within the lining of the uterus.

A cell derived from an early embryo that is capable of differentiating into any of the adult cell types.

A developmental pathway in which an offspring goes through radical changes in body form as it matures.

The innermost embryonic tissue layer, which gives rise to structures such as the lining of the digestive and respiratory tracts, and the liver and pancreas.

A type of stem cell produced from nonstem cells by the insertion of a specific set of genes that cause the cells to become capable of unlimited cell division and to be able to be differentiated into many different cell types, possibly any cell type of the body.

In the embryonic development of reptiles (including birds) and mammals, one of the chorion (functions in gas exchange), amnion (provision of the watery environment needed for development), allantois (waste storage), or yolk sac (storage of the yolk).

The process by which a group of cells causes other cells to differentiate into a specific tissue type.

A cluster of symptoms, including mental retardation and physical abnormalities, that occur in infants born to mothers who consumed large amounts of alcoholic beverages during pregnancy.

In human embryonic development, the cluster of cells, on the inside of the blastocyst, that will develop into the embryo.

The later stages of mammalian embryonic development (after the second month for humans), when the developing animal has come to resemble the adult of the species.

labor	organogenesis
lactation	placenta
larva (plural, larvae)	stem cell
mammary gland	therapeutic cloning
mesoderm	yolk
metamorphosis	yolk sac
morula	zygote

The process by which the layers of the gastrula (endoderm, ectoderm, mesoderm) rearrange to form organs.

A series of contractions of the uterus that result in birth.

In mammals, a structure formed by a complex interweaving of the uterine lining and the embryonic membranes, especially the chorion; functions in gas, nutrient, and waste exchange between embryonic and maternal circulatory systems, and also secretes the hormones estrogen and progesterone, which are essential to maintaining pregnancy.

The secretion of milk from the mammary glands.

An undifferentiated cell that is capable of dividing and giving rise to one or more distinct types of differentiated cell(s).

An immature form of an organism with indirect development before metamorphosis into its adult form; includes the caterpillars of moths and butterflies and the maggots of flies.

The production of a clone for medical purposes. Typically, the nucleus from one of a patient's own cells would be inserted into an egg whose nucleus had been removed; the resulting cell would divide and produce embryonic stem cells that would be compatible with the patient's tissues and therefore would not be rejected by the patient's immune system.

A milk-producing gland used by female mammals to nourish their young.

Protein-rich or lipid-rich substances contained in eggs that provide food for the developing embryo.

The middle embryonic tissue layer, lying between the endoderm and ectoderm, and normally the last to develop; gives rise to structures such as muscles, the skeleton, the circulatory system, and the kidneys.

One of the embryonic membranes of reptiles (including birds) and mammals. In reptiles, the yolk sac is a membrane surrounding the yolk in the egg; in mammals, it forms part of the umbilical cord and the digestive tract but does not contain yolk.

In animals with indirect development, a radical change in body form from larva to sexually mature adult, as seen in amphibians (tadpole to frog) and insects (caterpillar to butterfly).

In sexual reproduction, a diploid cell (the fertilized egg) formed by the fusion of two haploid gametes.

In animals, an embryonic stage during cleavage, when the embryo consists of a solid ball of cells.

SELF TEST

1. Which organism would you expect to have a larval stage during development?
   a. birds
   b. kangaroos
   c. butterflies
   d. worms

2. Which of the following animals utilizes indirect development?
   a. duck
   b. iguana
   c. ostrich
   d. frog

3. From the information presented in this chapter, the reptiles represent which major step in the evolution of vertebrates?
   a. Reptiles are the first group of organisms to display the gastrulation stage of development.
   b. Reptiles are the first group of organisms to have scaly skin.
   c. Reptiles are the first group to display direct development.
   d. Reptiles are the first group to have an amniotic egg.

4. In animal development, the process of cleavage can BEST be described as
   a. meiosis.
   b. mitosis.
   c. differentiation.
   d. fertilization.

5. Which of the following stages of development is defined by the three embryonic tissue layers (ectoderm, mesoderm, and endoderm)?
   a. gastrula
   b. zygote
   c. morula
   d. blastula

6. Women who want to have children are advised to take prenatal vitamins even before they become pregnant because folic acid helps prevent neurological disorders. This is important in light of the information from this chapter because
   a. the neural tube develops before a woman would know she is pregnant.
   b. the folic acid would keep gastrulation from taking place.
   c. cleavage cannot occur unless there is folic acid, and cleavage would take place before a woman would know she is pregnant.
   d. one stage in the development cycle cannot take place without another, and folic acid is required.

7. To prove that most cells in the human body contain sufficient DNA to produce a clone, DNA should be harvested from a(n) _____ cell.
   a. sperm
   b. brain
   c. egg
   d. skin
   e. both b and d

8. You would expect the structures depicted in the figure below to form in which organism?
   a. tuna
   b. kangaroo
   c. cardinal
   d. raccoon

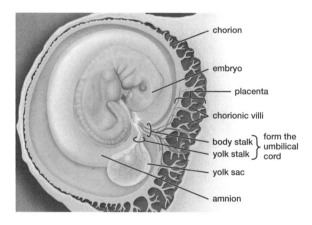

9. After the end of the _____ month of development, the embryo is called a fetus.
   a. first
   b. second
   c. fifth
   d. eighth

10. Which of the following would NOT provide evidence that aging is controlled by DNA?
    a. The number of mitotic divisions that a cell can undergo is directly proportional to the shortening of the telomeres.
    b. An increased ability to repair DNA (e.g., by combating the deleterious effects of reactive oxygen species) is directly proportional to increased life span for a cell.
    c. Some "immortal" cancer cell lines have reproduced for decades while maintaining genetically identical cells, indicating that they have found a way to bypass the regulatory mechanisms in the life span of a cell.
    d. As we age we see a decrease in elasticity of skin due to a reduction in mitotic replacement of skin cells.

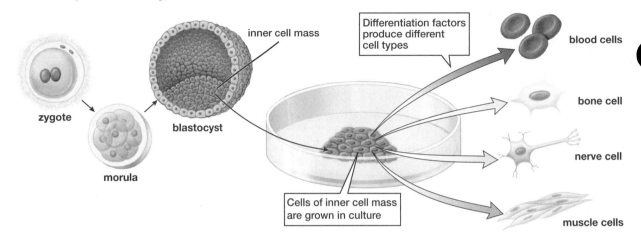

11. The process depicted in the figure above is used to harvest
    a. adult stem cells.
    b. embryonic stem cells.
    c. induced pluripotent stem cells.
    d. fetal stem cells.

12. In human embryonic development, the neural tube eventually becomes part of the
    a. digestive tract.
    b. liver.
    c. heart muscle.
    d. brain and spinal cord.

13. Which organism would not have three germ layers?
    a. jellyfish
    b. butterfly
    c. lizard
    d. eagle

14. Having legs where antennae are supposed to be would be a result of
    a. induction.
    b. errors in organogenesis.
    c. mutations in homeobox genes.
    d. mutations in embryonic stem cells.

15. Which of the following stages of development represents the earliest time that the human fetus would be viable if born?
    a. 8 weeks
    b. 12 weeks
    c. 5 months
    d. 7 months

16. Which of the following statements about reptilian extraembryonic membranes is NOT true?
    a. The chorion helps regulate the exchange of water between the environment and the embryo.
    b. The amnion encloses the embryo in fluid.
    c. The allantois carries blood between the embryo and the placenta.
    d. The yolk sac contains food for use by the embryo.

17. Which structure would implant into the endometrium?
    a. blastula
    b. morula
    c. blastocyst
    d. embryo

18. Developmentally speaking, what problems would occur in an embryo if gastrulation did NOT happen?
    a. The embryo could not form a small, solid ball of cells.
    b. Cleavage would not occur.
    c. The embryo could not form a hollow, spherical ball of cells.
    d. Organs would not form.

19. The first stage of cleavage is
    a. gastrulation.
    b. the blastula stage.
    c. the morula.
    d. the zygote.

20. Which of the following germinal layers will form the skin and the nervous system in development?
    a. mesoderm
    b. endoderm
    c. ectoderm
    d. periderm

21. The brain will develop from
    a. ectoderm.
    b. endoderm.
    c. mesoderm.
    d. periderm.

22. If an insect requires a specific temperature to become sexually active and El Niño has kept the temperature below that, then the insect would
    a. develop abnormally because of the relationship between temperature and development.
    b. develop earlier to make up for the lost time waiting to develop by others of its kind.

c. delay sexual activity until the proper temperature is reached in its environment.

d. migrate to a nearby habitat where the temperature is optimum.

23. Cells constantly receive chemical messages from other cells of the body. These messages can alter developmental life by changing
a. cell membrane function.
b. carbohydrate and fat metabolism.
c. the transcription of genes and thus of proteins made by the cell.
d. the permeability of the membrane to glucose or oxygen.

24. The fate of embryonic cells can be determined by chemical messages received from other embryonic cells using many methods EXCEPT which of the following?
a. The cells can undergo the process of induction.
b. Specific surface proteins recognize chemical trails or pathways, causing the cells that bear them to migrate along these pathways.
c. Cells may adopt different developmental fates, depending on their response to a concentration gradient of a regulatory substance.
d. The cells can be modified by differentiation and organogenesis.

25. If induction within a developing embryo stopped, _____ would be MOST affected.
a. cleavage
b. blastula formation
c. gastrulation
d. organogenesis
e. both c and d

26. Which of the following hormones is released by the developing embryo to ensure the maintenance of the corpus luteum in the ovary?
a. human chorionic gonadotropin (hCG)
b. luteinizing hormone (LH)
c. estrogen
d. follicle-stimulating hormone (FSH)

27. All of the following are true about the fetal umbilical vein EXCEPT that it
a. carries nutrient-poor blood.
b. carries blood from the fetus to the placenta.
c. is located in the umbilical cord.
d. carries oxygen-rich blood.

28. If a woman were unable to produce milk due to an endocrine gland disorder, she might consider prescribed injections of _____ to alleviate the problem.
a. oxytocin
b. prolactin
c. growth hormone
d. luteinizing hormone

29. Colostrum is particularly beneficial to a newborn baby because it
a. is especially rich in fat.
b. provides antibodies to the baby that bolster its immune system.
c. is especially rich in lactose.
d. is high in protein.
e. both b and d.

30. Which of the following would NOT cross the placenta?
a. alcohol
b. glucose
c. HIV
d. blood cells

ANSWER KEY

1. c
2. d
3. d
4. b
5. a
6. a
7. e
8. d
9. b
10. d
11. b
12. d
13. a
14. c
15. d
16. c
17. c
18. d
19. c
20. c
21. a
22. c
23. c
24. d
25. e
26. a
27. b
28. b
29. e
30. d

CHAPTER 43 PLANT ANATOMY AND NUTRIENT TRANSPORT

OUTLINE

Section 43.1 What Challenges Are Faced by All Life on Earth?

All life on Earth requires energy, water, nutrients, gas exchange, and movement of these materials within the organism. All life must also grow, develop, and reproduce.

Section 43.2 How Are Plant Bodies Organized?

Roots and **shoots** are the two major components of plants (**Figure 43-1**).

Roots function to anchor the plant to the soil, absorb water and minerals, transport nutrients and hormones, store food, produce hormones, and absorb nutrients generated by soil fungi and bacteria.

Shoots are usually above ground and consist of the stems, flowers, and leaves. Photosynthesis and reproduction take place in the shoot portion of the plant.

Flowering plants are divided into two groups: **monocots** and **dicots** (**Figure 43-2**). The number of cotyledons or seed leaves is what determines which plants belong to which group.

Section 43.3 How Do Plants Grow?

The two general cell types that make up plants are **meristem cells** and **differentiated cells**. Meristem cells are those that undergo mitosis for growth. Differentiated cells are those that go on to become specialized cells for specific functions.

Apical meristems are found on the tips of roots and shoots and are responsible for **primary growth**, resulting in length and height (**Figures 43-8 and 43-13**).

Lateral meristems, or **cambium**, are responsible for **secondary growth** resulting in an increase in diameter of the roots and shoots (**Figure 43-8**).

Section 43.4 What Are the Tissues and Cell Types of Plants?

Three **tissue systems** are found in plants: **dermal tissue**, **ground tissue**, and **vascular tissue** (Table 43-1).

Dermal tissue provides a cover for the plant. It is composed of **epidermis** found on the roots, stems, and leaves of young plants. The epidermis is replaced by **periderm** in the roots and stems of older plants. Periderm is composed of waterproof **cork cells**.

The ground tissue system consists of three tissue types (**Figure 43-3**). **Parenchyma** tissue cells are living and thin walled and perform many functions. **Collenchyma** tissue cells are living; have thicker, but flexible, walls; and act to support the plant. **Sclerenchyma** tissue cells are dead; have thick, rigid walls; and support the plant body.

The vascular tissue system is composed of two conducting tissues: **xylem** and **phloem**. Xylem conducts water and minerals from the roots to the rest of the plant. It is composed of sclerenchyma fibers, **tracheids**, and **vessel elements** (**Figure 43-4**). Phloem conducts water, sugars, amino acids, and hormones throughout the plant body. It is composed of sclerenchyma fibers, **sieve-tube elements**, and **companion cells** (**Figures 43-5**).

Section 43.5 What Are the Structures and Functions of Leaves?

Leaves are the major photosynthetic structures of most plants. They are composed of a broad, flat **blade** that is connected to the stem by a **petiole** (**Figure 43-6**).

The leaf epidermis consists of a layer of nonphotosynthetic cells, a **cuticle**, and **stomata** (with **guard cells**) that function to reduce water loss while allowing CO_2 entry. The **mesophyll** (beneath the epidermis) consists of photosynthetic parenchyma cells, air spaces, and **vascular bundles** (**Figure 43-6**).

Leaves have been modified due to the amount of light, temperature, and the availability of water (**Figure 43-7**).

Section 43.6 What Are the Structures and Functions of Stems?

Stems elevate and support the plant body (**Figure 43-8**). The stem epidermis, like that of leaves, reduces water loss while allowing CO_2 to enter. The stem **cortex** and **pith** function to support the stem, store food, and (in some cases) photosynthesize. Stem vascular tissues transport water, dissolved nutrients, and hormones.

As a stem elongates during primary growth, apical meristem cells are left behind at **nodes** and give rise to **leaf primordia** and **lateral buds** (**Figure 43-8**). Under hormonal influence, lateral buds will develop into branches.

Stem secondary growth occurs from **vascular cambium** and **cork cambium** cell divisions (**Figures 43-8** and **43-9**). The vascular cambium produces secondary xylem and secondary phloem, increasing stem diameter. The cork cambium produces waterproof cork cells that cover the outside of the stem (**Figure 43-10**).

Stems are modified for a variety of functions: storage, sprouting of new individuals, adhering and climbing, and protection (**Figure 43-11**).

Section 43.7 What Are the Structures and Functions of Roots?

Roots anchor the plant, absorb nutrients, and store food. **Fibrous roots** and **taproots** are usually specific to dicots and monocots (**Figure 43-12**). Primary growth causes roots to elongate and branch, forming structures consisting of an outer epidermis, a cortex, and an inner vascular cylinder. The root tip consists of apical meristem cells protected by a **root cap** (**Figure 43-13**).

Cells of the root epidermis absorb water and minerals from the soil, often through the use of **root hairs** (**Figure 43-14**). The root cortex is primarily responsible for the storage of sugars in the form of starch, although endodermis cells control the movement of water and minerals from the soil into the vascular cylinder (**Figure 43-17**). The root vascular cylinder contains conducting tissues that receive water and minerals from the cortex endodermis and then transport them to the rest of the plant.

Section 43.8 How Do Plants Acquire Nutrients?

Most minerals are actively transported into root hair cells and then diffuse through the cortex to the **pericycle** cells just inside the vascular cylinder. The pericycle cells actively transport minerals to the extracellular fluid of the vascular cylinder, where they then diffuse into the tracheids and vessel elements of xylem (**Figure 43-17**).

Fungal **mycorrhizae** help convert inaccessible soil minerals into forms that the root can absorb, whereas the fungus obtains sugars, amino acids, and vitamins from the plant (**Figure 43-19**).

Legumes have formed a symbiotic relationship with nitrogen-fixing bacteria, which allow them to obtain nitrogen from normally inaccessible atmospheric sources (**Figure 43-20**).

Section 43.9 How Do Plants Move Water and Minerals from Roots to Leaves?

The **cohesion–tension theory** explains how water moves from the roots to the leaves (**Figure 43-21**). When water evaporates from the leaves through **stomata** (**Figures 43-22** and **43-23**), water within the xylem enters the leaves to replace the lost water.

Cohesion of water molecules to each other (and the xylem walls) creates a "chain" of water molecules that runs down the length of the xylem plant roots.

Transpiration of water from the leaf creates "tension" that pulls the chain of water molecules toward the leaf and also draws water into the vascular cylinder of the root from surrounding endodermis cells.

Section 43.10 How Do Plants Transport Sugars?

The **pressure-flow theory** explains how sugar moves in phloem (**Figure 43-25**).

Sugar is produced in **source cells** (e.g., leaves) and is actively transported into phloem tubes, which causes the movement of water (by osmosis) into the phloem from adjacent xylem tubes.

The increase in water volume causes an increase in fluid pressure that drives fluid movement (and dissolved sugars) into regions of lower pressure.

Cells in sugar **sinks** (e.g., fruit or roots) actively transport sugars out of phloem tubes, with water following by osmosis. Thus, a fluid pressure gradient is established between source cells and sink cells that perpetuates sugar movement.

FLASH CARDS

To use the flash cards, tear the page from the book and cut along the dashed lines. The key term appears on one side of the flash card, and its definition appears on the opposite side.

annual ring	cohesion
apical meristem	cohesion-tension theory
bark	collenchyma
blade	companion cell
branch root	cork cambium
cambium (plural, cambia)	cork cell
Casparian strip	cortex

The tendency of the molecules of a substance to stick together.

A pattern of alternating light (early) and dark (late) xylem in woody stems and roots; formed as a result of the unequal availability of water in different seasons of the year, normally spring and summer.

A model for the transport of water in xylem; water is pulled up the xylem tubes, powered by the force of evaporation of water from the leaves (producing tension) and held together by hydrogen bonds between nearby water molecules (cohesion).

The cluster of meristem cells at the tip of a shoot or root (or one of their branches).

An elongated plant cell type, with thickened, flexible cell walls, that is alive at maturity and that supports the plant body.

The outer layer of a woody stem, consisting of phloem, cork cambium, and cork cells.

A cell adjacent to a sieve-tube element in phloem; involved in the control and nutrition of the sieve-tube element.

The flat part of a leaf.

A lateral meristem in woody roots and stems that gives rise to cork cells.

A root that arises as a branch of a preexisting root; occurs through divisions of pericycle cells and subsequent differentiation of the daughter cells.

A protective cell of the bark of woody stems and roots; at maturity, cork cells are dead, with thick, waterproof cell walls.

A lateral meristem, parallel to the long axis of roots and stems, that causes secondary growth of woody plant stems and roots. See *cork cambium; vascular cambium.*

The part of a primary root or stem located between the epidermis and the vascular cylinder.

A waxy, waterproof band, located in the cell walls between endodermal cells in a root, that prevents the movement of water and minerals into and out of the vascular cylinder through the extracellular space.

cuticle	fibrous root system
dermal tissue system	flower bud
dicot	ground tissue system
differentiated cell	guard cell
endodermis	heartwood
epidermal tissue	internode
epidermis	lateral bud

A root system, commonly found in monocots, that is characterized by many roots of approximately the same size arising from the base of the stem.

A waxy or fatty coating on the surfaces of aboveground epidermal cells of many land plants; aids in the retention of water.

A cluster of meristem cells (a bud) that forms a flower.

A plant tissue system that makes up the outer covering of the plant body.

A plant tissue system, consisting of parenchyma, collenchyma, and sclerenchyma cells, that makes up the bulk of a leaf or young stem, excluding vascular or dermal tissues. Most ground tissue cells function in photosynthesis, support, or carbohydrate storage.

Short for dicotyledon; a type of flowering plant characterized by embryos with two cotyledons, or seed leaves, that are usually modified for food storage.

One of a pair of specialized epidermal cells that surrounds the central opening of a stoma in the epidermis of a leaf or young stem, and regulates the size of the opening.

A mature cell specialized for a specific function; in plants, differentiated cells normally do not divide.

Older secondary xylem that usually no longer conducts water or minerals, but that contributes to the strength of a tree trunk.

The innermost layer of small, close-fitting cells of the cortex of a root that form a ring around the vascular cylinder; see also *Casparian strip*.

The part of a stem between two nodes.

Dermal tissue in plants that forms the epidermis, the outermost cell layer that covers leaves, young stems, and young roots.

A cluster of meristem cells at the node of a stem; under appropriate conditions, it grows into a branch.

In plants, the outermost layer of cells of a leaf, young root, or young stem.

lateral meristem

leaf

leaf primordium (plural, primordia)

legume

meristem cell

mesophyll

mineral

monocot

mycorrhiza (plural, mycorrhizae)

nitrogen fixation

nitrogen-fixing bacterium

node

nodule

nutrient

Short for monocotyledon; a type of flowering plant characterized by embryos with one seed leaf, or cotyledon.

A meristem tissue that forms cylinders parallel to the long axis of roots and stems; normally located between the primary xylem and primary phloem (vascular cambium) and just outside the phloem (cork cambium); also called *cambium*.

A symbiotic association between a fungus and the roots of a land plant that facilitates mineral extraction and absorption.

An outgrowth of a stem, normally flattened and photosynthetic.

The process that combines atmospheric nitrogen with hydrogen to form ammonia (NH_3).

A cluster of dividing cells, surrounding a terminal or lateral bud, that develops into a leaf.

A bacterium that possesses the ability to remove nitrogen (N_2) from the atmosphere and combine it with hydrogen to produce ammonia (NH_3).

A member of a family of plants characterized by root swellings in which nitrogen-fixing bacteria are housed; includes soybeans, lupines, alfalfa, and clover.

In plants, a region of a stem at which the petiole of a leaf is attached; usually, a lateral bud is also found at a node.

An undifferentiated cell that remains capable of cell division throughout the life of a plant.

A swelling on the root of a legume or other plant that consists of cortex cells inhabited by nitrogen-fixing bacteria.

Loosely packed, usually photosynthetic cells located beneath the epidermis of a leaf.

A substance acquired from the environment and needed for the survival, growth, and development of an organism.

An inorganic substance, especially one in rocks or soil.

parenchyma	plasmodesma (plural, plasmodesmata)
pericycle	pressure-flow theory
periderm	primary growth
petiole	root
phloem	root cap
pit	root hair
pith	root pressure

A cell-to-cell junction in plants that connects the cytoplasm of adjacent cells.

A plant cell type that is alive at maturity, normally with thin cell walls, that carries out most of the metabolism of a plant. Most dividing meristem cells in a plant are parenchyma.

A model for the transport of sugars in phloem; the movement of sugars into a phloem sieve tube causes water to enter the tube by osmosis, whereas the movement of sugars out of another part of the same sieve tube causes water to leave by osmosis. The resulting pressure gradient causes the bulk movement of water and dissolved sugars from the end of the sieve tube into which sugar is transported (a source) toward the end of the sieve tube from which sugar is removed (a sink).

The outermost layer of cells of the vascular cylinder of a root.

Growth in length and development of the initial structures of plant roots and shoots; results from cell division of apical meristems and differentiation of the daughter cells.

The outer cell layers of roots and stems that have undergone secondary growth; consists primarily of cork cambium and cork cells.

The part of the plant body, normally underground, that provides anchorage, absorbs water and dissolved nutrients and transports them to the stem, produces some hormones, and in some plants serves as a storage site for carbohydrates.

The stalk that connects the blade of a leaf to the stem.

A cluster of cells at the tip of a growing root, derived from the apical meristem; protects the growing tip from damage as it burrows through the soil.

A conducting tissue of vascular plants that transports a concentrated solution of sugars (primarily sucrose) and other organic molecules up and down the plant.

A fine projection from an epidermal cell of a young root that increases the absorptive surface area of the root.

An area in the cell walls between two plant cells where the two cells are separated only by a thin, porous cell wall.

Pressure within a root caused by the transport of minerals into the vascular cylinder, accompanied by the entry of water by osmosis.

Cells forming the center of a root or stem.

root system	sink
sapwood	source
sclerenchyma	stem
secondary growth	stoma (plural, stomata)
shoot system	taproot system
sieve plate	terminal bud
sieve-tube element	tissue

In plants, any structure that uses up sugars or converts sugars to starch, and toward which phloem fluids will flow.

All of the roots of a plant.

In plants, any structure that actively synthesizes sugar, and away from which phloem fluid will be transported.

Young secondary xylem that transports water and minerals in a tree trunk.

The portion of the plant body, normally located above ground, that bears leaves and reproductive structures such as flowers and fruit.

A plant cell type with thick, hardened cell walls that normally dies as the last stage of differentiation; may support or protect the plant body.

An adjustable opening in the epidermis of a leaf or young stem, consisting of a pair of guard cells surrounding a central pore; it regulates the diffusion of carbon dioxide, oxygen, and water into and out of the leaf or stem.

Growth in the diameter and strength of a stem or root due to cell division in lateral meristems and differentiation of their daughter cells.

A root system, commonly found in dicots, that consists of a long, thick main root and many smaller lateral roots that grow from the main root.

All the parts of a vascular plant exclusive of the root; normally above ground. Consists of stem, leaves, buds, and (in season) flowers and fruits; functions include photosynthesis, transport of materials, reproduction, and hormone synthesis.

Meristem tissue and surrounding leaf primordia that are located at the tip of a plant shoot or a branch.

In plants, a structure between two adjacent sieve-tube elements in phloem, where holes formed in the cell walls interconnect the cytoplasm of the sieve-tube elements.

A group of (normally similar) cells that together carry out a specific function; a tissue may also include extracellular material produced by its cells.

In phloem, one of the cells of a sieve tube.

tissue system	vascular tissue system
tracheid	vein
transpiration	vessel
vascular bundle	vessel element
vascular cambium	xylem
vascular cylinder	

A plant tissue system consisting of xylem (which transports water and minerals from root to shoot) and phloem (which transports water and sugars throughout the plant).

A group of two or more tissues that together perform a specific function.

In plants, a vascular bundle in a leaf.

An elongated cell type in xylem, with tapered ends that overlap the tapered ends of other tracheids, forming tubes that transport water and minerals. Pits in the cell walls of tracheids allow easy movement of water and minerals into and out of the cells, including from one tracheid to the next.

In plants, a tube of xylem composed of vertically stacked vessel elements with heavily perforated or missing end walls, leaving a continuous, uninterrupted hollow cylinder.

The evaporation of water through the stomata of a leaf or young stem.

One of the cells of a xylem vessel; elongated, dead at maturity, with thick lateral cell walls for support but with end walls that are either heavily perforated or missing.

A strand of xylem and phloem in leaves and stems; in leaves, commonly called a *vein*.

A conducting tissue of vascular plants that transports water and minerals from root to shoot.

A lateral meristem that is located between the xylem and phloem of a woody root or stem and that gives rise to secondary xylem and phloem.

The centrally located conducting tissue of a young root; consists of primary xylem and phloem, surrounded by a layer of pericycle cells.

SELF TEST

1. When you were a 3-foot-tall child, you drove a nail in a tree trunk at about eye level to hang a toy on. Now it is 10 years later and you are 6 feet tall. The tree has grown 1 foot a year. How high will the nail be on the side of the tree?
   a. 3 feet
   b. 6 feet
   c. 9 feet
   d. 12 feet

2. Primary growth, but NOT secondary growth,
   a. is produced by only lateral meristems.
   b. is produced by either lateral or apical meristems.
   c. increases the length of stems and roots.
   d. increases the diameter of stems and roots.
   e. occurs in stems but not in roots.

3. The water seen on the tips of the leaves in this figure is a result of
   a. transpiration.
   b. early morning dew.
   c. root pressure.
   d. rain drops.

4. If you found a leaf on the path you were hiking, how would you be able to tell if you were holding a monocot or dicot leaf?
   a. If the leaf came from a plant with a tap root, it would be a monocot.
   b. If the leaf had parallel venation it would be a monocot.
   c. You would have to find the plant the leaf came from and see if the vascular bundles were scattered in the stem. If so, then the leaf would be from a dicot plant.
   d. There is no way to tell if a plant is a monocot or dicot from the leaves. You have to have flowers or stems to make a determination.

5. Which main type of ground tissue would you be ingesting if you ate a carrot?
   a. parenchyma
   b. collenchyma
   c. sclerenchyma
   d. equal parts of a, b, and c

6. Which tissue in the aboveground portion of a plant body contains the least densely packed cells? What is the function of the air spaces between the cells?
   a. collenchyma; source of oxygen for photosynthesis
   b. pith; helps with support
   c. xylem; allows pressure for transport
   d. spongy and palisade parenchyma; source of carbon dioxide for photosynthesis and oxygen for cellular respiration

7. Runners of strawberry plants are actually
   a. stems.
   b. leaves.
   c. roots.
   d. flowers.

8. Why would one NOT expect to find root hairs in the area between the region of cell elongation and the root cap?
   a. There are no dermal cells in this area.
   b. The root cap is pushed down between soil particles by cells undergoing elongation. Root hairs would be torn off. Root hairs are formed later in development in the mature region.
   c. There is no vascular cylinder in these regions.
   d. These regions of the root do not require water.

9. The majority of water and dissolved minerals absorbed by a plant is first taken up by the
   a. root hairs.
   b. cortex.
   c. endodermis.
   d. vascular tissue of the root.

10. Which of the following does NOT drive the movement of water from the soil into the central root?
    a. transpiration
    b. active transport
    c. osmosis
    d. turgor pressure

11. Which of the following structures would not be found in bark?
    a. cork
    b. cork cambium
    c. xylem
    d. phloem

12. The cohesion–tension mechanism of water movement occurs when
    a. the buildup of water pressure in the roots forces water up the stem through the xylem.
    b. the water is loaded into vessel elements by active transport at the source and unloaded at the sink to ensure a constant movement of water through the xylem.
    c. the transpiration of water from the leaf surface pulls water up the xylem as a result of hydrogen bonding between water molecules.
    d. the stem contracts and decreases its diameter to force the water molecules through the tracheids and vessels.

13. While visiting your aunt in Michigan, one morning you find a beaver gnawing on a birch tree. You scare it off but not before it has removed a strip of bark all the way around the tree. Despite erecting a fence around the tree to prevent additional damage, the tree dies after a few months. The first event that would occur would be the death of the
    a. leaves of the tree because they stopped receiving water from the soil.
    b. leaves of the tree because they could no longer produce sugars by photosynthesis.
    c. roots of the tree because they would no longer be able to absorb water from the soil.
    d. roots of the tree because they would no longer be able to receive carbohydrates from the leaves.

14. Maple syrup is produced from fluid taken from sugar maple trees in the late winter or very early spring. These trees are tapped only during this time of the year because this is
    a. when roots serve as a source, and young buds are the sink.
    b. the time of year when roots are a sink.
    c. the only time when sugars are transported.
    d. when cold temperatures allow sapwood to transport fluids.

15. Pressure-flow theory explains sugar movement in phloem from _____, where sugars are actively made, toward _____, where sugars are used or converted to starch.
    a. sink; source
    b. sink; sink
    c. source; sink
    d. source; source

16. When you were a child you could "hug" the maple tree in your front yard. What type of growth would explain why your children cannot reach all the way around the same tree?
    a. primary growth
    b. secondary growth
    c. meristematic growth
    d. cork growth

17. The lateral bud can give rise to
    a. leaves and flowers.
    b. roots and branches.
    c. branches and flowers.
    d. leaves and roots.

18. The portion of the stem where the leaves are attached is called the _____; the portion of the stem that is free of leaves is called the _____.
    a. junction; interjunction
    b. node; internode
    c. bud site; interbud site
    d. pith; interpith

19. In a tree that is several years old, it is usually impossible to find any of the primary tissues that were outside the vascular cambium in the primary plant body. Those tissues
    a. were absorbed by the vascular cambium.
    b. became part of the secondary phloem.
    c. fragmented and fell off the outside of the stem.
    d. are now in the center of the stem.

20. In the phloem tissue, the cells that conduct water, sugars, amino acids, and hormones throughout the plant are called
    a. sieve-tube elements.
    b. companion cells.
    c. tracheids.
    d. vessel elements.

21. A very waxy leaf would suggest to you that
    a. the plant is adapted for an area that receives a substantial amount of rainfall.
    b. the plant is adapted to prevent the loss of water in the leaves.
    c. the plant is adapted to store water in the leaves for future use by the plant for photosynthesis.
    d. the plant is going to lose its leaves soon.

22. The primary function of the vascular tissue system in plants is to
    a. transport water, minerals, and sugars throughout the plant.
    b. store mineral and food reserves for the plant.
    c. cover and protect the outer surfaces of the plant.
    d. perform photosynthesis within the plant body.

23. If you were looking at a cross section of a fallen tree and noticed six annual rings that were very close to each other and four that were much wider, you could say that
    a. the tree had six years of very rapid growth.
    b. the tree had four years of nutrient loss.
    c. there is nothing special about tree rings other than to determine age of the tree.
    d. there were probably six consecutive years of drought in the tree's history.

24. Farmers who alternate soybeans in one planting season and wheat in the next planting season are taking advantage of
    a. the nitrogen-fixing ability of the bacteria associated with legumes.
    b. the less expensive wheat seed.
    c. the ability of wheat to live in hotter, drier seasons than soybeans.
    d. the short growing season of legumes.

25. The evaporation and loss of water vapor through the leaves of a plant is called
    a. photosynthesis.
    b. respiration.
    c. cohesion.
    d. transpiration.

26. Which type of leaf adaptation would you expect to find in a plant adapted for low light?
    a. Leaves that are needle-like in shape.
    b. Leaves that are very broad.
    c. Leaves that are narrow and covered with a cuticle.
    d. Leaves that are very thick and heavy.

27. Which of the tissues represented in the figure below would be responsible for photosynthesis?
    a. parenchyma
    b. collenchyma
    c. sclerenchyma
    d. ground tissue

28. If a hollow needle were inserted into the phloem of a plant
    a. air would be drawn into the cell.
    b. sugary sap would be pushed out of the cell.
    c. pure water would be pushed out of the cell.
    d. nothing would happen.

29. How do plants overcome the force of gravity and make water flow upward?
    a. photosynthesis in the leaves
    b. hydrogen bonds between water molecules
    c. transpiration and the hydrogen bonding of water molecules
    d. cohesion

30. Which of the following would NOT be a method that plants would use to balance the need to photosynthesize, which requires carbon dioxide from the atmosphere, with the water loss that comes from transpiration?
    a. regulating the opening and closing of the stomata
    b. regulating the potassium ion concentration within the guard cells
    c. increasing potassium concentrations when photosynthesis outstrips cellular respiration because of the influence of low carbon dioxide levels under these conditions.
    d. regulating the rate of cohesion as transpiration occurs

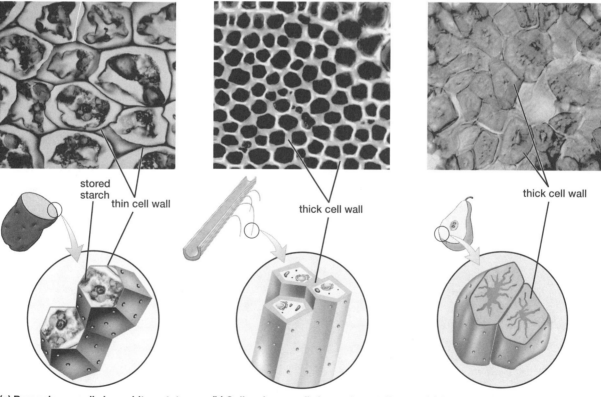

(a) Parenchyma cells in a white potato   (b) Collenchyma cells in a celery stalk   (c) Sclerenchyma cells in a pear

ANSWER KEY

1. a		16. b	
2. c		17. c	
3. c		18. b	
4. b		19. c	
5. a		20. a	
6. d		21. b	
7. a		22. a	
8. b		23. d	
9. a		24. a	
10. d		25. d	
11. c		26. b	
12. c		27. a	
13. d		28. b	
14. a		29. c	
15. c		30. d	

CHAPTER 44 PLANT REPRODUCTION AND DEVELOPMENT

OUTLINE

Section 44.1 How Do Plants Reproduce?

Plants can reproduce either asexually or sexually. Asexual reproduction takes place by mitotic divisions, which lead to the development of a new plant.

The sexual life cycle of plants involves an **alternation of generations** (**Figure 44-1**) between a multi-cellular diploid form (**sporophyte** generation) and a multicellular haploid form (**gametophyte** generation). The gametophyte stage is the dominant stage in the mosses and liverworts. The sporophyte stage is the dominant stage for all other groups of plants. Angiosperms and gymnosperms do not require water for the dispersal of sperm.

In gymnosperms and angiosperms, a male gametophyte is surrounded by a watertight coat, forming a microscopic **pollen grain** that can withstand desiccation and be dispersed by the wind or animals (**Figure 44-1**).

The female gametophyte remains moist within the flower of a plant. Upon fertilization by pollen, the egg is enclosed in a drought-resistant seed that contains a food reserve for the embryonic plant.

Section 44.2 What Is the Function and Structure of the Flower?

Flowers are the reproductive structures of angiosperms, most of which lure animals that directly pollinate them.

Complete flowers have four parts: **sepals**, **petals**, **stamens**, and **carpels** (**Figure 44-2**). Sepals form the outer protective covering of the flower bud and are found at the base of a flower. Petals can be brightly colored and attract pollinators to the flower. A stamen consists of a **filament** bearing an **anther** that produces **pollen grains**. The **carpel** consists of an **ovary** (in which female gameto-phytes develop) and a **style**, at the tip of which is a sticky **stigma** to which pollen adheres during **pollination**.

Pollen (**Figure 44-4**) contains the male gametophyte that develops within the anther of the sporophyte plant (**Figure 44-8**). Diploid **microspore mother cells** meiotically form four haploid **microspores** that meiotically produce a haploid male gametophyte. The male gametophyte usually consists of a **tube cell** and a **generative cell**, the latter of which forms two sperm cells.

The ovary contains the female gametophyte (**ovule**) that will form egg cells (**Figure 44-7**). Within the ovule, a diploid **megaspore mother cell** meiotically forms four haploid **megaspores**, of which only one survives. The surviving megaspore mitotically produces eight haploid nuclei, which eventually form seven haploid cells. The central cell is larger than the other six and has two nuclei. One of the cells near the ovule opening will become the egg cell.

Pollination leads to fertilization of the egg cell and occurs when a pollen grain lands on a stigma (**Figure 44-8**). When this happens, the pollen tube cell grows through the style to the female gametophyte. The generative cell forms two sperm cells that travel down the style (through the tube cell). One of these sperm cells will fertilize the egg cell, forming a diploid zygote that will become an embryo. The other sperm cell will fertilize the central cell, producing a triploid cell that will form the **endosperm** of the seed.

Section 44.3 How Do Fruits and Seeds Develop?

Fruit develops from the ovary and is modified to encourage seed dispersion (**Figure 44-9**). Many fruits are sweet to encourage animal consumption, have hooks so they can be carried and dispersed on animals, or have wings for wind dispersal.

The **seed** contains an embryo, consisting of an embryonic root and embryonic shoot, the latter of which contains the **cotyledon** (one in monocots, two in dicots; **Figure 44-10**). Cotyledons absorb food from the endosperm and transfer it to the growing embryo.

Section 44.4 How Do Seeds Germinate and Grow?

Seed **germination** requires warmth and moisture but may also require periods of drying, cold, or **seed coat** weathering. If these conditions are not met, the seed may remain dormant for an extended period of time.

During germination (**Figure 44-11**), the root emerges from the seed and absorbs water and nutrients. The embryonic shoot consists of a **hypocotyl** and **epicotyl**, which dicots use to disrupt the soil and form a path for the root apical meristem. Monocots use a **coleoptile** that encloses the shoot, allowing it to push aside the soil as the shoot emerges.

Cotyledons supply the developing plant with food energy as it grows (**Figure 44-12**).

Section 44.5 How Do Plants and Their Pollinators Interact?

Plants and their animal pollinators and seed dispersers have coevolved over time, acting as agents of natural selection for one another. Flowers attract animals with scent, food, or appealing colors (**Figures 44-13, 44-14, and 44-15**). Other flowers deceive pollinators with sexual attractants (**Figure 44-16**). Some plants and pollinators are codependent (**Figure 44-17**), with some plants providing nurseries for their pollinators.

Section 44.6 How Do Fruits Help to Disperse Seeds?

Fruits can disperse seeds by shotgun dispersal, being carried on wind or water (**Figure 44-19**), clinging to animal fur (**Figure 44-20**), or enticing an animal to eat the fruit (thus carrying the seed in its digestive tract; **Figure 44-21**).

FLASH CARDS

To use the flash cards, tear the page from the book and cut along the dashed lines. The key term appears on one side of the flash card, and its definition appears on the opposite side.

alternation of generations	double fertilization
anther	egg
carpel	endosperm
coleoptile	epicotyl
complete flower	filament
cotyledon	flower
dormancy	fruit

In flowering plants, the fusion of two sperm nuclei with the nuclei of two cells of the female gametophyte. One sperm nucleus fuses with the egg to form the zygote; the second sperm nucleus fuses with the two haploid nuclei of the central cell to form a triploid endosperm cell.

A life cycle, typical of plants, in which a diploid sporophyte (spore-producing) generation alternates with a haploid gametophyte (gamete-producing) generation.

The haploid female gamete, usually large and nonmotile; contains food reserves for the developing embryo.

The uppermost part of the stamen, in which pollen develops.

A triploid food storage tissue in the seeds of flowering plants that nourishes the developing plant embryo.

The female reproductive structure of a flower, composed of stigma, style, and ovary.

The part of the embryonic shoot located above the attachment point of the cotyledons but below the tip of the shoot.

A sheath surrounding the shoot in monocot seedlings that protects the shoot from abrasion by soil particles during germination.

In flowers, the stalk of a stamen, which bears an anther at its tip.

A flower that has all four floral parts (sepals, petals, stamens, and carpels).

The reproductive structure of an angiosperm plant.

A leaflike structure within a seed that absorbs food molecules from the endosperm and transfers them to the growing embryo; also called *seed leaf*.

In flowering plants, the ripened ovary (plus, in some cases, other parts of the flower), which contains the seeds.

A state in which an organism does not grow or develop; usually marked by lowered metabolic activity and resistance to adverse environmental conditions.

gametophyte	megaspore
generative cell	megaspore mother cell
germination	microspore
hypocotyl	microspore mother cell
imperfect flower	ovary
incomplete flower	ovule
integument	petal

A haploid cell formed by meiotic cell division from a diploid megaspore mother cell; through mitotic cell division and differentiation, it develops into the female gametophyte.

The multicellular haploid stage in the life cycle of plants; produces gametes (sperm and eggs).

A diploid cell, within the ovule of a flowering plant, that undergoes meiotic cell division to produce four haploid megaspores.

In flowering plants, one of the haploid cells of a pollen grain; undergoes mitotic cell division to form two sperm cells.

A haploid cell formed by meiotic cell division from a microspore mother cell; through mitotic cell division and differentiation, it develops into the male gametophyte.

The growth and development of a seed, spore, or pollen grain.

A diploid cell contained within an anther of a flowering plant; undergoes meiotic cell division to produce four haploid microspores.

The part of the embryonic shoot located below the attachment point of the cotyledons but above the root.

In flowering plants, a structure at the base of the carpel that contains one or more ovules and develops into the fruit.

A flower that is missing either stamens or carpels.

A structure within the ovary of a flower, inside which the female gametophyte develops; after fertilization, develops into the seed.

A flower that is missing one of the four floral parts (sepals, petals, stamens, or carpels).

Part of a flower, typically brightly colored and fragrant, that attracts potential animal pollinators.

In plants, the outer layers of cells of the ovule that surround the female gametophyte; develops into the seed coat.

pollen grain	sporophyte
pollination	stamen
seed	stigma
seed coat	style
sepal	tube cell
spore	zygote

The multicellular diploid stage of the plant life cycle; produces haploid, asexual spores through meiotic cell division.

The male gametophyte of a seed plant.

The male reproductive structure of a flower, consisting of a filament and an anther, in which pollen grains develop.

In flowering plants, when pollen grains land on the stigma of a flower of the same species; in conifers, when pollen grains land within the pollen chamber of a female cone of the same species.

The pollen-capturing tip of a carpel.

The reproductive structure of a seed plant, protected by a seed coat; contains an embryonic plant and a supply of food for it.

A stalk connecting the stigma of a carpel with the ovary at its base.

The thin, tough, and waterproof outermost covering of a seed, formed from the integuments of the ovule.

The outermost cell of a pollen grain, containing the sperm. When the pollen grain germinates, the tube cell produces a tube penetrating through the tissues of the carpel, from the stigma, through the style, and to the opening of an ovule in the ovary.

One of the group of modified leaves that surrounds and protects a flower bud; in dicots, usually develops into a green, leaflike structure when the flower blooms; in monocots, usually similar to a petal.

In sexual reproduction, a diploid cell (the fertilized egg) formed by the fusion of two haploid gametes.

A reproductive cell capable of developing into an adult without fusing with another cell; in the alternation-of-generations life cycle of plants, a haploid cell that is produced by meiotic cell division and then undergoes repeated mitotic cell divisions and differentiation of daughter cells to produce the gametophyte (the multicellular, haploid stage of the life cycle).

SELF TEST

1. In the life cycle of a plant, spores are produced by the
   a. gametes.
   b. sporophyte.
   c. eggs.
   d. gametophyte.

2. All plant life cycles consist of two unique stages, one in which the chromosomal number of cells is haploid and the other in which the chromosomal number of cells is diploid. Which haploid plant cell undergoes repeated mitotic divisions to produce a haploid adult organism?
   a. sporophyte
   b. gametophyte
   c. spore
   d. gamete

3. Which of the following structures is used to attract pollinators?
   a. sepals
   b. petals
   c. anthers
   d. pistols

4. The embryo sac is contained within the _____, which is located inside the _____.
   a. anther; stamen
   b. stigma; carpel
   c. ovule; ovary
   d. ovary; ovule

5. Where would you expect to find pollen?
   a. within the stamens
   b. within the megaspore mother cell
   c. within the microspores
   d. within the generative cell

6. Which of the following is true of double fertilization?
   a. One sperm fuses with the egg to form a zygote, and a second sperm fuses with the polar nucleus to form the endosperm.
   b. One sperm fuses with the tube cell to form the embryo, and a second sperm fuses with a generative cell to form the endosperm.
   c. The first fertilization event results in the production of a zygote, and a second fertilization event occurs several days or weeks later to produce a second zygote, which degenerates to the endosperm.
   d. Double fertilization is a rare event in flowering plants that results in the production of two identical embryos.

7. The endosperm is
   a. haploid.
   b. diploid.
   c. triploid.
   d. tetraploid.

8. A pollen grain has three different nuclei. The tube cell nucleus is one. Of the other two nuclei, one will fertilize the egg and the other
   a. will fertilize a synergid.
   b. will direct the growth of the pollen tube.
   c. is used as a spare.
   d. will unite with the polar nuclei.

9. Fruit flesh is derived from the _____ and functions to _____.
   a. ovary; supply nutrients to the growing embryo
   b. endosperm; protect the developing embryo
   c. ovule; supply nutrients to the growing embryo
   d. ovary; aid in seed dispersal

10. Many "vegetables" such as squash and tomatoes are actually fruits because they
    a. are too sweet to be true vegetables.
    b. are too brightly colored to be vegetables.
    c. develop from the ovary of the flower.
    d. develop from the ovule in a flower.

11. If a fire is hot enough to kill all of the existing plants in an area, new plant life is usually established in the area within a year. The first plants are commonly grasses and annual dicots. How can the rapid establishment of these plants be explained?
    a. Since the first plants are non-woody, they grow quickly. Trees would take much longer than one year to grow.
    b. The seeds of the plants required sterile soil in order to sprout.
    c. Seeds sprout quickly due to primary succession.
    d. The seeds of these plants were near the surface of the soil at the time of the fire and were protected by their seed coat and dormancy. Their dormancy was broken by their new exposure to the light when the covering plants were burned away.

12. What would NOT be considered an advantage that dormancy (characterized by decreased metabolic activity and resistance to adverse environmental conditions) would provide to many recently matured seeds?
    a. It provides an enforced delay before germination can begin upon maturation of the seed.
    b. It can provide sufficient time for the dispersal of seeds from the parent plant by weather and/or animals.
    c. It can provide an enforced delay until environmental conditions that are favorable for the growth of the seed exist.
    d. It can provide extra food resources to the embryo required for germination.

13. What are some of the mechanisms that plants have developed to protect growth of structures during germination?
    a. covering the delicate tips of structures growing through soil with protective caps
    b. extending the period of dormancy
    c. allowing the delicate tip of a growing shoot to "trail in the wake" of another structure that is "armored" against the abrasive effects of soil particles
    d. both a and c
    e. both b and c

14. The first leaflike structures to emerge from the seed are
    a. true leaves.
    b. not true leaves but cotyledons.
    c. not true leaves but part of the coleoptile.
    d. not true leaves but sepals.

15. Which group of plants requires a watery environment for the sperm to reach the egg?
    a. mosses
    b. pines
    c. flowering plants
    d. All plants require water for sperm movement.

16. When you eat fruit, what structure are you consuming?
    a. ovary
    b. stamen
    c. ovule
    d. seed

17. In which environment would you be less likely to find seeds that utilize the strategy of dormancy?
    a. tropical forests
    b. savannas
    c. deciduous forests
    d. chaparral

18. Which of the following is the outermost structure of the flower?
    a. stamen
    b. stigma
    c. sepals
    d. ovary

19. Which of the following is specialized to receive pollen?
    a. stamen
    b. stigma
    c. sepals
    d. ovary

20. Which of the following is NOT part of the carpel?
    a. anther
    b. style
    c. ovary
    d. stigma

21. Before it is carried to the flower of another plant, the pollen grain is created within the _____, which is part of the _____.
    a. anther; stamen
    b. stigma; carpel
    c. ovule; ovary
    d. ovary; ovule

22. The seeds depicted in this figure are adapted to disperse by
    a. the digestive system of mammals.
    b. wind.
    c. water.
    d. being attached to the fur of mammals.

(a) Dandelion fruits   (b) Maple fruits

23. What happens to the megaspore mother cell contained inside an ovary?
    a. Nothing. The megaspore is eventually fertilized by a pollen grain.
    b. The megaspore cell develops into a series of layers called *integuments*.
    c. The megaspore cell divides meiotically to produce four haploid megaspores.
    d. The megaspore later becomes the embryo producing a seed.

24. A seed contains each of the following EXCEPT a(n)
    a. embryonic plant.
    b. food source.
    c. seed coat.
    d. stoma.

26. Cotyledons are
    a. the dicot equivalent of sepals in monocots.
    b. resorbed shortly after the shoot emerges.
    c. the structures responsible for dormancy in seeds.
    d. the structures that attract pollinators.

27. The endosperm of the seed functions to
    a. provide nutrients to the growing embryo.
    b. develop into the new plant.
    c. provide protection for the growing embryo.
    d. aid in seed dispersal.

28. The function of the cotyledons in most dicot seeds is to
    a. store food for the embryo and growing seedling.
    b. protect the embryo.
    c. photosynthesize to provide food for the dormant embryo.
    d. help with seed dispersal.

29. When a bee gets nectar from a flower, the bee picks up pollen from the _____ and carries it to another flower.
    a. stigma
    b. ovary
    c. sepals
    d. anthers
    e. filament

25. How would you expect the plant shown in the photograph above to be pollinated?
    a. wind
    b. bees
    c. bats
    d. birds

30. Plants that have separate male and female flowers are
    a. complete flowers.
    b. incomplete flowers.
    c. all lacking sepals.
    d. found only in plants that occupy arid environments.

ANSWER KEY

1. b	16. a
2. c	17. a
3. b	18. c
4. c	19. b
5. c	20. a
6. a	21. a
7. c	22. b
8. d	23. c
9. d	24. d
10. c	25. a
11. d	26. b
12. d	27. a
13. d	28. a
14. b	29. d
15. a	30. b

CHAPTER 45 PLANT RESPONSES TO THE ENVIRONMENT

OUTLINE

Section 45.1 What Are Some Major Plant Hormones?

Plant **hormones** are chemicals produced by cells in one location and transported to other parts of the plant body, where they exert specific effects.

The six major plant hormone categories are **auxins**, **gibberellins**, **cytokinins**, **ethylene**, **abscisic acids**, and **florigens**. The functions of these hormones are listed in **Table 45.1**.

Section 45.2 How Do Hormones Regulate Plant Life Cycles?

Abscisic acid maintains seed dormancy and must be washed away by moisture before growth occurs (**Figure 45-1**).

Gibberellin stimulates germination, initiating the synthesis of enzymes that digest food reserves.

Auxin stimulates seedling shoot growth toward light (**phototropism**) and root growth against gravity (**gravitropism**). This is accomplished by the stimulation of cell elongation because of the accumulation of auxin along the cell layers of the shoot/root (**Figure 45-2**). Young plants may sense gravity because of the accumulation of plastids along the interior of cell surfaces (**Figure 45-2**).

The shape of a mature plant is determined by the interplay between auxin and cytokinin. The ratio of auxin (which inhibits lateral bud formation and is produced by the shoot tip) to cytokinin (which stimulates lateral bud formation and is produced by the root tip) controls lateral bud (and branch) formation (**Figure 45-7**). The phenomenon of the shoot tip dominating the activity of lateral buds is referred to as **apical dominance** (**Figure 45-6**).

The timing of flowering is typically controlled by the duration of darkness (**Figure 45-8**), which may stimulate the production of florigens.

Plants can detect light and darkness by changes in a leaf pigment called **phytochrome** (**Figure 45-9**). Phytochrome influences plant responses to light, including flowering, seedling elongation, leaf growth, chlorophyll synthesis, and straightening of the epicotyl or hypocotyl hook in dicots.

Developing seeds produce auxin and/or gibberellin in the surrounding ovary tissues, which stimulates fruit production (**Figures 45-10 and 45-11**).

As a seed matures, a surge in auxin concentration stimulates ethylene production, which causes fruit ripening (**Figure 45-11**). During ripening, starches are converted into sugars, the fruit softens, and colors brighten.

Temperate plants prepare themselves for winter by undergoing senescence, due primarily to a fall in auxin and cytokinin levels. This process of rapid aging results in the formation of abscission layers at fruit and leaf petioles (**Figure 45-12**). Simple sugars are transported to the roots for winter storage and plant buds become dormant, a condition maintained by high levels of abscisic acid.

Section 45.3 How Do Plants Communicate and Capture Prey?

Some plants release volatile chemicals when attacked. These chemicals attract insect predators that can remove the threat (**Figure 45-13**).

Injured plants release volatile chemicals that signal neighboring plants to produce substances (e.g., salicylic acid) that protect them from predation or infection.

Some plants, such as the Venus flytrap and bladderwort, can move rapidly to capture prey. In touch-sensitive plants such as the Venus flytrap, movement occurs when touch sensors that stimulate ion flow are triggered, resulting in the rapid loss of water in specialized cells that cause petiole movement. Movement of the sensitive plant is a result of **thigmotropism** (**Figure 45-14**). Bladderworts generate lower pressure in their bladders, which suck in prey when bristles around the bladder openings are triggered (**Figure 45-16**).

FLASH CARDS

To use the flash cards, tear the page from the book and cut along the dashed lines. The key term appears on one side of the flash card, and its definition appears on the opposite side.

abscisic acid	ethylene
abscission layer	florigen
apical dominance	gibberellin
auxin	gravitropism
biological clock	hormone
cytokinin	long-day plant
day-neutral plant	phototropism

A plant hormone that promotes the ripening of some fruits and the dropping of leaves and fruit; promotes senescence of leaves.

A plant hormone that generally inhibits the action of other hormones, enforcing dormancy in seeds and buds and causing the closing of stomata.

One of a group of plant hormones that may stimulate or inhibit flowering in response to day length.

A layer of thin-walled cells, located at the base of the petiole of a leaf, that produces an enzyme that digests the cell walls holding the leaf to the stem, allowing the leaf to fall off.

One of a group of plant hormones that stimulates seed germination, fruit development, flowering, and cell division and elongation in stems.

The phenomenon whereby a growing shoot tip inhibits the sprouting of lateral buds.

Growth with respect to the direction of gravity.

A plant hormone that influences many plant functions, including phototropism, gravitropism, apical dominance, and root branching.

A chemical that is secreted by one group of cells and transported to other cells, whose activity is influenced by reception of the hormone.

A metabolic timekeeping mechanism found in most organisms, whereby the organism measures the approximate length of a day (24 hours) even without external environmental cues such as light and darkness.

A plant that will flower only if the length of uninterrupted darkness is shorter than a species-specific critical period; also called a *short-night plant*.

A group of plant hormones that promotes cell division, fruit development, and the sprouting of lateral buds; also delays the senescence of plant parts, especially leaves.

Growth with respect to the direction of light.

A plant in which flowering occurs as soon as the plant has undergone sufficient growth and development, regardless of day length.

phytochrome

short-day plant

plant hormone

thigmotropism

senescence

A plant that will flower only if the length of uninterrupted darkness exceeds a species-specific critical period; also called a *long-night plant*.

Growth in response to touch.

A light-sensitive plant pigment that mediates many plant responses to light, including flowering, stem elongation, and seed germination.

A chemical produced by specific plant cells that influences the growth, development, or metabolic activity of other cells, typically some distance away in the plant body.

In plants, a specific aging process, typically including deterioration and the dropping of leaves and flowers.

SELF TEST

1. The downward growth of roots into the ground is an example of
   a. phototropism.
   b. gravitropism.
   c. abscission.
   d. senescence.

2. Which hormone is responsible for stimulation of cell division and cell differentiation as lateral buds are released from dormancy?
   a. cytokinin
   b. auxin
   c. ethylene
   d. abscisic acid

3. Long-day plants flower specifically when the
   a. light period is less than some critical length.
   b. light period is greater than some critical length.
   c. dark period is less than some critical length.
   d. dark period is greater than some critical length.

4. Gravitropism in roots is a response to _____ and to the hormone_____.
   a. gravity; auxin
   b. Earth's rotation; cytokinin
   c. altitude; ethylene
   d. the biological clock; phytochrome

5. Rapid stem elongation and larger fruit size are promoted by
   a. auxins.
   b. gibberellins.
   c. cytokinins.
   d. abscisic acid.

6. During the hot dry summers, some plants prevent water loss by transpiration by producing
   a. auxins.
   b. florigens.
   c. abscisic acid.
   d. cytokinins.

7. Within a plant cell that has statoliths concentrated at the bottom of the cell, the highest concentration of auxin would be found
   a. at the upper end of the cell.
   b. in the middle of the cell.
   c. at the bottom of the cell.
   d. evenly distributed throughout the cell.

8. Karrikins trigger
   a. phototropism.
   b. gravitropism.
   c. fruit production.
   d. germination.

9. In the cycle of photoperiodism, far-red light is absorbed
   a. by both forms of phytochrome.
   b. by neither form of phytochrome.
   c. only by the red form of phytochrome.
   d. only by the far-red form of phytochrome.

10. When the phytochrome in the red form is exposed to red light, it will
    a. be converted into the phytochrome in the far-red form.
    b. be broken down into a physiologically inactive compound.
    c. stay in the phytochrome red form.
    d. reflect the red light.

11. Flowering in a long-day plant can be produced by an
    a. uninterrupted light period of 12 hours and an uninterrupted dark period of 12 hours.
    b. uninterrupted light period of 16 hours and an uninterrupted dark period of 8 hours.
    c. uninterrupted light period of 8 hours and an uninterrupted dark period of 16 hours.
    d. keeping the plant in 24-hour light cycles for two consecutive days.

12. Pinching the tips of your chrysanthemums to make them bushy to fill out your flower bed causes the plant to stop production of
    a. auxin.
    b. gibberellins.
    c. cytokinin.
    d. florigen.

13. The green bags sold in stores to help keep your fruit "fresh" longer block _____ to keep the fruit from ripening too quickly.
    a. ethylene
    b. florigen
    c. auxin
    d. abscisic acid

14. In the fall, the increased production of ethylene promotes
    a. germination in cool season plants.
    b. senescence.
    c. apical dominance.
    d. flowering of short-day plants.

15. When a plant is injured by being fed on by insects, its close neighbors may not suffer the same fate because
    a. the neighboring plants begin to produce chemicals that make them taste bitter to the insect so the insects will not eat those plants.
    b. the injured plant produces toxins that kill the insects.
    c. the neighboring plants produce chemicals that act as pesticides after receiving the chemical signals sent by the injured plant.
    d. the insects are unable to eat enough to have an effect on the neighboring plants.

16. One rotten apple spoils the whole barrel because the rotting apple produces
    a. auxin.
    b. cytokinin.
    c. gibberellin.
    d. ethylene.

17. One of the possible consequences for a plant with a defective gene for the production of abscisic acid would be
    a. an increased rate of senescence of the plant's tissues.
    b. an increased rate of lateral root development.
    c. the development of fruit without fertilization.
    d. a decreased dormancy period for the plant's seeds.

18. If the tip of a plant shoot were removed, the lateral buds near the top of the shoot will
    a. die from the lack of an important resource.
    b. remain dormant.
    c. begin to grow into branches or flowering stalks.
    d. develop into roots instead of stem branches.

19. In experiments with plant growth, scientists have found that a plant tip will bend toward the light because
    a. when the shoot tip is not exposed to light, it will not produce auxin.
    b. auxin will move to the darker side of the stem and cause more cell elongation there.
    c. auxin moves to the lighter side of the stem, which causes elongation to the cells there.
    d. statoliths move in response to the UV radiation.

20. Predatory plants utilize the _____ found in their prey.
    a. ATP
    b. ethylene
    c. nitrogen
    d. glucose

21. Sensitive plants and sundew move due to
    a. the movement of sensory triggers that stimulate ATP production.
    b. thigmotropism.
    c. sensory receptors that detect chemicals released from the organism in contact with the plant.
    d. electrical signals transmitted by the organism in contact with the plant.

22. If the source of cytokinin were eliminated during the development of a plant, the lateral buds would
    a. remain dormant.
    b. produce branches.
    c. produce roots instead of branches.
    d. produce flowering stalks instead of branches.

23. Photoperiodism is a response to
    a. light and dark periods.
    b. humidity.
    c. temperature.
    d. carbon dioxide levels.

24. Why do the coleoptiles of oats bend toward the light?
    a. Oat coleoptiles exhibit negative phototropism.
    b. Gibberellins stimulate growth on the illuminated side of the shoot.
    c. Auxins are found at highest concentrations and stimulate growth on the darker side of the shoot.
    d. Changes in turgor pressure in cells on the darker side of the shoot cause the shoot to bend toward the light.

25. If a plant were strapped upside down to the wall of a rapidly rotating drum (like the ones you might ride on at an amusement park), which of the following would you expect to happen?
    a. The artificial G forces would cause the roots to grow toward the middle of the drum and the stem to grow away from the middle.
    b. The artificial G forces would cause the roots to grow away from the middle of the drum and the stem to grow toward the middle.
    c. The roots and stem would curl around such that the roots would grow toward Earth and the stem would grow toward the sky.
    d. The roots and stem would grow in the direction they were oriented because they would no longer be influenced by Earth's gravitational forces.

26. The plant in this photo illustrates an example of
    a. gravitropism.
    b. thigmotropism.
    c. apical dominance.
    d. phototropism.

27. How does auxin control the direction of root growth so that roots grow down into the ground, rather than up out of the ground?
    a. Roots respond to light penetrating the soil in a manner known as *positive phototropism.*
    b. Auxin accumulates along the side of the growing root farthest from the ground surface. This inhibits cell elongation on the lower side of the root, and the root bends and grows downward.
    c. Auxin stimulates a negative gravitropism.
    d. Gibberellins are produced that respond to the pull of gravity, causing the roots to move down instead of up.

28. What is believed to be responsible for the detection of gravity in roots and shoots?
    a. Plastids settling onto the lowest surface of the root/shoot (i.e., farthest away from the surface of the soil) initiate a series of steps that end with auxin triggering an unequal elongation in the root/shoot.
    b. Gibberellin is able to detect the direction of gravity.
    c. The movement of root hairs detects gravity.
    d. Auxin production causes cell elongation and the bending of the shoots away from gravity and lengthening of the roots toward gravity.

29. What is responsible for ensuring that the relative growth rate of the roots does not overwhelm the growth rate of the shoots and vice versa?
    a. auxin
    b. cytokinin
    c. apical dominance
    d. the ratio of auxin to cytokinin

30. The abscission layer shown in the figure below is where
    a. the leaf will be lost.
    b. the new leaf will be formed.
    c. the auxins are concentrated to promote the growth of the leaf.
    d. the florigens are concentrated to promote the blooms of flowers.

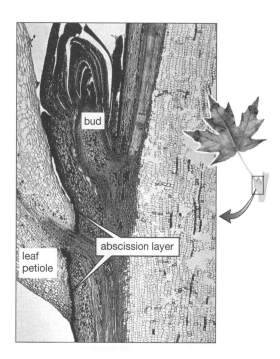

ANSWER KEY

1. b		16. d	
2. a		17. d	
3. c		18. c	
4. a		19. b	
5. b		20. c	
6. c		21. b	
7. c		22. a	
8. d		23. a	
9. d		24. e	
10. a		25. b	
11. b		26. b	
12. a		27. b	
13. a		28. d	
14. b		29. d	
15. a		30. a	